Frontiers in Parasitology

(Volume 2)

(Water-borne Protozoa in Humans)

Edited by

Fabrizio Bruschi

Università di Pisa, Pisa, Italy

Bentham Science Publishers Ltd.
Executive Suite Y - 2
PO Box 7917, Saif Zone
Sharjah, U.A.E.
Email: subscriptions@benthamscience.org

BENTHAM
SCIENCE

CONTENTS

FOREWORD

Parasitology continues to be a very exciting and stimulating field of study and has regained global attention and interest especially due to the data obtained from the recent 2010 and 2013 Global Burden of Disease studies that highlight how parasitic infections continue to be amongst the leading causes for the highest disabilities worldwide. In addition, even though amazing strides have been achieved towards the Millennium Development Goals, the UN member states and the world are now faced with much broader and more audacious post-2015 goals as part of the new Sustainable Development Goals to be achieved by 2030. For these objectives to be achieved it is very important not only to focus and monitor available and ongoing control implementation programs globally, but also complement these efforts with novel research leading to new and affordable health innovations.

Therefore, when Professor Fabrizio Bruschi invited me to write the foreword for this book titled **"Frontiers in Parasitology: Water-borne Protozoa in Humans"** it immediately captured my interest and excitement to know that with this book physicians, health professionals and scientists around the world will rapidly gain access to the most up to date advances and innovation about the epidemiology, pathobiology, laboratory and clinical diagnosis, and available prevention and treatment measures for this very important group of infectious and tropical diseases. It is well known that protozoan infections are important causes of diarrheal and other gastrointestinal diseases in humans including the major causes for travellers' diarrhea. In addition, some of the protozoans can affect the central nervous system and even cause keratitis.

After reading this book the hope is that it will raise awareness of the importance of water-borne protozoan infections as global public health problems, while at the same time generate new ideas, new networks and partnerships and bring new translational discoveries from the bench into the clinic ultimately leading to an improvement in health delivery mechanisms benefiting global strategies.

Indeed, and of no surprise if one were to search for a good compilation of recent articles in the last 10 years that summarizes the global distribution and burden in this field of study, there are very limited publications that provide a comprehensive review of the global health impact of waterborne parasitic protozoan infections worldwide. Furthermore, for most if not all the water-borne diseases, effective drug treatments and vaccines are not yet available and diagnostic tools are not highly sensitive or reliable especially for point of care diagnostics in resource-poor settings. One of the most recent reviews on water-borne protozoa was written by Baldursson and Karanis and published in Water Research in 2011, and shows that in Australia, Europe and North America at least 199 outbreaks of these diseases occurred and were reported between 2004-2010. However, the authors also highlight that the countries that most likely have the greatest populations afflicted by water-borne parasitic infectious and other tropical diseases lack strong surveillance systems, leading to a chronic under-reporting or even no reporting of their public health burden.

This book, then, aims to fill this gap in knowledge as well as provide a comprehensive overview of major parasitic water-borne protozoan diseases. It includes an impressive list of authors and co-authors internationally recognized in the field of parasitology, covering in a compilation of 8 chapters the new and recent insights for waterborne-protozoa including *Blastocystis*, *Cryptosporidium, Cyclospora, Cystoisospora, Entamoeba and other pathogenic intestinal amoebae, free living amoebae, Giardia and Microsporidia.*

Maria Elena Bottazzi
Associate Dean, National School of Tropical Medicine
Professor, Pediatrics, Section of Pediatric Tropical Medicine & Molecular Virology & Microbiology
Baylor College of Medicine and Texas Children's Hospital, Houston, Texas, USA

Distinguished Professor
Department of Biology, Baylor University, Waco, Texas, USA

Deputy Director, Sabin Vaccine Institute Product Development Partnership
Director for Product Development, Sabin Vaccine Institute and Texas Children's Hospital
Center for Vaccine Development, Texas, USA

Editor in Chief, Current Tropical Medicine Reports, Springer, US

PREFACE

Water is essential for living organisms on Earth, but it can also spread toxic substances and pathogens for humans and animals, particularly in low income Countries. At global level, waterborne pathogen exposure in fact correlates significantly with low socio-economical conditions.

During the recent past, human development, population growth, extreme weather events as well as natural calamities due to climate changes have affected both the quality and quantity of water resources, increasing the occurrence of water-borne infectious diseases. These are a major cause of morbidity and mortality, with an estimate of 4.0% of global deaths and 5.7% of the global disease burden (in terms of Disability Adjusted Life Years) attributable to a complex of factors like water, sanitation, and hygiene conditions. Water-borne pathogens such as viruses, bacteria and protozoa are responsible for diarrheal diseases which affect, sometimes in a serious manner, in particular children, worldwide.

This E-book is designed to inform the reader on water-borne protozoa infections in humans.

The audience is represented by medical students, parasitologists, clinical microbiologists, infectivologists, researchers, environmental technicians, people who are engaged in water purification and control and public health personnel. Each chapter addresses the history, morphology and life cycle, global epidemiology and risk factors, immunology and immunopathology, symptoms, diagnosis and detection methods, treatment and perspectives of control for major waterborne parasites transmissible with water, either used for drinking or for recreational activities.

In addition to the parasites considered in this book, others can be transmitted by water, like *Toxoplasma gondii* for example, but were not considered in this book.

The Editor is grateful to David Di Cave, Marco Lalle, Simonetta Mattiucci, Walhul Khan, Una Ryan, Chuck Sterling, Magda Azab, Nicola Coppola for valuable reviewing process.

Fabrizio Bruschi
Università di Pisa
Pisa
Italy

List of Contributors

Claudia Ledda	Department of Experimental Medicine, Division of Microbiology and Clinical Microbiology, II Università di Napoli, Naples, Italy
Eduardo R. Cobo	Gastrointestinal Research Group, University of Calgary, Calgary, Alberta, Canada
Fabio Tosini	Department of Infectious, Parasitic and Immunomediated Diseases, Istituto Superiore di Sanità, Rome, Italy
Gianluigi Franci	Department of Experimental Medicine, Division of Microbiology and Clinical Microbiology, II Università di Napoli, Naples, Italy
Ildebrando Patamia	Department of Biomedical and Biotechnological Sciences, University of Catania, Catania, Italy
Kris Chadee	Gastrointestinal Research Group, University of Calgary, Calgary, Alberta, Canada
Marco Prato	Department of Biomedical and Biotechnological Sciences, University of Catania, Catania, Italy
Maria Elena Della Pepa	Department of Experimental Medicine, Division of Microbiology and Clinical Microbiology, II Università di Napoli, Naples, Italy
Marilena Galdiero	Department of Experimental Medicine, Division of Microbiology and Clinical Microbiology, II Università di Napoli, Naples, Italy
Massimiliano Galdiero	Department of Experimental Medicine, Division of Microbiology and Clinical Microbiology, II Università di Napoli, Naples, Italy
Nadia Aly E-Dib	Faculty of Medicine, Cairo University, Cairo, Egypt
Settimia Alfano	Department of Experimental Medicine, Division of Microbiology and Clinical Microbiology, II Università di Napoli, Naples, Italy
Showgy Maayeh	Department of Cell and Molecular Biology, Science for Life Laboratory, Uppsala University, Uppsala, Sweden
Simone M. Cacciò	Department of Infectious, Parasitic and Immunomediated Diseases, Istituto Superiore di Sanità, Rome, Italy
Staffan Svärd	Department of Cell and Molecular Biology, Science for Life Laboratory, Uppsala University, Uppsala, Sweden

Thomas Wetzel Clínica Alemana School of Medicine,Universidad del Desarrollo, Santiago, Chile

Vinu Khanna Department of Microbiology, Kasturba Medical College, Manipal, Karnataka, India

Ynes R. Ortega Center for Food Safety, University of Georgia, Griffin, Georgia, USA

Frontiers in Parasitology

Frontiers in Parasitology

Volume # 2

Water-borne Protozoa in Humans

Editor: Fabrizio Bruschi

ISSN (Print): 2542-4211

eISSN (Online): 2542-422X

eISBN (Online): 978-1-68108-433-6

ISBN (Print): 978-1-68108-434-3

The figure shown on the front cover of this book is kindly supplied by Dr. Massimiliano Galdiero, Department of Experimental Medicine, Division of Microbiology and Clinical Microbiology, Second University of Naples, Italy. The cover figure shows the sexual and asexual replication of Cystoisospora belli (see chapter 4 of this book).

CHAPTER 1

Blastocystis spp.

Ildebrando Patamia and **Marco Giuseppe Prato***

Department of Biomedical and Biotechnological Sciences, University of Catania, Catania, Italy

Abstract: *Blastocystis* spp. are intestinal parasitic protozoa of humans and many other organisms Using molecular approaches the taxonomy of *Blastocystis* was determined. At present, this parasite is believed to belong to the subclass of Stramenopiles (or Heterokonts), subgroup Opalinata. *Blastocystis* is the only Stramenopile known to be a human parasite. For this reason it was classified first as *Blastocystis* hominis, but in the light of new knowledge, the term of *Blastocystis* spp. is more appropriate In this chapter epidemiology, life cycle and transmission, morphology and structure, laboratory diagnosis including microscopy, culture and molecular techniques are reviewed. Finally, clinical aspects, including the association with irritable bowel syndrome and therapeutic approaches are considered.

Keywords: *Blastocystis* spp., Diagnostic method, Epidemiology, Human infection, Immunology, Molecular epidemiology, Molecular subtypes, Pathogenesis, Treatment.

INTRODUCTION

Blastocystis spp. are single-celled protozoa, intestinal parasites of humans and numerous other organisms such as mammals, birds, amphibians, reptiles, fish, arthropods, annelids, insects and mollusks [1 - 5]. For years, the taxonomy of *Blastocystis* was very controversial and through the use of molecular methods it has been possible to define the phylogenetic characteristics of the parasite. Today it is believed to belong to the subclass of Stramenopiles (or Heterokonts), subgroup Opalinata [6 - 8]. *Blastocystis* is the only Stramenopile known to be a human parasite, for this reason Brumpt suggested the name of *Blastocystis* hominis [9], which today has been substituted, in the light of new acquisitions, by *Blastocystis* spp.

* **Corresponding author Marco Giuseppe Prato:** Department of Biomedical and Biotechnological Sciences, University of Catania, Catania, Italy; Tel: (+39) 0953782773; E-mail: prato.marco1@gmail.com

Fabrizio Bruschi (Ed.)

EPIDEMIOLOGY

The geographic distribution of *Blastocystis* spp. is very wide, above all in developing countries, and it is estimated that about 1 billion subjects are colonized [10, 11]. Even if the epidemiology of this parasite has not yet been well defined [12, 13] it is probably the most widespread eukaryote organism in humans [14].

Table 1. Prevalence of *Blastocystis* infections in different countries.

Prevalence (%)	Country	Reference
2.1	Turkey	[15]
3.7	China	[16]
7	Mexico	[10]
13.5	Thailand	[11]
13.6	Italy	[17]
14.2	Turkey	[18]
14.7	India	[19]
15	Germany	[20]
20	USA	[21]
22.9	Argentina	[22]
26.5	Brazil	[23]
26.6	Libya	[24]
36.4	Colombia	[25]
40.9	Brazil	[26]
52.3	Malaysa	[27]

The prevalence of *Blastocystis* spp. is clearly higher in developing countries, approximately between 30% and 50% [28, 29]. Immigrants, adopted children and people who work in close contact with animals, have a greater probability of contracting an infection caused by *Blastocystis* spp [30, 31]. Scientific evidence suggests that patients with AIDS and severe immunosuppression, have higher risk of infection by *Blastocystis* spp [32 - 35].

The prevalence of *Blastocystis* spp. in HIV/AIDS patients is estimated at 8% in India [35], and 38% in Germany [33]. A recent study evaluated it at 72.4% in Jakarta, Indonesia [34].

Fig. (1)

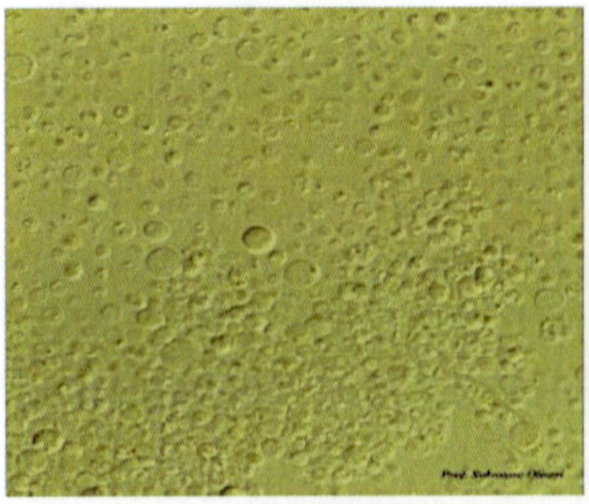

Fig. (2)

Fig. (3)

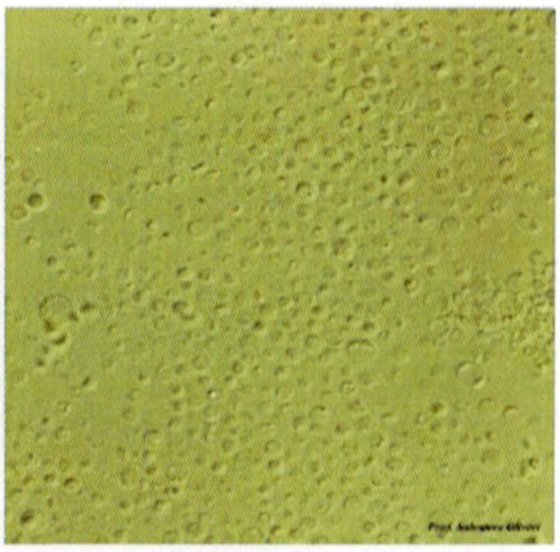

Fig. (1-3). Cystic form of *Blastocystis spp.* from laboratory culture on Boeck and Drbohlav's egg enriched medium. Optical microscopy, x400 magnification.

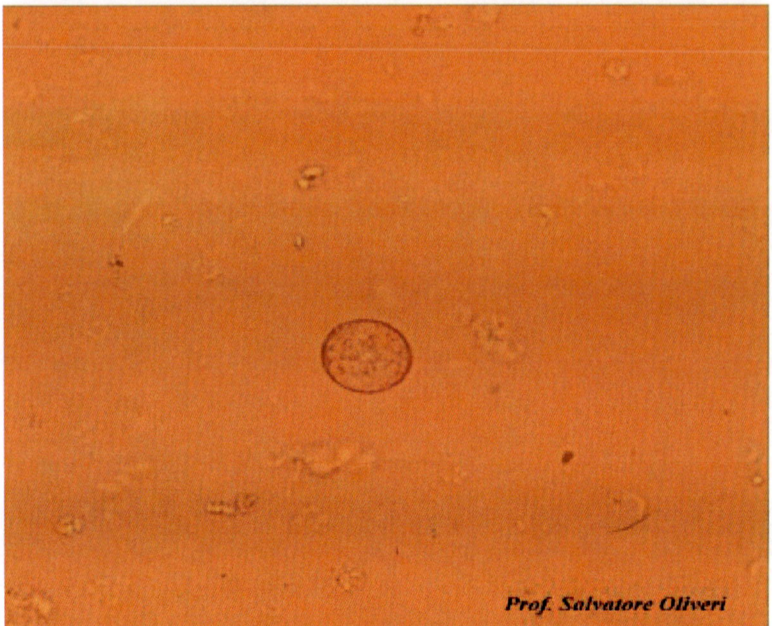

Fig. (4). Detail of a Cystic form of *Blastocystis spp.* from culture. Optical microscopy, x1000 magnification.

MORPHOLOGY AND STRUCTURE

Blastocystis spp. is an extremely polymorphic organism of which four principle forms have been described, in fecal samples or in *in vitro* cultures: vacuolar, granular, amoeboid and cystic [1, 35 - 37].

Vacuolar Form

The vacuolar form is the most frequently identified in axenic cultures and fresh feces of infected patients. Under brightfield illumination this form is seen as a roundish mass between 2 and 200 µm in size, with an average diameter of 4-45 μm [1], however, gigantic forms up to 400 μm, also have been described in axenic cultures [38].

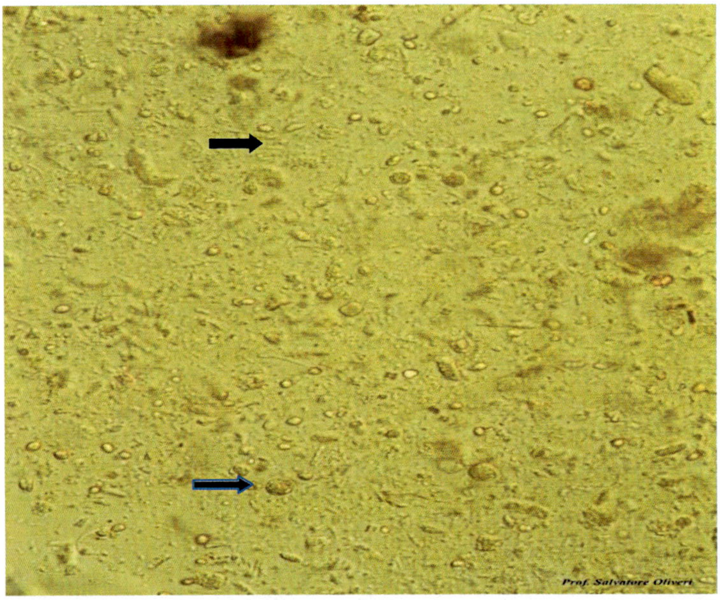

Fig. (5). Arrows show cystic form of *Blastocystis spp.* from fresh stool under Lugol coloration.

A large central vacuole containing fine granulations [39], whose function has not yet been fully clarified, occupies about 90% of the organism. Its lipid [40] and carbohydrate [41] content has suggested its possible function as a reserve [42].

The vacuole is surrounded by a thin cytoplasmatic layer containing 4 (or more) spherical or ovoid nuclei with a diameter of 1-2 μm [43], and Golgi apparatus and mitochondria-like organelles (MLO) with numerous highly electron-dense crests. The vacuolar form has a thick superficial fibril covering made up of mainly of carbohydrates, whose function has not been completely investigated. It has been postulated that the structure could be an aid to capturing, during phagocytosis of bacteria [44, 45] and in protection from osmotic shock [46]. It would also be an important protective barrier for membrane proteins, shielding them from the host's immune system [47]. The vacuolar forms reproduce quickly by binary fission [48].

Granular Form

The granular form is generally larger than the vacuolar form with dimensions between 12 and 25 μm, though still remaining similar to it. What distinguishes the two forms is essentially the shape of the granules that, in the granular form, can be differentiated morphologically. Studies carried out with the electron microscope, have identified different typologies of granules: lipidic and metabolic. The existence of reproductive granules, linked to the asexual multiplication of *Blastocystis* spp., has been reported in several studies [49, 50], but not confirmed yet. In fact, the granules are heterogeneous and inclusions have been described such as myelin like, lipidic drops, small vesicles or crystalline formations.

Amoeboid Form

The amoeboid form, which is smaller than vacuolar and granular forms (2.6 to 7.8 um in diameter), is rarely observed in axenic cultures or fresh feces, they have characteristics similar to the previously described form, but also have two or more pseudopodia that make them apparently similar to amebae. Dunn and coll [39]. described amoeboid forms of *Blastocystis* spp. containing phagocytized bacteria and digestive vacuoles similar to lysosomes. This led to the hypothesis that a nutritional role could be attributed to this form.

Cystic Forms

The cysts are the form of the parasite most easily seen in fresh feces of colonized subjects. They were first described by Mehlhorn (1988) from feces of patients with AIDS. The dimensions of the cysts are variable, but generally smaller than those of the granular and vacuolar forms. They have a diameter between 2 and 5 μm [51] or between 5 and 10 μm according to some authors [52]. They are generally roundish in shape, the cysts have a multi-layer protective wall with an external fibrillar layer, which encloses a thin cytoplasm with numerous mitochondria, granules of glycogen, small vacuoles and nuclei, containing chromatin consolidations, whose number can vary from one to four [53, 54].

It has been hypothesized that the cysts are a resistant form of the parasite. In fact, they are believed to be able to survive for more than thirty days at 25°C and up to two months at 4°C [55]. Resistant to osmotic shock and aerobic environments, they are still sensitive to heat and common disinfectants [51], although some authors report a possible resistance to chloride at concentrations normally used in drinking water [56].

Today there seems to be an agreement in defining the cysts as the only forms responsible for the transmission of *Blastocystis* spp [35, 51, 55, 57].

The life cycle of *Blastocystis* spp. is still uncertain, even if numerous theories have been made. Singh *et al.* (1995) found the presence of two cystic forms of the parasite: one thin-walled and the other thick-walled. The former is considered responsible for the passage of the parasite into the external environment by the fecal-oral route; the latter is implicated in mechanisms of auto infestation. In the host's intestine, the parasite would reproduce by schizogony generating the pre-cystic form(stage) that evolve into thick-walled cysts; after their rupture, they liberate the daughter vacuolar form.

Genetic Variability

The isolates of *Blastocystis* spp. from both humans and animals are morphologically indistinguishable.

However, a genetic variability has been observed that has led some authors to substitute the previous denomination of *Blastocystis hominis* with the current *Blastocystis* spp. The conserved and variable regions of the gene encoding the small subunit ribosomal RNA gene (SSU rRNA) are the target of PCR reactions [8].

The molecular methods most commonly used to study *Blastocystis* spp. are:

- PCR-RFLP (Restriction Fragment Length Polymorphism) [58, 59]
- PCR and dideoxysequencing [60, 61]
- PCR-STS (Sequence Tagged Site) [61 - 63]
- PCR fingerprinting [64, 65]

Based on the data obtained from the analysis of SSU-rRNA, to date 17 subtypes (ST) have been identified [66]. 90% of the isolates obtained from humans belong to the subtypes between ST1 and ST4,the subtypes from ST5 to ST9 are, instead, less frequent. Subtypes ST10 - ST17 are exclusive to animals [67 - 69].

LIFE CYCLE AND TRANSMISSION

The method of transmission of *Blastocystis* spp. remains unclear also if we presume that it can take place by the direct fecal-oral route or by means of environmental contamination with the cystic form of the parasite, present in the feces of the individual affected, the only one considered transmissible, from individual to individual [62, 71, 82, 83].

The transmission of *Blastocystis* spp. through the use of untreated water, above all in the presence of poor hygiene conditions, has been repeatedly reported [56, 76, 84].

The ample distribution of *Blastocystis* spp. in animals has suggested the hypothesis that some STs are not host-specific and that cross-infectivity between various animal species could be taken into consideration in this regard human infection can be considered as an anthropozoonosis originating from a reservoir made up of numerous animal species.

Numerous life cycle model have been proposed for *Blastocystis* spp. and the differences hypothesized are probably due to the common opinion that the parasite can have different reproductive cycles [36, 70].

1. The cysts are swallowed by the human or animal host.
2. The cysts transform into the vacuolar form in the intestine which multiply by binary fission possibly also evolving into amoeboid organular forms.
3. The vacuolar forms begin to encyst in the host's intestine. The forms that develop in the intermediate stages of the process, can be surrounded by a thick fibrillar layer that later is lost during the passage to the external environment.
4. The cysts, once excreted with the feces, are the forms responsible for the transmission of the parasite.

As regards the transmission route, numerous studies describe water contaminated by cysts as a potential transmission route of the parasite, in the same way as numerous intestinal parasitosis caused by other protozoa [51, 73 - 76]. Ascertaining the existence of STs common to both humans and animals has allowed the integration of the classic route mediated by contaminated water with that linked to the zoonotic transmission of the parasite [77].

Transmission by the fecal-oral route has been suggested by numerous studies developed with traditional and molecular methods [61, 78] and experimentally demonstrated in rats [55].

LABORATORY DIAGNOSIS

Microscopy

As for other intestinal protozoa it is possible to look for *Blastocystis* spp. by observing fecal samples under an optical microscope. The observation can be carried out using fresh feces but generally after concentration of the material.

The formalin-ethyl acetate concentration technique (FECT) is most frequently used in numerous studies report a modest sensitivity to this method compared to other techniques such as xenic *in vitro* culture (XIVC) and permanent trichrome staining of feces fixed in sodium acetate-acetic acid-formalin (SAF-PST) [79]. Moreover, various authors have reported that the processes of centrifugation can damage or destroy the vacuolar, multi-vacuolar and granular forms of the parasite, making identification difficult [80].

A density gradient (FICOLL) is also used [81]. Confounding factors in the microscopic identification of cysts of *Blastocystis* spp. could be fecal leucocytes, oocysts of Cryptosporidium spp. and, above all, cysts of E.histolytica, E.hartmanni, and Endolimax nana. The observation of a good number of fecal samples is necessary to find the presence even of few parasites, as it is always recommended in traditional microscopy diagnostics. To set up microscope samples the following are used: temporary staining like that one with solution of MIF, Bailenger, Lugol and permanent staining such as: Gomori trichrome, iron-ematoxilin, Ziehl-Neelsen, Gram or Giemsa. All the stains permanent staining, except for Giemsa, generally includes fixing of the sample with different compounds such as ethanol, formalin, SAF (sodium acetate-acetic acid-formalin), PVA (polyvinyl alcohol). Microscopic observations are carried out at 400x magnification and at 1000x using immersion oil.

Culture

It is possible to obtain axenic cultures of *Blastocystis* spp. on specific media such as Jones medium (Jones, 1946) enriched with 10% horse serum [72] and Boeck and Drbohlav's egg enriched medium. Antibiotics are added to the media to inhibit bacterial growth that could interfere with the growth of the parasite. The cultures are incubated at 37°C and subcultured into fresh media every 3 days. After growth, it is possible to observe *Blastocystis* spp. in the form of compact masses, similar to bacterial colonies. Culturing improves the detection sensitivity but it is not routinely used in diagnostic labs, respect to microscopic observations, even if greater sensitivity has been reported [81].

Molecular Techniques

Over the last few years the diagnostic approach to the patient affected by intestinal parasitic infections, especially protozoonosis, has undergone a rapid evolution thanks to the introduction of molecular methods in the clinical microbiology laboratory. Molecular methods have been widely used also in the study and diagnosis of *Blastocystis* spp. Using these techniques, just a little less rapid in the diagnosis of an FECT, but certainly quicker than culture, it is possible to use samples qualitatively unsuitable to be used for a culture examination, giving a quick response and greatly increasing the sensitivity and the specificity of the diagnostic test. A commercial molecular kit for the detection of *Blastocystis* spp. is available (Clongen Laboratories, LLC).

CLINICAL ASPECTS

The clinical significance of one of the most widespread single-cell intestinal parasites in the world, *Blastocystis* spp., remains uncertain. Numerous epidemiological and clinical studies have been made to evaluate the prevalence in various populations and study the role of the parasite as an etiological agent of intestinal and extra-intestinal pathologies.

As regards the clinical aspects, it should be noted that the knowledge of the period of incubation is an essential prerequisite for a clear understanding of the onset of any pathology and the relative importance of the various routes of transmission. However, for *Blastocystis* spp., neither period of incubation or duration of infection have been described. Human infection with *Blastocystis* spp. is called "blastocystosis". There are a variety of signs and symptoms associated with blastocystosis, from non-specific intestinal symptoms to cutaneous manifestations.

Intestinal Symptoms

Blastocystosis manifests, most frequently with diarrhea and abdominal pain [72]. Other non-specific gastrointestinal symptoms can include abdominal distension, flatulence, anorexia, weight loss, nausea, vomiting, constipation, and dysentery [85, 86].

As regards the severity of the manifestations, there have been reported cases ranging from chronic diarrhea to acute gastroenteritis. Some researchers saw a possible association between severity of the pathology and the number of parasites found in fecal samples [83, 87] or the finding of amoebic forms of *Blastocystis* spp.in symptomatic patents.The association was also demonstrated by the elevated protease activity was in isolates that had higher percentages of amoebic forms [88].

Other studies have evaluated the possible correlation existing between ST infections and severity of clinical manifestations. It has been repeatedly reported how genotypic variability can play a role in the pathogenicity of *Blastocystis* spp [89]. Subtype 3 is most frequently isolated from patients with gastrointestinal disorders [90]. Rats have been experimentally infected with isolates from symptomatic and asymptomatic patients. It is interesting to note that the pathological manifestations, evaluated by the authors as moderate and severe, were detected only in the experimental animals infected with isolates from symptomatic patients, and that differences in severity were observed among the various STs of *Blastocystis*, thus suggesting the existence of particularly virulent strains. Pathophysiological variability between different *B. hominis* genotypes was evaluated in experimentally infected rats. In symptomatic isolates, subtypes 1, 3, and 4 were often detected. Subtype 3 was the commonest in humans [91].

Infection of immunocompetent BALB/c rats has shown the presence of inflammatory infiltrates only in the intestine of some of them, suggesting the idea that host factors play a relevant role in the evolution of the infection [72]. As regards the observations in humans, endoscopically removed samples from symptomatic patients showed that bioptic *Blastocystis* spp. do not invade the mucosa of the colon. Therefore it can be possible that the intestinal symptoms, correlated with the presence of the parasite, could be linked to alterations of permeability of the epithelial barrier at the level of the tight cellular junctions [92].

Irritable Bowel Syndrome (IBS)

Irritable bowel syndrome is described as a functional gastrointestinal disorder characterized by abdominal pain associated with bowel variations. Notwithstanding the fact that numerous studies correlate the clinical manifestations referable to irritable bowel syndrome with the presence of *Blastocystis* spp., [93, 94] its possible role in the pathogenesis of the syndrome is still debated. Since the 1990s, researchers have, reported that serum of patients with IBS show elevated levels of IgG antibodies for *Blastocystis* spp. compared to healthy controls [95, 96].

In the pathogenesis of IBS, a mechanism linked to a low degree of inflammation by means of chronic exposure to *Blastocystis* spp. antigen [97] has been proposed. Recently, the production by the ST7 strain of *Blastocystis* spp. of two cysteine-proteases able to induce rho kinase-mediated response on the intestinal epithelial barrier [69, 98].

Cysteine proteases induce pro-inlammatory cytokine interleukin-8 (IL-8) and degrade human immunoglobulin A (IgA) are important enzymes for host invasion and infection and are well recognized as virulence factors in pathogenic protozoa, too [132].

Extra-Intestinal Manifestations

Cutaneous Lesions

Infections from *Blastocystis* spp. have been indicated in numerous reports as the possible etiology of cutaneous lesions: acute or chronic rash [99], palmoplantar itching [100] or diffuse angioedema [101], with the finding of peripheral hypereosinophilia [102]. The stimulation, by parasite antigens, ofTh2 cells leads to the production of interleukins (IL-3, IL-4, IL- 5, IL-13) that mediate allergic IgE reactions [103]. It has also been postulated that antigens of *Blastocystis* spp. are able to activate the Complement cascade inducing the production of C3a and C5a that, because of their interaction with mastocytes and basophils, would activate the production of histamine with consequent cutaneous alterations [104]. An association has also been proposed between acute rash and the presence of *Blastocystis* spp. ST 3 in the amoeboid form, mediated by carbohydrates of the parasite membrane [105].

Other Extra-Intestinal Manifestations

A research group [106] has reported a case of severe hypoalbuminemia and anasarca with the finding of peripheral hypereosinophilia and the presence of *Blastocystis* spp. in the feces as the only parasite. The authors correlated the pathological manifestations with the presence of *Blastocystis* spp. based on the favorable response of the patient to the therapy with metronidazole and to the absence of other known causes. They also suggested that the conditions of the subject under observation, affected by a psychiatric pathology and in poor general condition, could have favored the severe manifestations attributed to *Blastocystis* spp. In the literature, there is only one previous report of a medical case similar to the previous described [107].

Another report in the literature describes the detection of *Blastocystis* spp. in synovial liquid drained from the knee joint of a patient affected by rheumatoid arthritis, in treatment with steroids and concomitant acute diarrhea. The presence of the cystic form of *Blastocystis* spp. was also detected with a parasitological examination of the feces. The administration of therapy with metronidazole has been correlated by some authors with the disappearance of diarrhea and the resolution of the articular phlogistic process [108]. The presence of *Blastocystis* spp. ST3 has recently been found in the contents of splenic cysts, in a 22 year old female patient complained of pain in the left hypochondrium [109].

High-risk Subjects

As in many parasitic infections (Cryptosporidium spp. Isospora spp., Cyclospora,), some groups of subjects appear more susceptible to contract infection with *Blastocystis* spp [110]. Over the last few years, the presence of the parasite has been frequently reported in subjects affected by cancer or with immune system compromised by HIV/AIDS.The prevalences reported in various studies range from about 10% [111, 112] to 13% [113], in subjects affected by cancer. Greater differences have been reported in studies on patients affected by AIDS. The prevalence, in the various studies, vary from 16% [114] to 38% [115]. A prevalence of 72.4% [34] in AIDS patients with diarrhea has been found. As regards the ST of *Blastocystis* spp. isolated from subjects affected by cancer or HIV/AIDS, the presence of ST 1, 2, 3, 4 has been reported. ST3 has been found with greater frequency, followed in order by ST4, ST1, ST2 [112]. A high incidence of isolation of *Blastocystis* is reported in renal transplantation in immunosuppressive therapy [116] and in children with nephrotic syndrome who are undergoing steroid therapy [73]. Subjects living in developing countries or coming from them, show a higher prevalence of *Blastocystis* spp. compared to the general population of developed countries [117, 118]. This has been correlated with poor hygiene, contact with animals, and consumption of contaminated food or water [117].

THERAPEUTIC APPROACH

The treatment of infections from *Blastocystis* spp., given the controversial pathogenesis, is carried out, in symptomatic patients, only when every other possible etiology has been excluded [119, 120]. In these cases, metronidazole is considered the therapy of choice [119, 121, 122]. Numerous trials have been made in the last few years on immune competent subjects affected by diarrhea with the finding of *Blastocystis* spp. in the parasitological examination of the feces.The drugs used in the treatment were: metronidazole [123, 124], cotrimoxazole [125] nitazoxanide [126]. The authors report a high percentage of eradication of the parasite and resolution of the intestinal symptoms. This would lead us to consider *Blastocystis* spp. responsible for the pathological manifestation [123]. The recent reports of therapeutic failure [122] would lead to the consideration of drug resistance of the parasite or, at least, a variable sensitivity of *Blastocystis* spp. to drugs [127]. Numerous other drugs have been used in the therapy of infections from *Blastocystis* spp., singularly or in association with metronidazole: cotrimoxazole [125], nitazoxanide [126, 128], paromomicine [129]. According to some authors these can be considered second-line drugs [73, 130], moreover, there have also been indications· of therapeutic failure [127]. Molecular identification of different ST has led to the possibility that the varying sensitivity to drugs is correlated with the ST [119].

Studies *in vitro* have shown different patterns of sensitivity of the parasite to drugs for the various ST [127, 131]. These studies would lead to consider metronidazole as only one of the possible therapeutic options and not necessarily a first line drug.

CONFLICT OF INTEREST

The authors confirm that they have no conflict of interest to declare for this publication.

ACKNOWLEDGEMENTS

The images are courtesy of Prof. Oliveri Salvatore.

REFERENCES

[1] Stenzel DJ, Boreham PF. *Blastocystis hominis* revisited. Clin Microbiol Rev 1996; 9(4): 563-84. [PMID: 8894352]

[2] König G, Müller HE. *Blastocystis hominis* in animals: incidence of four serogroups. Zentralbl Bakteriol 1997; 286(3): 435-40. [http://dx.doi.org/10.1016/S0934-8840(97)80105-3] [PMID: 9361389]

[3] Belova LM, Krylov MV. [The distribution of *Blastocystis* according to different systematic groups of hosts]. Parazitologia 1998; 32(3): 268-76. [PMID: 9702803]

[4] Yoshikawa H, Wu Z, Howe J, Hashimoto T, Geok-Choo N, Tan KS. Ultrastructural and phylogenetic studies on *Blastocystis* isolates from cockroaches. J Eukaryot Microbiol 2007; 54(1): 33-7. [http://dx.doi.org/10.1111/j.1550-7408.2006.00141.x] [PMID: 17300516]

[5] Finding of *Blastocystis sp.* in bivalves of the genus Donax. Rev Peru Biol 2007; 2.

[6] Sina M, Ad J, *et al.* The revised classification of eukaryotes. Eukaryot Microbiol 2013; 60: 321.

[7] Margulis L, Chapman MJ. Kingdoms and Domains: An Illustrated Guide to the Phyla of Life on Earth. 4th ed., Amsterdam Academic Press Elsevier 2009.

[8] Tan KS, Singh M, Yap EH. Recent advances in *Blastocystis hominis* research: hot spots in terra incognita. Int J Parasitol 2002; 32(7): 789-804. [http://dx.doi.org/10.1016/S0020-7519(02)00005-X] [PMID: 12062550]

[9] Sheehan DJ, Raucher BG, McKitrick JC. Association of Blastocystis hominis with signs and symptoms of human disease. J Clin Microbiol 1986; 24(4): 548-50. [PMID: 3771743]

[10] Wang KX, Li CP, Wang J, Cui YB. Epidemiological survey of *Blastocystis hominis* in Huainan City, Anhui Province, China. World J Gastroenterol 2002; 8(5): 928-32. [http://dx.doi.org/10.3748/wjg.v8.i5.928] [PMID: 12378644]

[11] Diaz E, Mondragon J, Ramirez E, Bernal R. Epidemiology and control of intestinal parasites with nitazoxanide in children in Mexico. Am J Trop Med Hyg 2003; 68(4): 384-5. [PMID: 12875284]

[12] Abe N, Wu Z, Yoshikawa H. Molecular characterization of *Blastocystis* isolates from primates. Vet Parasitol 2003; 113(3-4): 321-5. [http://dx.doi.org/10.1016/S0304-4017(03)00081-5] [PMID: 12719144]

[13] Yan Y, Su S, Lai R, *et al.* Genetic variability of *Blastocystis hominis* isolates in China. Parasitol Res 2006; 99(5): 597-601.
[http://dx.doi.org/10.1007/s00436-006-0186-z] [PMID: 16688468]

[14] Souppart L, Sanciu G, Cian A, *et al.* Molecular epidemiology of human *Blastocystis* isolates in France. Parasitol Res 2009; 105(2): 413-21.
[http://dx.doi.org/10.1007/s00436-009-1398-9] [PMID: 19290540]

[15] Alfellani M. The significance of '*Blastocystis*' in different hosts. PhD thesis, London School of Hygiene & Tropical Medicine 2012.

[16] Köksal F, Başlanti I, Samasti M. [A retrospective evaluation of the prevalence of intestinal parasites in Istanbul, Turkey]. Turkiye Parazitol Derg 2010; 34(3): 166-71.
[PMID: 20954117]

[17] Yaicharoen R, Ngrenngarmlert W, Wongjindanon N, Sripochang S, Kiatfuengfoo R. Infection of *Blastocystis hominis* in primary schoolchildren from Nakhon Pathom province, Thailand. Trop Biomed 2006; 23(1): 117-22.
[PMID: 17041560]

[18] Ostan I, Kilimcioğlu AA, Girginkardeşler N, Ozyurt BC, Limoncu ME, Ok UZ. Health inequities: lower socio-economic conditions and higher incidences of intestinal parasites. BMC Public Health 2007; 7: 342-9.
[http://dx.doi.org/10.1186/1471-2458-7-342] [PMID: 18042287]

[19] Rayan P, Verghese S, McDonnell PA. Geographical location and age affects the incidence of parasitic infestations in school children. Indian J Pathol Microbiol 2010; 53(3): 498-502.
[http://dx.doi.org/10.4103/0377-4929.68292] [PMID: 20699511]

[20] Paschke C, Apelt N, Fleischmann E, *et al.* Controlled study on enteropathogens in travellers returning from the tropics with and without diarrhoea. Clin Microbiol Infect 2011; 17(8): 1194-200.
[http://dx.doi.org/10.1111/j.1469-0691.2010.03414.x] [PMID: 21054662]

[21] Amin OM. Seasonal prevalence of intestinal parasites in the United States during 2000. Am J Trop Med Hyg 2002; 66(6): 799-803.
[PMID: 12224595]

[22] Minvielle MC, Pezzani BC, Cordoba MA, De Luca MM, Apezteguia MC, Basualdo JA. Epidemiological survey of Giardia spp. and *Blastocystis hominis* in an Argentinian rural community. Korean J Parasitol 2004; 42(3): 121-7.
[http://dx.doi.org/10.3347/kjp.2004.42.3.121] [PMID: 15381860]

[23] Nascimento SA, Moitinho MdaL. *Blastocystis hominis* and other intestinal parasites in a community of Pitanga City, Paraná State, Brazil. Rev Inst Med Trop Sao Paulo 2005; 47(4): 213-7.
[http://dx.doi.org/10.1590/S0036-46652005000400007] [PMID: 16138204]

[24] Al-Fellani MA, Khan AH, Al-Gazoui RM, Zaid MK, Al-Ferjani MA. Prevalence and Clinical Features of *Blastocystis hominis* Infection among Patients in Sebha, Libya. Sultan Qaboos Univ Med J 2007; 7(1): 35-40.
[PMID: 21654943]

[25] Londoño AL, Mejía S, Gómez-Marín JE. [Prevalence and risk factors associated with intestinal parasitism in preschool children from the urban area of Calarcá, Colombia]. Rev Salud Publica (Bogota) 2009; 11(1): 72-81.
[http://dx.doi.org/10.1590/S0124-00642009000100008] [PMID: 19721981]

[26] Aguiar JI, Gonçalves AQ, Sodré FC, *et al.* Intestinal protozoa and helminths among Terena Indians in the State of Mato Grosso do Sul: high prevalence of *Blastocystis hominis*. Rev Soc Bras Med Trop 2007; 40(6): 631-4.
[http://dx.doi.org/10.1590/S0037-86822007000600006] [PMID: 18200414]

[27] Noor Azian MY, San YM, Gan CC, *et al.* Prevalence of intestinal protozoa in an aborigine community in Pahang, Malaysia. Trop Biomed 2007; 24(1): 55-62.
[PMID: 17568378]

[28] Ashford RW, Atkinson EA. Epidemiology of *Blastocystis hominis* infection in Papua New Guinea: age-prevalence and associations with other parasites. Ann Trop Med Parasitol 1992; 86(2): 129-36.
[http://dx.doi.org/10.1080/00034983.1992.11812642] [PMID: 1417203]

[29] Torres P, Miranda JC, Flores L, *et al.* [Blastocystosis and other intestinal protozoan infections in human riverside communities of the Valdivia River basin, Chile]. Rev Inst Med Trop Sao Paulo 1992; 34(6): 557-64.
[http://dx.doi.org/10.1590/S0036-46651992000600010] [PMID: 1342125]

[30] Tan KS. *Blastocystis* in humans and animals: new insights using modern methodologies. Vet Parasitol 2004; 126(1-2): 121-44.
[http://dx.doi.org/10.1016/j.vetpar.2004.09.017] [PMID: 15567582]

[31] Parkar U, Traub RJ, Vitali S, *et al.* Molecular characterization of *Blastocystis* isolates from zoo animals and their animal-keepers. Vet Parasitol 2010; 169(1-2): 8-17.
[http://dx.doi.org/10.1016/j.vetpar.2009.12.032] [PMID: 20089360]

[32] Brites C, Barberino MG, Bastos MA, Silva N, Silva N. *Blastocystis hominis* as a potential cause of diarrhea in AIDS patients: a report of six cases in Bahia, Brazil. Braz J Infect Dis 1997; 1(2): 91-4.
[PMID: 11107245]

[33] Prasad KN, Nag VL, Dhole TN, Ayyagari A. Identification of enteric pathogens in HIV-positive patients with diarrhoea in northern India. J Health Popul Nutr 2000; 18(1): 23-6.
[PMID: 11014766]

[34] Kurniawan A, Karyadi T, Dwintasari SW, *et al.* Intestinal parasitic infections in HIV/AIDS patients presenting with diarrhoea in Jakarta, Indonesia. Trans R Soc Trop Med Hyg 2009; 103(9): 892-8.
[http://dx.doi.org/10.1016/j.trstmh.2009.02.017] [PMID: 19327806]

[35] Albrecht H, Stellbrink HJ, Koperski K, Greten H. *Blastocystis hominis* in human immunodeficiency virus-related diarrhea. Scand J Gastroenterol 1995; 30(9): 909-14.
[http://dx.doi.org/10.3109/00365529509101600] [PMID: 8578192]

[36] Tan TC, Suresh KG, Smith HV. Phenotypic and genotypic characterisation of *Blastocystis hominis* isolates implicates subtype 3 as a subtype with pathogenic potential. Parasitol Res 2008; 104(1): 85-93.
[http://dx.doi.org/10.1007/s00436-008-1163-5] [PMID: 18795333]

[37] Suresh K, Venilla GD, Tan TC, Rohela M. *In vivo* encystation of *Blastocystis hominis*. Parasitol Res 2009; 104(6): 1373-80.
[http://dx.doi.org/10.1007/s00436-009-1340-1] [PMID: 19238443]

[38] Zierdt CH. *Blastocystis hominis*past and future. Clin Microbiol Rev 1991; 4(1): 61-79.
[http://dx.doi.org/10.1128/CMR.4.1.61] [PMID: 2004348]

[39] Dunn LA, Boreham PF, Stenzel DJ. Ultrastructural variation of *Blastocystis hominis* stocks in culture. Int J Parasitol 1989; 19(1): 43-56.
[http://dx.doi.org/10.1016/0020-7519(89)90020-9] [PMID: 2707962]

[40] Yoshikawa H, Satoh J, Enose Y. Light and electron microscopic localization of lipids in *Blastocystis hominis*. J Electron Microsc (Tokyo) 1995; 44(2): 100-3.
[PMID: 7544390]

[41] Yoshikawa H, Kuwayama N, Enose Y. Histochemical detection of carbohydrates of *Blastocystis hominis*. J Eukaryot Microbiol 1995; 42(1): 70-4.
[http://dx.doi.org/10.1111/j.1550-7408.1995.tb01542.x] [PMID: 7537145]

[42] Kozubsky LE, Archelli S. Acta Bioquim Clin Latinoam 2010; 44: 371-6.

[43] Tan KS, Ibrahim M, Ng GC, *et al.* Exposure of *Blastocystis* species to a cytotoxic monoclonal antibody. Parasitol Res 2001; 87(7): 534-8.
[http://dx.doi.org/10.1007/s004360000365] [PMID: 11484849]

[44] Zaman V, Howe J, Ng M. Observations on the surface coat of *Blastocystis hominis*. Parasitol Res 1997; 83(7): 731-3.
[http://dx.doi.org/10.1007/s004360050329] [PMID: 9272568]

[45] Zaman V, Howe J, Ng M, Goh TK. Scanning electron microscopy of the surface coat of *Blastocystis hominis*. Parasitol Res 1999; 85(12): 974-6.
[http://dx.doi.org/10.1007/s004360050668] [PMID: 10599919]

[46] Cassidy MF, Stenzel DJ, Boreham PF. Electron microscopy of surface structures of Blastocystis sp. from different hosts. Parasitol Res 1994; 80(6): 505-11.
[http://dx.doi.org/10.1007/BF00932698] [PMID: 7809001]

[47] Tan SW, Ho LC, Moe KT, *et al.* Production and characterization of murine monoclonal antibodies to *Blastocystis hominis*. Int J Parasitol 1996; 26(4): 375-81.
[http://dx.doi.org/10.1016/0020-7519(96)00016-1] [PMID: 8773525]

[48] Zierdt CH, Rude WS, Bull BS. Protozoan characteristics of *Blastocystis hominis*. Am J Clin Pathol 1967; 48(5): 495-501.
[http://dx.doi.org/10.1093/ajcp/48.5.495] [PMID: 6058384]

[49] Tan HK, Zierdt CH. Ultrastructure of *Blastocystis hominis*. Z Parasitenkd 1973; 42(4): 315-24.
[http://dx.doi.org/10.1007/BF00328892] [PMID: 4360331]

[50] Singh M, Suresh K, Ho LC, Ng GC, Yap EH. Elucidation of the life cycle of the intestinal protozoan *Blastocystis hominis*. Parasitol Res 1995; 81(5): 446-50.
[http://dx.doi.org/10.1007/BF00931510] [PMID: 7501648]

[51] Moe KT, Singh M, Howe J, *et al.* Observations on the ultrastructure and viability of the cystic stage of *Blastocystis hominis* from human feces. Parasitol Res 1996; 82(5): 439-44.
[http://dx.doi.org/10.1007/s004360050142] [PMID: 8738284]

[52] Mehlhorn H. Blastocystis hominis, Brumpt 1912: are there different stages or species? Parasitol Res 1988; 74(4): 393-5.
[http://dx.doi.org/10.1007/BF00539464] [PMID: 3387411]

[53] Yoshikawa H. Genomic polymorphism among *Blastocystis* isolates and development of PCR-based identication of zoonotic isolates. J Eukaryot Microbiol 2003; 50(710–711): 383-5.

[54] Abou El Naga IF, Negm AY. Morphology, histochemistry and infectivity of *Blastocystis hominis* cyst. J Egypt Soc Parasitol 2001; 31(2): 627-35.
[PMID: 11478461]

[55] Yoshikawa H, Yoshida K, Nakajima A, Yamanari K, Iwatani S. Parasitol Res 2004; 101: 43-51.

[56] Zaki M, Zaman V, Sheikh NA. Resistance of *blastocystis hominis* cysts to chlorine. J Pak Med Assoc 1996; 46(8): 178-9.
[PMID: 8936976]

[57] Suresh K, Howe J, Chong SY, *et al.* Ultrastructural changes during *in vitro* encystment of *Blastocystis hominis.* Parasitol Res 1994; 80(4): 327-35. [http://dx.doi.org/10.1007/BF02351875] [PMID: 8073021]

[58] Böhm-Gloning B, Knobloch J, Walderich B. Five subgroups of Blastocystis hominis from symptomatic and asymptomatic patients revealed by restriction site analysis of PCR-amplified 16S-like rDNA. Trop Med Int Health 1997; 2(8): 771-8. [http://dx.doi.org/10.1046/j.1365-3156.1997.d01-383.x] [PMID: 9294547]

[59] Rivera WL, Tan MA. Molecular characterization of *Blastocystis* isolates in the Philippines by riboprinting. Parasitol Res 2005; 96(4): 253-7. [http://dx.doi.org/10.1007/s00436-005-1344-4] [PMID: 15886995]

[60] Abe N. Molecular and phylogenetic analysis of *Blastocystis* isolates from various hosts. Vet Parasitol 2004; 120(3): 235-42. [http://dx.doi.org/10.1016/j.vetpar.2004.01.003] [PMID: 15041098]

[61] Yoshikawa H, Wu Z, Kimata I, *et al.* Polymerase chain reaction-based genotype classification among human *Blastocystis hominis* populations isolated from different countries. Parasitol Res 2004; 92(1): 22-9. [http://dx.doi.org/10.1007/s00436-003-0995-2] [PMID: 14598169]

[62] Abe N, Wu Z, Yoshikawa H. Zoonotic genotypes of Blastocystis hominis detected in cattle and pigs by PCR with diagnostic primers and restriction fragment length polymorphism analysis of the small subunit ribosomal RNA gene. Parasitol Res 2003; 90(2): 124-8. [PMID: 12756546]

[63] Yan Y, Su S, Lai R, *et al.* Genetic variability of *Blastocystis hominis* isolates in China. Parasitol Res 2006; 99(5): 597-601. [http://dx.doi.org/10.1007/s00436-006-0186-z] [PMID: 16688468]

[64] Yoshikawa H, Nagono I, Yap EH, Singh M, Takahashi Y. DNA polymorphism revealed by arbitrary primers polymerase chain reaction among *Blastocystis* strains isolated from humans, a chicken, and a reptile. J Eukaryot Microbiol 1996; 43(2): 127-30. [http://dx.doi.org/10.1111/j.1550-7408.1996.tb04492.x] [PMID: 8720942]

[65] Tan TC, Suresh KG, Thong KL, Smith HV. PCR fingerprinting of *Blastocystis* isolated from symptomatic and asymptomatic human hosts. Parasitol Res 2006; 99(4): 459-65. [http://dx.doi.org/10.1007/s00436-006-0177-0] [PMID: 16628457]

[66] Stensvold CR. *Blastocystis*: Genetic diversity and molecular methods for diagnosis and epidemiology. Trop Parasitol 2013; 3(1): 26-34. [http://dx.doi.org/10.4103/2229-5070.113896] [PMID: 23961438]

[67] Malheiros AF, Stensvold CR, Clark CG, Braga GB, Shaw JJ. Short report: Molecular characterization of *Blastocystis* obtained from members of the indigenous Tapirapé ethnic group from the Brazilian Amazon region, Brazil. Am J Trop Med Hyg 2011; 85(6): 1050-3. [http://dx.doi.org/10.4269/ajtmh.2011.11-0481] [PMID: 22144442]

[68] Alfellani MA, Stensvold CR, Vidal-Lapiedra A, Onuoha ES, Fagbenro-Beyioku AF, Clark CG. Variable geographic distribution of *Blastocystis* subtypes and its potential implications. Acta Trop 2013; 126(1): 11-8. [http://dx.doi.org/10.1016/j.actatropica.2012.12.011] [PMID: 23290980]

[69] Wawrzyniak I, Texier C, Poirier P, *et al.* Characterization of two cysteine proteases secreted by *Blastocystis* ST7, a human intestinal parasite. Parasitol Int 2012; 61(3): 437-42. [http://dx.doi.org/10.1016/j.parint.2012.02.007] [PMID: 22402106]

[70] Govind SK, Khairul AA, Smith HV. Multiple reproductive processes in *Blastocystis*. Trends Parasitol 2002; 18(12): 528.
[http://dx.doi.org/10.1016/S1471-4922(02)02402-9] [PMID: 12482534]

[71] Moe KT, Singh M, Howe J, *et al.* Experimental *Blastocystis hominis* infection in laboratory mice. Parasitol Res 1997; 83(4): 319-25.
[http://dx.doi.org/10.1007/s004360050256] [PMID: 9134552]

[72] Tan K. New Insights on Classification, Identification, and Clinical Relevance of Blastocystis spp. Clinic Microbiolgy Reviews 2008; pp. 639-65.

[73] Nimri LF. Evidence of an epidemic of *Blastocystis hominis* infections in preschool children in northern Jordan. J Clin Microbiol 1993; 31(10): 2706-8.
[PMID: 8253970]

[74] Khalifa AM, El Temsahy MM, Abou El Naga IF. Effect of ozone on the viability of some protozoa in drinking water. J Egypt Soc Parasitol 2001; 31(2): 603-16.
[PMID: 11478459]

[75] Waikagul J, Krudsood S, Radomyos P, *et al.* A cross-sectional study of intestinal parasitic infections among schoolchildren in Nan Province, Northern Thailand. Southeast Asian J Trop Med Public Health 2002; 33(2): 218-23.
[PMID: 12236415]

[76] Leelayoova S, Siripattanapipong S, Thathaisong U, *et al.* Drinking water: a possible source of *Blastocystis* spp. subtype 1 infection in schoolchildren of a rural community in central Thailand. Am J Trop Med Hyg 2008; 79(3): 401-6.
[PMID: 18784233]

[77] Lee LI, Chye TT, Karmacharya BM, Govind SK. *Blastocystis* sp.: waterborne zoonotic organism, a possibility? Parasit Vectors 2012; 5: 130.
[http://dx.doi.org/10.1186/1756-3305-5-130] [PMID: 22741573]

[78] Tanizaki A, Yoshikawa H, Iwatani S, Kimata I. Infectivity of *Blastocystis* isolates from chickens, quails and geese in chickens. Parasitol Res 2005; 96(1): 57-61.
[http://dx.doi.org/10.1007/s00436-005-1326-6] [PMID: 15772866]

[79] Stensvold CR, Arendrup MC, Jespersgaard C, Mølbak K, Nielsen HV. Detecting *Blastocystis* using parasitologic and DNA-based methods: a comparative study. Diagn Microbiol Infect Dis 2007; 59(3): 303-7.
[http://dx.doi.org/10.1016/j.diagmicrobio.2007.06.003] [PMID: 17913433]

[80] Miller RA, Minshew BH. *Blastocystis hominis*: an organism in search of a disease. Rev Infect Dis 1988; 10(5): 930-8.
[http://dx.doi.org/10.1093/clinids/10.5.930] [PMID: 3055191]

[81] Zaman V, Khan KZ. A concentration technique for obtaining viable cysts of *blastocystis hominis* from faeces. J Pak Med Assoc 1994; 44(9): 220-1.
[PMID: 7799511]

[82] Leelayoova S, Rangsin R, Taamasri P, Naaglor T, Thathaisong U, Mungthin M. Evidence of waterborne transmission of *Blastocystis hominis*. Am J Trop Med Hyg 2004; 70(6): 658-62.
[PMID: 15211009]

[83] Yan Y, Su S, Ye J, *et al.* *Blastocystis* sp. subtype 5: a possibly zoonotic genotype. Parasitol Res 2007; 101(6): 1527-32.
[http://dx.doi.org/10.1007/s00436-007-0672-y] [PMID: 17665214]

[84] Taamasri P, Mungthin M, Rangsin R, Tongupprakarn B, Areekul W. Transmission of intestinal blastocystosis related to the quality of drinking water in Southeast Asian. J Trop Med Public Health 2000; 31: 112-7.

[85] Stensvold CR, Lewis HC, Hammerum AM, *et al.* Blastocystis: unravelling potential risk factors and clinical significance of a common but neglected parasite. Epidemiol Infect 2009; 137(11): 1655-63. [http://dx.doi.org/10.1017/S0950268809002672] [PMID: 19393117]

[86] Kaya S, Cetin ES, Aridoğan BC, Arikan S, Demirci M. Pathogenicity of *Blastocystis hominis*, a clinical reevaluation. Turkiye Parazitol Derg 2007; 31(3): 184-7. [PMID: 17918055]

[87] Moghaddam DD, Ghadirian E, Azami M. Blastocystis hominis and the evaluation of efficacy of metronidazole and trimethoprim/sulfamethoxazole. Parasitol Res 2005; 96(4): 273-5. [http://dx.doi.org/10.1007/s00436-005-1363-1] [PMID: 15915364]

[88] Rajamanikam A, Govind SK. Amoebic forms of Blastocystis spp. - evidence for a pathogenic role. Parasit Vectors 2013; 6(1): 295.

[89] Hu KC, Lin CC, Wang TE, Liu CY, Chen MJ, Chang WH. Amoebic liver abscess or is it? Gut 2008; 57(5): 627-683, 683. [http://dx.doi.org/10.1136/gut.2006.119149] [PMID: 18408101]

[90] Vassalos CM, Spanakos G, Vassalou E, Papadopoulou C, Vakalis N. Differences in clinical significance and morphologic features of Blastocystis sp subtype 3. Am J Clin Pathol 2010; 133(2): 251-8. [http://dx.doi.org/10.1309/AJCPDOWQSL6E8DMN] [PMID: 20093234]

[91] Hussein EM, Hussein AM, Eida MM, Atwa MM. Pathophysiological variability of different genotypes of human *Blastocystis hominis* Egyptian isolates in experimentally infected rats. Parasitol Res 2008; 102(5): 853-60. [http://dx.doi.org/10.1007/s00436-007-0833-z] [PMID: 18193282]

[92] Piche T, Barbara G, Aubert P, *et al.* Impaired intestinal barrier integrity in the colon of patients with irritable bowel syndrome: involvement of soluble mediators. Gut 2009; 58(2): 196-201. [http://dx.doi.org/10.1136/gut.2007.140806] [PMID: 18824556]

[93] Cekin AH, Cekin Y, Adakan Y, Tasdemir E, Koclar FG, Yolcular BO. Blastocystosis in patients with gastrointestinal symptoms: a case-control study. BMC Gastroenterol 2012; 12: 122. [http://dx.doi.org/10.1186/1471-230X-12-122] [PMID: 22963003]

[94] Poirier P, Wawrzyniak I, Vivarès CP, Delbac F, El Alaoui H. New insights into *Blastocystis* spp.: a potential link with irritable bowel syndrome. PLoS Pathog 2012; 8(3): e1002545. [http://dx.doi.org/10.1371/journal.ppat.1002545] [PMID: 22438803]

[95] Hussain R, Jaferi W, Zuberi S, *et al.* Significantly increased IgG2 subclass antibody levels to *Blastocystis hominis* in patients with irritable bowel syndrome. Am J Trop Med Hyg 1997; 56(3): 301-6. [PMID: 9129532]

[96] Mahmoud MS, Saleh WA. Secretory and humoral antibody responses to *Blastocystis hominis* in symptomatic and asymptomatic human infections. J Egypt Soc Parasitol 2003; 33(1): 13-30. [PMID: 12739797]

[97] Stark D, van Hal S, Marriott D, Ellis J, Harkness J. Irritable bowel syndrome: a review on the role of intestinal protozoa and the importance of their detection and diagnosis. Int J Parasitol 2007; 37(1): 11-20. [http://dx.doi.org/10.1016/j.ijpara.2006.09.009] [PMID: 17070814]

[98] Mirza H, Wu Z, Kidwai F, Tan KS. A metronidazole-resistant isolate of *Blastocystis* spp. is susceptible to nitric oxide and downregulates intestinal epithelial inducible nitric oxide synthase by a novel parasite survival mechanism. Infect Immun 2011; 79(12): 5019-26. [http://dx.doi.org/10.1128/IAI.05632-11] [PMID: 21930763]

[99] Dina AM. J Health Sci 2012.

[100] Kick G, Rueff F, Przybilla B. Palmoplantar pruritus subsiding after *Blastocystis hominis* eradication. Acta Derm Venereol 2002; 82(1): 60. [http://dx.doi.org/10.1080/000155502753600948] [PMID: 12013204]

[101] Micheloud D, Jensen J, Fernandez-Cruz E, Carbone J. [Chronic angioedema and blastocystis hominis infection]. Rev Gastroenterol Peru 2007; 27(2): 191-3. [PMID: 17712397]

[102] Valsecchi R, Leghissa P, Greco V. Cutaneous lesions in *Blastocystis hominis* infection. Acta Derm Venereol 2004; 84(4): 322-3. [http://dx.doi.org/10.1080/00015550410025949] [PMID: 15339085]

[103] Pasqui AL, Savini E, Saletti M, Guzzo C, Puccetti L, Auteri A. Chronic urticaria and *blastocystis hominis* infection: a case report. Eur Rev Med Pharmacol Sci 2004; 8(3): 117-20. [PMID: 15368795]

[104] Kozubsky LE, Archelli S. Some considerations about *Blastocystis* sp, a controversial parasite. Acta Bioquim Clin Latinoam 2010; 44(3): 371-6.

[105] Katsarou-Katsari A, Vassalos CM, Tzanetou K, Spanakos G, Papadopoulou C, Vakalis N. Acute urticaria associated with amoeboid forms of Blastocystis sp. subtype 3. Acta Derm Venereol 2008; 88(1): 80-1. [http://dx.doi.org/10.2340/00015555-0338] [PMID: 18176765]

[106] Nassir E, Awad J, Abel AB, Khoury J, Shay M, Lejbkowicz F. *Blastocystis hominis* as a cause of hypoalbuminemia and anasarca. Eur J Clin Microbiol Infect Dis 2004; 23(5): 399-402. [http://dx.doi.org/10.1007/s10096-004-1122-8] [PMID: 15112065]

[107] Levy Y, George J, Shoenfeld Y. Severe *Blastocystis hominis* in an elderly man. J Infect 1996; 33(1): 57-9. [http://dx.doi.org/10.1016/S0163-4453(96)92833-9] [PMID: 8842998]

[108] Lee MG, *et al.* A survey of *Blastocystis* in domestic chickens. Ann Rheum Dis 1990; 49: 192-3. [http://dx.doi.org/10.1136/ard.49.3.192] [PMID: 2322029]

[109] Santos HL, Sodré FC, de Macedo HW. *Blastocystis* sp. in splenic cysts: causative agent or accidental association? A unique case report. Parasit Vectors 2014; 7: 207. [http://dx.doi.org/10.1186/1756-3305-7-207] [PMID: 24779380]

[110] Cirioni O, Giacometti A, Drenaggi D, Ancarani F, Scalise G. Prevalence and clinical relevance of *Blastocystis hominis* in diverse patient cohorts. Eur J Epidemiol 1999; 15(4): 389-93. [http://dx.doi.org/10.1023/A:1007551218671] [PMID: 10414382]

[111] Koltas S, Ozcan K, Tannverdi S, Paydas S, Baslamish F. The prevalence of *Blastocystis hominis* in immunosuppressed patients. Ann Med Sci 1999; 8: 117-9.

[112] Tan TC, Ong SC, Suresh KG. Genetic variability of *Blastocystis* sp. isolates obtained from cancer and HIV/AIDS patients. Parasitol Res 2009; 105(5): 1283-6. [http://dx.doi.org/10.1007/s00436-009-1551-5] [PMID: 19603182]

[113] Taşova Y, Sahin B, Koltaş S, Paydaş S. Clinical significance and frequency of Blastocystis hominis in Turkish patients with hematological malignancy. Acta Med Okayama 2000; 54(3): 133-6. [PMID: 10925738]

[114] Junod C. [*Blastocystis hominis*: a common commensal in the colon. Study of prevalence in different populations of Paris]. Presse Med 1995; 24(36): 1684-8.
[PMID: 8545396]

[115] Albrecht H, Stellbrink HJ, Koperski K, Greten H. *Blastocystis hominis* in human immunodeficiency virus-related diarrhea. Scand J Gastroenterol 1995; 30(9): 909-14.
[http://dx.doi.org/10.3109/00365529509101600] [PMID: 8578192]

[116] Noureldin MS, Shaltout AA, El Hamshary EM, Ali ME. Opportunistic intestinal protozoal infections in immunocompromised children. J Egypt Soc Parasitol 1999; 29(3): 951-61.
[PMID: 12561933]

[117] Cheng HS, Haung ZF, Lan WH, Kuo TC, Shin JW. Epidemiology of *Blastocystis hominis* and other intestinal parasites in a Vietnamese female immigrant population in southern Taiwan. Kaohsiung J Med Sci 2006; 22(4): 166-70.
[http://dx.doi.org/10.1016/S1607-551X(09)70302-X] [PMID: 16679297]

[118] Sohail MR, Fischer PR. *Blastocystis hominis* and travelers. Travel Med Infect Dis 2005; 3(1): 33-8.
[http://dx.doi.org/10.1016/j.tmaid.2004.06.001] [PMID: 17292002]

[119] Stensvold CR, Smith HV, Nagel R, Olsen KE, Traub RJ. Eradication of *Blastocystis* carriage with antimicrobials: reality or delusion? J Clin Gastroenterol 2010; 44(2): 85-90.
[http://dx.doi.org/10.1097/MCG.0b013e3181bb86ba] [PMID: 19834337]

[120] Coyle CM, Varughese J, Weiss LM, Tanowitz HB. *Blastocystis*: to treat or not to treat.... Clin Infect Dis 2012; 54(1): 105-10.
[http://dx.doi.org/10.1093/cid/cir810] [PMID: 22075794]

[121] Cassano N, Scoppio BM, Loviglio MC, Vena GA. Remission of delayed pressure urticaria after eradication of *Blastocystis hominis.* Acta Derm Venereol 2005; 85(4): 357-8.
[PMID: 16191864]

[122] Moghaddam DD, Ghadirian E, Azami M. *Blastocystis hominis* and the evaluation of efficacy of metronidazole and trimethoprim/sulfamethoxazole. Parasitol Res 2005; 96(4): 273-5.
[http://dx.doi.org/10.1007/s00436-005-1363-1] [PMID: 15915364]

[123] Nigro L, Larocca L, Massarelli L, *et al.* A placebo-controlled treatment trial of *Blastocystis hominis* infection with metronidazole. J Travel Med 2003; 10(2): 128-30.
[http://dx.doi.org/10.2310/7060.2003.31714] [PMID: 12650658]

[124] Kaya S, Cetin ES, Aridoğan BC, Arikan S, Demirci M. Pathogenicity of *Blastocystis hominis*, a clinical reevaluation. Turkiye Parazitol Derg 2007; 31(3): 184-7.
[PMID: 17918055]

[125] Ok UZ, Girginkardeşler N, Balcioğlu C, Ertan P, Pirildar T, Kilimcioğlu AA. Effect of trimethoprim-sulfamethaxazole in *Blastocystis hominis* infection. Am J Gastroenterol 1999; 94(11): 3245-7.
[http://dx.doi.org/10.1111/j.1572-0241.1999.01529.x] [PMID: 10566723]

[126] Rossignol JF, Kabil SM, Said M, Samir H, Younis AM. Effect of nitazoxanide in persistent diarrhea and enteritis associated with *Blastocystis hominis.* Clin Gastroenterol Hepatol 2005; 3(10): 987-91.
[http://dx.doi.org/10.1016/S1542-3565(05)00427-1] [PMID: 16234044]

[127] Mirza H, Teo JD, Upcroft J, Tan KS. A rapid, high-throughput viability assay for Blastocystis spp. reveals metronidazole resistance and extensive subtype-dependent variations in drug susceptibilities. Antimicrob Agents Chemother 2011; 55(2): 637-48.
[http://dx.doi.org/10.1128/AAC.00900-10] [PMID: 21098237]

[128] Cimerman S, Ladeira MC, Iuliano WA. [Blastocystosis: nitazoxanide as a new therapeutic option]. Rev Soc Bras Med Trop 2003; 36(3): 415-7.
[PMID: 12908045]

[129] van Hellemond JJ, Molhoek N, Koelewijn R, Wismans PJ, van Genderen PJ. Is paromomycin the drug of choice for eradication of *Blastocystis* in adults? J Infect Chemother 2013; 19(3): 545-8. [http://dx.doi.org/10.1007/s10156-012-0496-2] [PMID: 23053509]

[130] Wawrzyniak I, Poirier P, Viscogliosi E, *et al.* Blastocystis, an unrecognized parasite: an overview of pathogenesis and diagnosis. Ther Adv Infect Dis 2013; 1(5): 167-78. [http://dx.doi.org/10.1177/2049936113504754] [PMID: 25165551]

[131] Mirza H, Wu Z, Kidwai F, Tan KS. A metronidazole-resistant isolate of *Blastocystis* spp. is susceptible to nitric oxide and downregulates intestinal epithelial inducible nitric oxide synthase by a novel parasite survival mechanism. Infect Immun 2011; 79(12): 5019-26. [http://dx.doi.org/10.1128/IAI.05632-11] [PMID: 21930763]

[132] Mirza H, Wu Z, Teo JD, Tan KS. Statin pleiotropy prevents rho kinase-mediated intestinal epithelial barrier compromise induced by *Blastocystis* cysteine proteases. Cell Microbiol 2012; 14(9): 1474-84.

CHAPTER 2

Cryptosporidium and Cryptosporidiosis

Simone M. Cacciò* and **Fabio Tosini**

Department of Infectious, Parasitic and Immunomediated Diseases, Istituto Superiore di Sanità, Rome, Italy

Abstract: *Cryptosporidium* spp. has emerged as an important cause of diarrheal disease in humans and animals, with a global distribution. The epidemiology of cryptosporidiosis is complex, and transmission occurs through both direct (human-to-human, animal-to-human) and indirect (through water and food) routes. In humans, the majority of infections are caused by *C. hominis* and *C. parvum*, but a number of other species have been recognized as human pathogens. The burden of disease is particularly high among young children in developing regions of the world, where cryptosporidiosis is one of the major causes of moderate-to-severe diarrhea and is associated with an increased risk of death. In developed countries, contamination of drinking water has caused several large waterborne outbreaks. Effective drug treatments and vaccines are not yet available, but the partial immunity after exposure suggests the potential for developing vaccines. Routine diagnostic methods for *Cryptosporidium* often have a low sensitivity, and those based on antigen or DNA detection, that greatly improve sensitivity, are underused. Recent advances in next-generation sequencing techniques will significantly improve our knowledge of the transmission of *Cryptosporidium*. However, increased funding will be essential to combat this important pathogen.

Keywords: Cryptosporidiosis, *Cryptosporidium*, Diagnostic methods, Epidemiology, Genomics, Genotyping, Human infection, Immunology, Intestinal disease, Molecular epidemiology, Pathogenesis, Risk factors, Transmission routes, Treatment.

INTRODUCTION

History

The first description of the parasite dates back to more than a century ago, with two publications by Ernest Edward Tyzzer, a medical parasitologist at Harvard University in Boston. Tyzzer gave the name *Cryptosporidium* because of the

* **Corresponding author Simone M. Cacciò:** Department of Infectious, Parasitic and Immunomediated Diseases, Istituto Superiore di Sanità, Rome, Italy; Tel: +39 06 4990 3016; Fax: +39 06 4990 3561; E-mail: simone.caccio@iss.it

Fabrizio Bruschi (Ed.)

absence of sporocysts within the oocysts, a characteristic of other coccidians. He first described the type species, *C. muris*, from the gastric glands of laboratory mice [1] and provided a more complete description of the life cycle in a subsequent publication [2]. Soon thereafter, Tyzzer described a second species, *C. parvum*, also from the laboratory mice, which differed from the type species because infection occurred in the small intestine and not in the stomach, and because of the smaller size of the oocysts [3]. Remarkably, these first three publications have defined most of what we currently know about the biology and life history of *C. muris* and *C. parvum* [4].

Research on *Cryptosporidium* then entered a long period of oblivion, and it was not until 1955 that its pathogenicity was demonstrated in turkeys [5], and, some twenty years later, in livestock [6]. In 1976, the first cases of human cryptosporidiosis were reported by two groups, and involved either children or immunosuppressed adults [7, 8].

Two events stand out for having significantly impacted our awareness of these parasites and the disease they cause. First, the HIV epidemic and the emergence of cryptosporidiosis as a potentially severe opportunistic infection in people living with AIDS was significant, not only because of the clinical implications, but because it motivated a substantial research effort and a desire to better understand these parasites [9]. A second notable event was the waterborne cryptosporidiosis outbreak in Milwaukee in 1993, which is still the largest waterborne outbreak ever recorded [10]. The importance of this outbreak is illustrated by the fact that the article reporting on the epidemiological investigation of the outbreak was cited over 600 times, more than any other paper dealing with this parasite.

Taxonomy

The taxonomy of the genus *Cryptosporidium*, like is the case for many other protozoa, is still in a state of flux. The genus *Cryptosporidium* belongs to the phylum Apicomplexa, but the exact position within the phylum has been a matter of debate. Traditionally, *Cryptosporidium* has been placed within the coccidia, since it shares with *Eimeria*, *Isospora* and *Cyclospora* the ability to develop in the gastrointestinal tract and to complete the entire life cycle within a single host. However, *Cryptosporidium* shows several peculiarities that separate it from any other coccidian, including its location within the host cell (intracellular, but extracytoplasmic), its ability to form a feeder organelle that facilitates the uptake of nutrients from the host cell, the presence of two morpho-functional types of oocysts (thick-walled and thin-walled), and its insensitivity to all anti-coccidial agents tested so far.

Besides these observations, molecular phylogenetic studies have challenged the inclusion of *Cryptosporidium* within the coccidia. Indeed, analysis of ribosomal and tubulin genes showed that *Cryptosporidium* is more closely related to the gregarines, a basal group within the phylum Apicomplexa [11], and this phylogenetic association was supported by genome comparison of *Cryptosporidium* and the gregarine *Ascogregarina taiwanensis* [12]. A clearer understanding of the correct taxonomic placement of the genus *Cryptosporidium* will probably require further genomic studies of gregarines.

At the species level, inherent difficulties have hampered the definition of an accurate taxonomy, not least the lack of consensus as to which definition of species can be applied to organisms like *Cryptosporidium*. However, it has been proposed that the naming of new species must be based on: a) morphometric studies of oocysts; b) genetic characterization, with sequences deposited in public databases; c) demonstration of natural and, whenever feasible, at least some experimental host specificity; and d) compliance with International Code of Zoological Nomenclature [13].

As a result of intense research, at present 27 species are regarded as valid; of these, 19 are from mammals, 3 from birds, 1 from amphibians, 2 from reptiles, and two from fish [14]. Besides species, there are at least 40 genotypes of still undefined taxonomic status. The majority of human infections are caused by two species, *C. hominis* and *C. parvum*, but many other species and genotypes have been associated with human cryptosporidiosis, although with a lower prevalence [15]. Table **1** lists the species/genotypes currently regarded as human pathogens.

Table 1. List of the *Cryptosporidium* species and genotypes currently recognized as human pathogens. Original references of species description can be found in [15].

Species or Genotypes	Main Hosts	Public Health Relevance
C. hominis	Human	High
C. parvum	Cattle, human	High
C. meleagridis	Turkey	Moderate
C. felis	Cat	Moderate
C. canis	Dog	Moderate
C. cuniculus	Rabbit	Moderate
C. ubiquitum	Sheep, wildlife	Moderate
C. viatorum	Human	Moderate
C. muris	Rodent	Minor
C. scrofarum	Pig	Minor

(Table 1) contd.....

Species or Genotypes	Main Hosts	Public Health Relevance
C. suis	Pig	Minor
C. andersoni	Cattle	Minor
C. bovis	Cattle	Minor
C. tyzzeri	Mouse	Minor
C. erinacei	Hedgehog	Minor
C. fayeri	Marsupial	Minor
Horse genotype	Horse	Minor
Skunk genotype	Skunk	Minor
Chipmunk genotype I	Chipmunk	Minor

Morphology and Life Cycle

The life cycle of *Cryptosporidium*, similar to that of other Apicomplexa, is complex, and involves both asexual and sexual phases [16] (Fig. **1**). The cycle starts by ingestion of infectious oocysts by a susceptible host. Once inside the body, infection begins with the release of four infectious sporozoites that, through an opening of the oocyst wall along a suture at one pole, leave the oocyst. This step is thought to be triggered by exposure to reducing conditions in the stomach and subsequently to pancreatic enzymes and bile salts in the small intestine. The sporozoites are motile stages that actively contact and ultimately invade the host cell; their structure is schematically depicted in Fig. (**2**). During the invasion process, the secretory organelles that comprise the apical complex (the single rhoptry, the micronemes and the dense granules; Fig. **2**) discharge their content in a coordinated manner to facilitate the invasion. The attached sporozoites become oval or spherical and vacuoles form near the anterior end. These vacuoles cluster together and surround the parasite, and then fuse with the membrane of the host cell to form the host-parasite interface. The parasite is thus contained into a parasitophorous vacuole, and during this process, a structure referred to as a feeder organelle matures from material released from the apical end of the sporozoite. The presence of the parasitophorous vacuole and of the feeder organelle keeps the parasite intracellular but extracytoplasmic, a feature unique to *Cryptosporidium*.

The internalized sporozoite, called a trophozoite, undergoes asexual multiplication (merogony) to generate meronts. In *C. parvum*, two type of metonts (Type I and II) are produced. Type I meronts have six or eight nuclei, each of which becomes incorporated into a merozoite, a stage similar to the sporozoite. The mature merozoite leaves the meront to infect another host cell, and develop into another Type I or into a Type II meront. Type II meronts produce four merozoites.

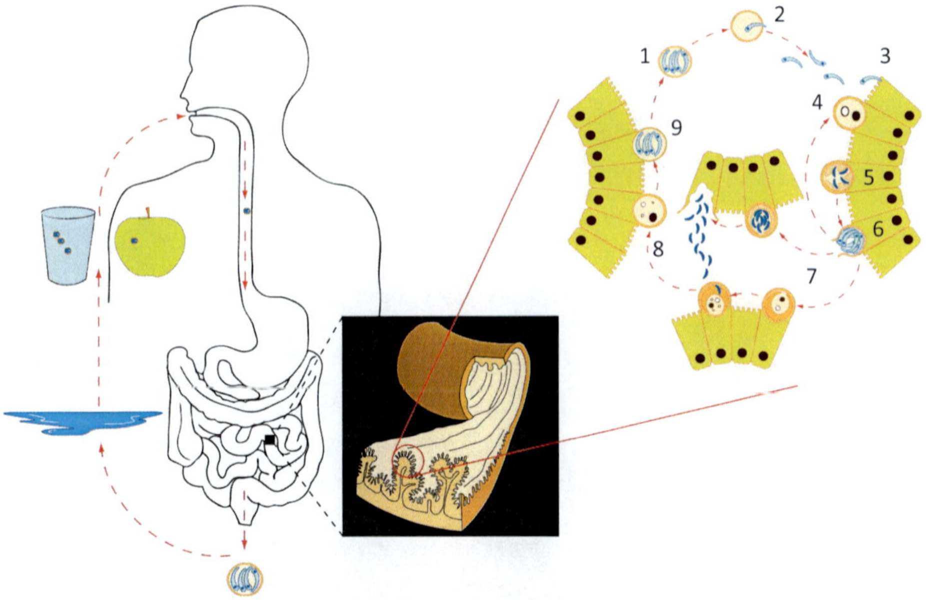

Fig. (1). Schematic representation of the life cycle of *Cryptosporidium*. After ingestion of oocysts (1), sporozoites emerge (2) and actively invade (3) host cells. The intracellular stage (4) of the parasite undergoes asexual multiplication to generate meronts (5,6). The sexual phase of the cycle involves the generation of macro- and micro-gamonts (7,8) and culminates with the production of new oocysts (9) that are passed with the feces of the host.

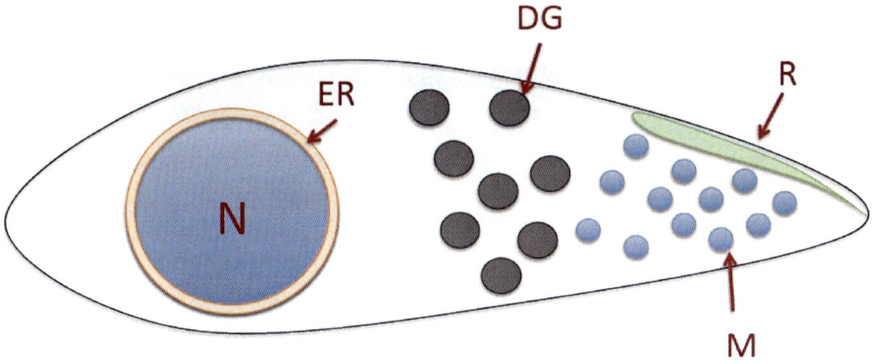

Fig. (2). Schematic representation of the invasive stage (sporozoite) of *Cryptosporidium*. The main cellular elements are indicated as follows: N, nucleus, R, rhoptry, M, micronemes, DG, dense granules, ER, endoplasmic reticulum.

The sexual stage is initiated by merozoites from Type II meronts that differentiate into either a male stage (microgamont) or a female stage (macrogamont). The microgamont becomes multinucleated and each nucleus is incorporated in the equivalent of a sperm cell, called a microgamete. The macrogamont remains uninucleate, representing the equivalent of an ovum. The exact mechanism by which microgametes detect macrogamonts is unknown, but penetration of the host cell membrane and the macrogamont membrane is necessary for fertilization to occur. Fertilized macrogamonts develop into oocysts and during the process, undergo meioisis, and generate four haploid sporozoites. *Cryptosporidium* can produce two types of oocysts: one with a thick wall, the environmentally resistant stage shed with the host feces, and another with a thin wall, which is responsible for autoinfection, a process in which oocysts excyst within the host to reinitiate the parasite life cycle.

GLOBAL EPIDEMIOLOGY AND RISK FACTORS

Parasites of the genus *Cryptosporidium* are ubiquitous, and the existence of both direct (human-to-human and animal-to-human) and indirect (through contaminated water and food) transmission routes accounts for the complex epidemiology of human cryptosporidiosis. This subject has been extensively reviewed in recent years [17, 18]; thus, in this section, we will briefly summarize some key elements.

Identity of the Species Causing Infection in Humans

As mentioned above, many *Cryptosporidium* species and genotypes are associated with human infection and only molecular methods allow their reliable identification. Therefore, our understanding of the biology and epidemiology of the parasite has been significantly improved with the development of molecular tools [19, 20]. Broadly, tools can be divided into those that are used to detect and differentiate *Cryptosporidium* species (genotyping tools), and those that allow subspecies variants to be identified (subtyping tools).

The application of genotyping tools to a large number of isolates collected in many regions of the world has demonstrated that about 90% of the infections in humans are due to *C. hominis* and *C. parvum*. The former is essentially adapted to the human host, although it has been detected in a few animals, whereas the latter is able to infect a number of mammals, is of particular importance in young ruminants, and has a well-established zoonotic potential [21].

The distribution of these two *Cryptosporidium* species varies geographically. In the United Kingdom, in other European countries and in New Zealand, *C. parvum* is responsible for slightly more infections than *C. hominis* [22]. In the Middle East, *C. parvum* is the dominant species in humans. In contrast, *C. hominis* is responsible for more infections than *C. parvum* in the United States, Australia,

China and Japan, as well as in most developing countries [17]. Other species play a role in human cryptosporidiosis, with *C. meleagridis*, *C. ubiquitum* and *C. cuniculus* being the most important [23]. Of note, infection with less common species is not limited to immunocompromised people, as originally believed, and, in specific settings, their prevalence can be as high as that of *C. parvum* [23].

There is high intra-species genetic variability, which is useful for the characterization of isolates to the level of subtypes, allowing finer molecular epidemiologic investigations. The glycoprotein 60 gene (GP60) is the most commonly used marker for subtyping isolates of *C. parvum* and *C. hominis* [22, 23], and, more recently, also for other human pathogens such as *C. meleagridis* [24] and *C. viatorum* [25]. Other highly polymorphic markers include a number of micro- and mini-satellites, usually found within coding sequences [26]. These markers have been useful to describe the population structure of different *Cryptosporidium* species and to investigate factors, including genetic recombination, involved in the diversification of natural parasite populations and in the emergence of virulent subtypes [27 - 29]. A relevant example is the identification of a *C. parvum* GP60 type (IIc) that has been found in humans but not in animals, supporting anthroponotic transmission and a possible ongoing mechanism of host adaptation [22]. Indeed, genome-wide comparison of zoonotic or anthroponotic *C. parvum* isolates has identified a number of candidate genes that may influence the host range [30].

Global Burden of Disease

Prior to the advent of molecular diagnostics methods, the rate of infection was estimated to be 1% among immunocompetent individuals in high-income countries, and 5–10% among those from low-resource settings [31]. More recent studies based on DNA and antigen detection methods have shown that as many as 15–25% of children with diarrhea are infected with *Cryptosporidium* [32], indicating that prevalence was underestimated due to the use of less sensitive methods.

Cryptosporidiosis is associated with prolonged diarrhea (7–14 days) and persistent diarrhea (≥14 days), and results in greater childhood morbidity and mortality than other causes [33]. Survivors often suffer persistent growth retardation and cognitive deficits, indicative of long-term sequelae caused by the infection [34]. Recently, the pathogenic role of *Cryptosporidium* in children has been even more firmly established by the Global Enteric Multicentre Study [35]. In this very large cohort study of children from seven sites in sub-Saharan Africa and South Asia, the authors aimed at assessing the causes, burden, clinical syndromes, and adverse outcomes of moderate-to-severe diarrhea. Among the main results, *Cryptosporidium* was found to be one of the four major contributors to moderate-to-severe diarrheal diseases during the first 2 years of life, at all sites and independently of the HIV status [35]. The authors showed that the risk of mortality increased 2-3 times in infected children aged between 11-23 months, further highlighting the public health importance of this parasite.

Human-to-Human Transmission

Cryptosporidium is easily transmitted among children and staff members in nurseries [36], day care centers [37], and schools [38]. Nosocomial infection is also well documented [39, 40], and secondary cases among roommates [41] and family members [42] have been described, further supporting the highly infectious capacity of the parasite.

In developing countries, the high prevalence of *C. hominis* and of anthroponotic subtypes of *C. parvum*, particularly in children, has been interpreted as an indication of the importance of person-to-person transmission [22]. Direct transmission among HIV-positive men who have sex with men occurs more frequently than in HIV-positive drug users [43]. Sexual contact represents a risk factor for fecal-oral parasite transmission: a case-control study in Australia has shown that men having more than one sexual partner are more likely to have diarrhea [44].

Zoonotic Transmission

The role played by animals in the transmission of cryptosporidiosis has been the subject of many investigations, and excellent reviews have been published [21, 45]. The large fraction of cases is attributed to *C. parvum*, undoubtedly the species of major zoonotic relevance, and the main reservoirs are livestock species, particularly young animals [21]. Convincing examples of zoonotic transmission comprises the numerous outbreaks of infection observed among veterinary students [46]. Epidemiological investigation of those outbreaks has identified a number of risk factors, with direct contact with animals having diarrhea being the most important [47].

Apart from *C. parvum*, many species of *Cryptosporidium* have animals as their major hosts but are also associated with human infections (Table **1**). In agreement with zoonotic potential, transmission of *C. meleagridis* from birds to humans has been described on a farm in Sweden [48]. Further support for zoonotic potential of this species was the demonstration of identical parasite genotypes recovered from birds and humans [49]. More recently, novel zoonotic species have been described, including *C. cuniculus*, a parasite of rabbits that has caused a waterborne outbreak of human cryptosporidiosis in the UK [50], and *C. erinacei*, a parasite of hedgehogs that has been recognized as a human pathogen [51].

Waterborne Transmission

Water plays a very important role in the global epidemiology of cryptosporidiosis, and is the most commonly reported vehicle of transmission in outbreaks. Waterborne cryptosporidiosis is a serious public health concern, particularly for populations at risk of severe infection (pregnant women, children, HIV-positive and transplanted patients). Despite the global distribution of *Cryptosporidium*, reports of waterborne outbreaks mostly originate from industrialized countries in the continents of Australia, Europe and North America [52]. The most extensive outbreak occurred in 1993 in Milwaukee, Wisconsin, USA, in which 400,000 persons were infected with *Cryptosporidium* oocysts by drinking water from a water treatment plant [10].

Historically, the first outbreak linked to drinking water was in 1984, at Braun Station, Texas, USA, and involved over 200 individuals, when sewage contaminated a groundwater supply [53]. In 1987, a larger outbreak affected an estimated 13,000 people at Carrollton, Georgia, USA, when mains water became contaminated from a surface supply [54]. At about the same time, outbreaks were recorded in the United Kingdom, when over 500 individuals acquired cryptosporidiosis by ingesting oocysts from drinking water [55]. These outbreaks showed that filtration and chlorine disinfection, which were assumed to control waterborne diseases in industrialized nations, were not always sufficient to protect people from becoming sick through the mains water supply, and prompted regulatory agencies to develop rules for public health protection through monitoring, removal or inactivation [56]. Because of improvements in drinking water quality, driven by both regulations and long-term investment of water companies, a substantial reduction in reported mains drinking water outbreaks has been observed. For example, only four outbreaks linked to public water supplies have been recorded in England and Wales since 2000 [56]; nevertheless, drinking water remains a risk factor for cryptosporidiosis in England and Wales. Furthermore, the recent waterborne outbreak in Östersund, Sweden, where 27,000 inhabitants (about 45%) were infected through drinking water [57], shows that outbreaks continue to occur, and further stresses the value of assessing microbial risks in raw water.

In recent years, outbreaks linked to recreational waters have increased in number, globally. Indeed, out of 120 reported waterborne *Cryptosporidium* outbreaks during the years 2004-2010, 46% were associated with drinking water and 54% with recreational waters [52]. Outbreaks have occurred in swimming pools, paddling or wading pools, water parks and fountains, and were caused by *C. hominis* and *C. parvum* [58].

Foodborne Transmission

Many outbreaks linked to food have been reported, but, often, investigation of relevant sources has been limited by the lack of molecular characterization and of national surveillance reporting [59]. Another important difficulty is that the implicated foods are generally short-shelf life products, which may not be available for testing after the cases of illness have been identified. Further, processing of large batches of products requires sampling procedures, but it is unclear how to optimize sampling for products where the distribution of the contamination is uneven.

Contamination with *Cryptosporidium* oocysts has been mostly investigated in vegetables, fruits and shellfish, as these foods are generally consumed raw and thus are more likely to represent vehicles of transmission [60]. In general, the level of contamination appears to be low, but detection methods show large variability in terms of recovery efficiency, reproducibility and applicability to different food matrices [61]. A recent study in Canada used both PCR and immunofluorescence to reveal *Cryptosporidium* spp. oocysts in 32 of 544 (5.9%) retail samples of a variety of salad blends and individual leafy greens; molecular analysis further showed *C. parvum* in 29 of 32 contaminated samples [62]. The lack of standard methods for the detection of *Cryptosporidium* in food makes comparison of results from different studies difficult. An International Organization for Standardization (ISO) method is in preparation for the detection of oocysts from soft fruit and leafy green vegetables, based on extensive testing from UK studies [63, 64].

A review of the literature data showed that *C. parvum* has been implicated more often than other species in foodborne cryptosporidiosis, and a role of animal contamination has been highlighted at least in some cases [59]. This parasite has been identified in outbreaks in Europe, the US and Japan, that were linked to the consumption of salads, beverages and raw liver. Despite considerable evidence shown that shellfish (oysters, clams, mussels) can concentrate oocysts in their body [65], to date there is no evidence of human infection linked to their consumption.

Airborne Transmission

Infection of the respiratory tract has been described in both animals and humans, particularly among immunocompromised individuals. In AIDS patients, infections involving the nasal mucosa, tracheobronchial tree and deep lung were reported. Unfortunately, these critically ill patients experienced fulminant disease, failed to respond to existing therapies, and succumbed [66]. While cryptosporidiosis probably disseminated from the intestinal tract in most cases, the lack of gastrointestinal involvement in some patients suggested other potential routes of transmission, including respiratory transmission.

For respiratory transmission to occur, the parasite must be able to establish an active life cycle within the respiratory tract. In agreement with this, various parasite forms lining the bronchial epithelium of lung sections, extracellular invasive forms and intramacrophagic oocysts, have been described. This suggests that *Cryptosporidium* may be capable of propagating within the human respiratory tract in the same way that it parasitizes the gastrointestinal epithelium [67].

The most compelling evidence for respiratory transmission has been provided by a study in Uganda, in which 35% of children diagnosed with intestinal cryptosporidiosis and cough, had sputum samples containing *Cryptosporidium* DNA, suggesting the presence of organisms in the respiratory tract [68]. Of note, samples from saliva were negative, indicating no contamination of sputum by gastrointestinal contents. Furthermore, 16 of 17 children were HIV negative, and 10 of 17 were normally nourished, thereby suggesting that respiratory cryptosporidiosis may occur in immunocompetent individuals. This study also provided information on the species involved, and showed that 13 of 17 (76%) children were infected with *C. hominis*, while the remaining four were infected with *C. parvum* [68]. Other evidence for airborne transmission has been recently reviewed [66]. The authors underlined that more investigations are needed to determine the role of the respiratory tract in disease transmission; in turn, this information will be necessary to establish specific guidelines for the prevention of airborne cryptosporidiosis.

Risk Factors

The majority of human cases is considered sporadic, but investigation of risk factors for sporadic cryptosporidiosis is complex. Indeed, our knowledge of risk factors for acquiring cryptosporidiosis mostly derives from outbreaks investigations. This has demonstrated that contact with infected animals, person-to-person transmission in households and care settings, consumption of contaminated foods and drinks, consumption of water from private and public supplies, exposure to recreational water in swimming pools or water parks, and travel to endemic countries [69]. Furthermore, risk factors are likely to differ for *C. parvum* and *C. hominis* and in different geographical settings, due to the difference in host range, transmission cycles and prevalence. This stresses the role of species or genotype identification when conducting epidemiologic studies. However, molecular typing has been used only in a few case-control studies, both from developed [70, 71] and developing countries [72, 73].

In industrialized countries, case-control studies have been conducted in the US [74], the UK [70, 71], and Australia [75]. In all these studies, multivariate analysis have demonstrated that contact with persons with diarrhea, particularly young children, and contact with cattle, especially calves, are the most significant risk factors. Moreover, travel abroad was significantly associated with an increased risk in the US and UK studies [70, 74]. The Australian study focused on endemic disease, thus history of travel abroad was not considered [75]. In addition, the US study identified swimming in fresh water as a risk factor, and the Australian study found swimming in a chlorinated swimming pool as a risk factor. Both the UK and Australian studies identified a dose dependent risk associated with drinking unboiled water, as was the case for a regional study in the UK [76]. The consumption of ice cream and raw vegetables was negatively associated with the risk of infection in the UK and US studies.

A more recent study from the UK [71] investigated the role of wider environmental and socioeconomic factors (*e.g.* water supply, socioeconomic status, land use, livestock densities and healthcare accessibility) in human cryptosporidiosis. By comparing 3,368 laboratory-confirmed cases to an equal number of controls, the authors found different risk factors for *C. hominis* and *C. parvum*. Indeed, for *C. hominis* cases the strongest risks factors were living in areas with many higher socioeconomic status individuals, living in areas with a high proportion of young children and living in urban areas. In contrast, agricultural land use surrounding the place of residence and the water supply were significant risk factors for *C. parvum* [71].

In comparison to industrialized countries, less is known about risk factors in developing and tropical countries, despite the higher prevalence of cryptosporidiosis in those areas. Taken together, the studies performed in developing countries have identified several risk factors for cryptosporidiosis: age <2 years, absence of breastfeeding, contact with pets, living in overcrowded conditions, low birth weight, male gender, malnourishment and co-infections [77].

GENOMICS

The sequencing of the genomes of *C. parvum*, *C. hominis* and *C. muris* has been a major advance in our understanding of the molecular biology of *Cryptosporidium* [78, 79]. The genomes of *C. parvum* and *C. hominis* are remarkably similar, displaying 95-97% DNA sequence identity and ~30% GC content, and are organized in eight chromosomes that are apparently completely collinear [80]. The highly compacted nature of the genome, which is comprised of only ~9.2 Mbp, is reflected in its high coding capacity (about 4000 genes that account for 2/3 of the genome). *Cryptosporidium* appears to have lost the apicoplastid, which is present in many other Apicomplexa but absent in the Gregarine [12]. Further, a degenerate 'mitosome' is present instead of a mitochondrion and both mitochondrial and nuclear genes encoding for mitochondrial proteins are absent, including those required for the TCA cycle, oxidative phosphorylation, and fatty acid oxidation (notably, these genes are present in the *C. muris* genome). Genes for *de novo* biosynthesis of amino acids, nucleotides, and sugars are also absent, as well as mechanisms for RNA splicing and gene silencing [78, 79]. The highly streamlined metabolic pathways imply that *Cryptosporidium* heavily relies on scavenging nutrients from the host, salvage rather than de novo biosynthesis, and glycolysis or substrate-level phosphorylation for energy production.

More recently, Next Generation Sequencing methodologies are being used to generate genome sequences at a lower cost and with much higher efficiency, which allows for the exploration of genetic differences between *C. parvum* and *C. hominis* and facilitates the identification of factors that can potentially contribute to differences in host specificity of the two closely related species [81]. Further database mining of an increased number of *Cryptosporidium* genomes will certainly assist in the discovery of new genes, biochemical pathways and protective antigens that may represent novel targets for chemotherapy.

PATHOPHYSIOLOGY AND SYMPTOMS

The majority of the studies concerning the pathogenesis of cryptosporidiosis are relative to *C. parvum* and *C. hominis* and their effects on the intestinal epithelium. Both species affect the small intestine and provoke comparable gastrointestinal diseases in terms of symptoms and extent.

Intestinal Pathology

The intestinal epithelium is the major mucosa of the body, and is composed of a unique layer of cells that form innumerable crypts and villi, increasing the surface exposed to the intestinal lumen. This provides an extended layer for nutrient acquisition, and for electrolyte and water exchange between the intestinal lumen and the underlying tissues. Intestinal epithelial cells exert an important role in protecting from potentially toxic contents and microbial organisms that pass through the gut. In addition, epithelial cells play an important role in segregating the microbial flora of the gut from the cells of the host immune system present underneath the epithelial layer. This intestinal barrier is fundamental to maintaining homeostasis between the commensal flora and the mucosal immune system [82].

Cryptosporidium infection alters in a drastic way the functions of enterocytes in the tract of intestine where the infection takes place. It appears that the main effect is relative to the permeability and/or to the secretive functions of enterocytes, since a characteristic trait of cryptosporidiosis is watery diarrhea. The mechanism causing the diarrhea, as well as the molecules involved, is not clearly understood, even though various hypothesis have been formulated. Some researchers have argued that the severity of cryptosporidiosis is largely dependent on the immunological status of the host (see below). Differences in the severity of disease have also been attributed to infection with either *C. parvum* or *C. hominis*, or even to differences among isolates of the same species (see above). Obviously, variable factors that influence the virulence of the different *Cryptosporidium* isolates are genetically based, although gene variants that determine different clinical pictures are not yet clearly identified. Thus, the severity of the disease is largely dependent on both host factors (such as immunological status), and parasite factors, including undetermined genetic differences among species and genotypes.

Cryptosporidiosis is associated with morphological changes in the intestinal epithelium and the underlying tissues that produce macroscopic effects, such as blunting of the intestinal villi. Histological evidence shows that cryptosporidiosis causes crypt hyperplasia and villus atrophy, with shortened or absent brush border microvilli [83]. Cryptosporidiosis also reduces the coupled glucose-NaCl transport and increases chloride secretion (Cl^-) [84, 85]. These alterations imply that the absorptive capacity of the villi is highly reduced and, hence, maintaining a proper fluid exchange is impaired, which accounts for the main clinical manifestations, like diarrhea and dehydration. Noteworthy, substance P, a neuropeptide that stimulates active secretion of Cl^- ion in the area of inflammation, is elevated in experimentally infected individuals and in AIDS patients with cryptosporidiosis and the level of substance P correlates positively to the severity of diarrhea [86].

Cryptosporidium damages the intestinal epithelial barrier that separates the microbial environment in the intestinal lumen from the mucosal immune system. This alteration results in the temporary disruption of the epithelial tight junctions at the site of the infection [85, 87, 88]. Whether damage is caused by parasite factors or an uncontrolled immune response is still unknown.

Epithelial cell death at the site of infection is an additional remarkable pathological sign of cryptosporidiosis. Both necrosis and apoptosis of enterocytes contribute to this phenomenon. The parasite delays the programmed cell death of the infected cells with an unclear mechanism, and then induces apoptosis of the host cell prior to its egress from the parasitophorous vacuole when the internal cycle is already completed [89, 90]. By contrast, the adjacent uninfected cells are subjected to apoptosis mediated by the Fas/FasL pathway [91].

In humans, diarrhea and abdominal pain represent the main symptoms of cryptosporidiosis, often in association with nausea, vomiting and fever. Others non-specific symptoms can be observed, such as headache, myalgia, malaise and anorexia [92]. In immunocompetent patients, cryptosporidiosis is usually self-limited and does not last more than 2-3 weeks [93]. In immunodeficient individuals such as AIDS patients, the disease can be prolonged (up to several months), leading to severe illness with considerable dehydration and wasting as consequence of the persistent diarrhea. In these cases, cryptosporidiosis represents an opportunistic infection with a high rate of mortality [94].

Atypical Cryptosporidiosis in Immunocompromised Subjects

Immunocompromised patients are prone to infections in atypical localizations, including the stomach, biliary tract and pancreas [95]. Infection of both the upper and lower respiratory tract has often been observed in immunocompromised individuals especially in AIDS patients [66]. Respiratory cryptosporidiosis provokes mild symptoms like inflammation of the nasal mucosa and other mucosae of the upper respiratory tract as well as cough and fever in the infections of the lower respiratory tract.

Recent studies indicated that *C. parvum* infection in immunosuppressed subjects may have a role in the development of cancer in the digestive tract. A zoonotic *C. parvum* isolate (subtype IIA15G2R1), that caused a fulminant cryptosporidiosis in a transplant recipient, was able to induce invasive gastrointestinal and biliary adenocarcinoma in SCID mice [96]. It was further demonstrated that low infective doses are sufficient to induce digestive adenocarcinoma in an immunosuppressed mouse model [97]. Very recently, it has been shown that the Wnt pathway is implicated in the onset of ileo-cecal adenocarcinoma in an immunodeficient mouse model [98]. The hypothesis that cryptosporidiosis increases the risk for some types of gastrointestinal cancer is of interest, even though the evidence for this correlation is limited to animal models with immunodeficiency.

Parasite Virulence Factors

As mentioned above, the major symptom of cryptosporidiosis is watery diarrhea, similar to that caused by other microbial pathogens such as *Vibrio cholera.* This fact stimulated the search for *Cryptosporidium* toxin-like factors that are able to act on the secretive host mechanisms in a way similar to bacterial toxins [99]. Pioneering studies demonstrated the electrical effect (chloride- and calcium-dependent), of supernatants from *C. parvum*-induced diarrhoic stools in an *"ex vivo"* and an *"in vitro"* model [100, 101]. However, the search for a *Cryptosporidium* entero-toxin has remained inconclusive and the *Limulus* amebocyte lysate (LAL) test, a classical assay to reveal the presence of toxins in a biological substance, gave no results when tested with *Cryptosporidium* extracts [102]. Finally, no genes encoding for candidate toxic proteins were found in the complete *Cryptosporidium* genomes [78, 79].

Possible *Cryptosporidium* pathogenicity and virulence factors that contribute to the establishment of an effective infection have been recently reviewed [103]. While solid correlations between parasite factors and the severity of disease are still lacking, it is plausible that proteins having a role in the early phases of infection could be of particular relevance. Therefore, intense research has focused on the proteins expressed at the sporozoite stage, and a list of those supposed to play an important role in host cell invasion is given in Table **2**. Some of these proteins belong to protein families highly conserved in eukaryotes, such as the elongation factor CpEF-1α [104] and the subtilisin CpSUB1 [105]. The glycoprotein gp40/15 [106] and the mucin Muc4 [107] are known to have a role in the binding to host cells. Some of the proteins secreted or associated to the sporozoite membrane were first identified for their immunogenicity [108 - 110]. In fact, free sporozoites are able to induce a strong and specific IgG response that overall is directed against proteins and other molecules exposed and accumulated at the apical end of the parasite. This capacity can be exploited to identify proteins specifically expressed at this stage [111]. Although, proteins secreted and/or exposed by sporozoites activate a specific humoral response, anti-*Cryptosporidium* antibodies (*e.g.* IgA and IgG) are not sufficient to counteract an ongoing infection nor to prevent new infections. Importantly, sporozoites release the contents of apical organelles in a coordinated way, similar to what has been observed for other apicomplexan species [112]. In particular, micronemes transport proteins to the parasite outer membrane and some of these proteins are also released in the external space during the sporozoite gliding on the epithelium surface. Among these secreted proteins, GP900, P23 and Cpa135, are also immunodominant sporozoite-antigens [109, 113, 114] that are discharged along the sporozoite trail toward the host cell. Hence, it is reasonable that these antigenic proteins may act as decoys for the host immune system since they can also activate proliferative responses of Th1 cells and other immune cells [115, 116]. The engagement of the host immune system in response to molecules released in the outer space can be an effective strategy for immune evasion, and may thus contribute to the virulence of the parasite.

Table 2. Antigenic proteins[1] expressed by *Cryptosporidium* sporozoites, with a putative role as virulence factors.

Protein	Biochemical and Molecular Features	Localization	Reference
Aminopeptidase N	Aminopeptidase	Sporozoite surface	[120, 121]
Cops-1(*C. parvum*) Chos-1 (*C. hominis*)	Glycosilated Telomerically encoded	Sporozoite surface, secreted	[103]
CP2	Transmembrane domain	Sporozoite surface	[122]
Cp20	Transmembrane domains	Sporozoite surface	[123]
Cp22.4.1	Zinc-finger domains	Unknown	[124]
Cp33	Unknown	Apical region	[125]
CpA135	Modular protein, LCCL domain	Apical region, secreted	[114]
CpATPase2	Heavy metal ATPase	Membrane associated	[126]
CpEF-1α	Elongation factor	Apical region, cytoskeleton	[104]
CpMIC1 (CpTSP8)	Thrombospondin (TSP) family	Sporozoite surface	[127]
CpMuc4	Mucin	Sporozoite surface, host cell binding	[107, 128]
CpMuc5	Mucin	Sporozoite surface	[129]
CpRom	Rhomboid protease	Sporozoite surface	[111]
CpSUB1	Subtilisin-like protease	Apical region	[105]
gp40/15 (also called Cp17, gp15/45/60 or S60)	Mature products by cleavage (gp40 and gp15), glycoprotein	Sporozoite surface, host cell binding	[106, 129]
GP900	Glycoprotein, polythreonine domain	Sporozoite surface, secreted	[109]
Hem A	Hemolytic activity	Unknown	[130]
HSP 70	Heat shock protein	Cytoplasmic	[131]
HSP 90	Heat shock protein	Unknown	[132]
LCFA, EC 6.2.1.3	long-chain fatty acyl-CoA synthetase	Apical region	[133]

(Table 2) contd.....

Protein	Biochemical and Molecular Features	Localization	Reference
P23	N-glycosylation site	Sporozoite surface, secreted	[134]
P30	Lectin	Intracellular, apical region	[135]
TrapC1	Thrombospondin (TSP) family	Sporozoite surface	[136]

[1]Only proteins with a corresponding, well defined DNA coding sequence were considered.

After the development of the parasitophorous vacuole, the host cell membrane envelopes the parasite; this is believed to protect the parasite from the host immune reaction and to limit the dispersion of molecules in the surrounding environment. Some of the proteins expressed in the sporozoites are also found in the internal stages and probably in the merozoites [113, 114, 117]. By contrast, little is known about proteins specific to the internal forms of the parasite. It is thought that the feeder organelle, an exclusive subcellular structure of *Cryptosporidium* spp., regulates the exchange of nutrients and other substances with the host cell. However, information about the molecules that pass through the host-parasite barrier is only hypothetical. In this respect, an ATP-binding cassette protein (CpABC) has been localized at the host-parasite boundary, where it probably mediates ion exchange between the parasite and the host cytoplasm [118]. It is plausible that various molecules cross the feeder organelle and the membrane layers that wrap the parasitophorous vacuole, but to date none of these molecules has been identified. The host cell is subjected to a drastic rearrangement of the cytoskeleton after parasite entry, due to the loss of brush border to accommodate the parasitophorous vacuole and the accumulation of actin that constitutes a plaque at the boundary of the host-parasite interface [119]. These relevant modifications are probably mediated by parasite factors, that are yet completely unknown. Thus, the elements that regulate the interactions between the parasite and the host cell represent important virulence factors that, if identified, will allow for a better understanding of the pathology of cryptosporidiosis.

IMMUNOLOGY AND IMMUNOPATHOLOGY

The immunological status of the host and other host factors constitute crucial elements that determine the extent and the severity of cryptosporidiosis. The fact that *Cryptosporidium* infections are self-limited in immunocompetent subjects indicates that hosts can activate effective immune defences against the parasite. On the other hand, it is well known that immune-deficient hosts are often subjected to a prolonged and severe diarrhea that can be life-threatening [137]. Depletion of $CD4^+$ cells in HIV-infected is a known factor of predisposition for severe cryptosporidiosis [138]. The maturity of the immune system also plays an important role in determining the susceptibility and the severity of the disease, as demonstrated by the fact that young animals and children are most susceptible to *Cryptosporidium* infections [35, 139]. However, the reasons for the high susceptibility of young hosts to cryptosporidiosis are rather unclear. Immune responses to *Cryptosporidium* infections have been recently reviewed in detail [140, 141]. Here, we discuss some relevant elements of the host immune system involved in the control of cryptosporidiosis and recent advances in understanding the complex interplay of factors involved in the innate response.

Innate Immune Response

Role of Mannose Binding Lectin

One of the basic defence mechanism to contain the invasion of microbial pathogens is represented by mannose binding lectin (MBL), a highly conserved protein that participates in the innate responses to *Cryptosporidium*. Members of the MBL protein family recognize various carbohydrates present on many microorganisms and represent the first level in the innate immune system that activates the complement pathway [142]. It has been reported that deficiency of serum MBL is associated with persistent diarrhea in AIDS patients and that this molecule can bind *C. parvum* sporozoites [143]. More recently, it has been shown that a MBL2 polymorphism is associated with a higher risk of recurrent cryptosporidiosis in young children [144]. MBL detected in the intestinal fluid of individuals with cryptosporidiosis is likely of serum origin [143].

Interaction of **Cryptosporidium** *with Epithelial Cells*

During the intracellular stages, *Cryptosporidium* spp. parasites have a peculiar localization since they assemble the parasitophorus vacuoles at the luminal side of the intestinal epithelial cells. Thus, in this particular niche, the parasites are completely wrapped by the host membrane, and are protected from the attack of the specialized cells of the immune system as well as from the circulating antibodies. By contrast, the epithelial cells in the infected area activate various protective mechanisms by the regulation of innate effectors and the stimulation of the adaptive cell response. Therefore, the cells of the epithelial barrier, once affected by *Cryptosporidium*, coordinate the fundamental defensive mechanisms to control the parasite and restore the intestinal homeostasis [82]. Overall, gastrointestinal epithelial cells provide the first defence against *Cryptosporidium* infection and they mobilize cells of the immune system to the infection site to activate adaptive immunity.

Epithelial cells respond to invasion by microbial pathogens by activating complex molecular pathways that involve many effectors both in the infected cells and in the surrounding uninfected cells. Pathogen associated molecular pattern molecules (PAMPs) can trigger the innate response *via* toll-like receptors (TLRs) on the surface of epithelial cells [82]. Thus, sporozoite invasion activates TLR4 and TLR2 pathways [145], although the *Cryptosporidium* PAMPs recognized by these receptors are unknown. None of the typical microbial molecules acting on these receptors (*e.g.* LPS, lipoproteins, phospholipids, *etc.*) has been described in *Cryptosporidium.* However, it is reasonable that molecules exposed on the sporozoite surface and/or substances released from the apical organelles (*e.g.* micronemes) prior to or during invasion of enterocytes may act as PAMPs and trigger the innate response of epithelial cells. *In vitro* experiments on cultured human cholangiocytes showed a post-transcriptional up-regulation of TLR4 in response to *C. parvum* infection [102]. Interestingly, inhibition of TLR4 expression mediated by the HIV TAT protein could account for the biliary cryptosporidiosis characteristic of HIV-infected individuals [146]. However, the role of TLR4 is crucial for the clearance of parasite in a mouse model of biliary cryptosporidiosis [147].

The activation of TLR4/NF-kB pathway constitutes the dominant response to *Cryptosporidium* invasion, by coordinating a network of various subordinate molecules. Thus, biliary epithelial cells respond to the direct interaction with *Cryptosporidium* by activating NF-kB, mediated by MyD88, which acts as an adaptor of TLR4 [144, 148]. The involvement of TLR4/NF-kB pathway has been demonstrated in a mouse model [149]. Therefore, *Cryptosporidium* invasion induces TLR4/NF-kB dependent responses as the first and principal host defence mechanism; once activated, NF-kB triggers a cascade of events by transcriptional up-regulation of pro-inflammatory genes [150]. This transactivation produces the expression of different classes of proteins such as adhesion molecules, antimicrobial peptides and inflammatory effectors (*e.g.* cytokines/ chemokines). Table **3** lists the molecules induced by anti-*Cryptosporidium* responses.

Table 3. Host factors dependent on the activation of TLR4/NFkB pathway and involved in the innate immune response to *Cryptosporidium* infection.

Cytokines	Chemokines	Apoptotic Genes	Regulatory Proteins	Antimicrobial Functions	Prostaglandins
IFN-γ [182, 183]	CXCL1 (GROα) [187]	FasL/ FasR [91]	CIS [166]	α-defensin [191]	PGE2 [190]
IL-15 [184]	IL-8 (CXCL8) [188]	Caspase 3 [89]	SOCS4 [167]	β-defensin [192]	PGF2α [190]

(Table 3) contd.....

Cytokines	Chemokines	Apoptotic Genes	Regulatory Proteins	Antimicrobial Functions	Prostaglandins
TNF-α [185]	CXCL10 [189]	Bcl-2 [89]	ICAM-1 (CD54) [170]	iNOS [171]	
IL-18 [186]	CXC3CL1 [172]	B7-H1 (PD-L1/CD274) [169]	KSRP (KHSRP) [171]		
	CCL20 [160]		HDACs [172]		
			SIRT1 [168]		

Cytokines, Chemokines and Effector Molecules

Upon *Cryptosporidium* infection, epithelial cells quickly initiate a series of innate immune reactions, including production of antimicrobial peptides (*e.g.* β-defensins and LL-37) and release of inflammatory cytokines/ chemokines, such as interleukin-8 (IL-8) [139, 151]. Production and secretion of antimicrobial peptides can kill *Cryptosporidium* or inhibit parasite growth. These cytokines/chemokines of epithelial cell origin can mobilize and activate immune effector cells (*e.g.* lymphocytes, macrophages and neutrophils) to the infection sites [152].

An important cytokine of the innate response is interferon-gamma (IFN-γ), which inhibits the development of the parasite and prevents invasion of the enterocytes if administered in human and murine cell lines [153]. Clear evidence of the role of IFN-γ has previously been reported both in humans [154, 155] and in experimental mouse infections [156, 157]. An important effect is determined by IL-15 expression: this cytokine activates NK cells to lyse *Cryptosporidium*-infected epithelial cells [158]. A recent study demonstrated that IL-18 plays an important role during infection, since IL-18 depleted SCID mice show an exacerbated form of cryptosporidiosis that is ameliorated by treatment with recombinant IL-18. Increased expression of IFN-γ and of some anti-microbial peptides was observed in IL-18-treated mice; hence, IL-18 stimulates epithelial cells to increase antimicrobial activity [159]. Recently it has been demonstrated that CCL20, a chemokine with antimicrobial activity, is diminished in *C. parvum* infection in neonatal mice, whereas oral administration of this chemokine reduced parasite load [160].

MicroRNAs

MicroRNAs (miRNAs) are small non-coding RNA (20-22 nucleotides) present in many eukaryotes with a regulatory function *via* RNA silencing and post-transcriptional regulation [161]; more than one thousand miRNAs have been identified in human [162].

Recently, it has been demonstrated that various host miRNAs are involved in the response to *Cryptosporidium*, and are implicated in the fine-tuning of the TLR-mediated inflammatory response [163]. The majority of the studies were based on experimental infection of cultured human cholangiocytes. In this model, *C. parvum* provoked reduced expression of primary let-7i and mature let-7 in a MyD88/NF-kappaB-dependent manner [145] by the formation of a NFkB p50-C/EBPbeta silencer complex that acted as a transcriptional repressor on the let-7i promoter [164]. In turn, this resulted in the up-regulation of TLR4 in infected cells regulated by let-7 [102]. Other evidence concerns the transactivation, by the NF-kB p65 subunit, of a cluster of miRNAs (mir-125b-1, mir-21, mir-30b, and mir-23b-27b-24-1) [165]. Cytokine-inducible Src homology 2-containing protein (CIS) and suppressors of cytokine signalling 4 (SOCS4) are down regulated by miR-98 and let-7 [166, 167]. MiRNA let-7i is also involved in the NAD-dependent deacetylase sirtuin-1 (SIRT1) protein that regulates multiple cellular processes, including inflammation. In fact, *C. parvum* infection reduces expression of let-7i that acts as post-transcriptional repressor of SIRT1 mRNA [168]. MiRNAs are also implied in the regulation of apoptotic genes by miR-513 that inhibits the expression of the B7-H1 protein [169]. Furthermore, miRNAs are involved in the regulation of adhesion proteins that mediate the adhesion of lymphocytes at the infection site, thus the expression of intercellular adhesion molecule-1 (ICAM-1), which increases in *C. parvum* infection, can be inhibited by miR-221 acting as post-transcriptional repressor [170]. Among the defence mechanisms controlled by miRNAs, is the NF-κB-dependent production of nitric oxide (NO) in host epithelial cells that acts by the stabilization of the inducible NO synthase (iNOS) mRNA. *C. parvum* infection induces iNOS expression and miR-27b participates in this process by suppressing KH-type splicing regulatory protein (KSRP) expression that, in turn, alters the stability of iNOS mRNA [171]. CX3CL1, a chemo-attractor of T cells and monocytes, is induced in *C. parvum* infection. Its activation is mediated by the suppression of transcription of miR-424 and miR-503 as a result of histone deacetylase (HDACs) activity, dependent on NF-kB signalling [172]. MiRNAs act as post-transcriptional repressors of most of the genes involved in the inflammatory response TLR4/NF-kB dependent. Overall, the studies on miRNAs are defining the regulatory mechanisms controlling innate responses and extend the knowledge to novel factors involved in the host responses to *Cryptosporidium* infections.

Exosomes

One of the most recent advances in understanding the cellular response to *Cryptosporidium* is the discovery that exosomes contribute to the host defences against the parasite. Exosomes are particular small vesicles (30-100 nm) assembled in cell compartments called multi-vesicular endosomes (MVEs) and secreted in the outer space by most cell types. These vesicles carry specific proteins, mRNAs and microRNAs and represent an effective intercellular-communication system [173]. In *Cryptosporidium* infection, the TLR4/NF-kB pathway also regulates luminal exosome release from the apical site of enterocytes and shuttling of antimicrobial peptides from the gastrointestinal epithelium. Exosomes carry antimicrobial peptides of epithelial origin, including cathelicidin-37 and beta-defensin 2. *Cryptosporidium parvum* sporozoites are bound by luminal exosomes that affect their viability and infectivity. Direct binding to the sporozoite surface is required for the anti-*C. parvum* activity of released exosomes [174]. *Cryptosporidium parvum* infected epithelial cells also release exosomes from the basolateral membrane and, differently from the apical exosomes secreted in the intestinal lumen, are probably involved in the antigen presentation since they carry MHC proteins and show anti-*Cryptosporidium* activity [174]. Thus, epithelial exosomes may constitute a new link between innate and adaptive immune responses.

Adaptive Immune Response

Cell Mediated Responses

IFN-γ plays a major role in inducing an effective cell-mediated immunity since it largely mediates $CD4^+$ cell responses [175]. The importance of $CD4^+$ T cells in limiting cryptosporidiosis is most evident in AIDS patients who are susceptible to prolonged, severe cryptosporidiosis [92]. The extent and the severity of the infection is strictly correlated with the $CD4^+$ count: individuals with less than 50 $CD4^+$ cells/mm^3 are at risk of contracting fulminant cryptosporidiosis, whereas those with more than 100 $CD4^+$ cells/mm^3 present with a less severe, self-limited disease [85]. Peripheral Blood Mononuclear Cells (PBMCs) from healthy donors with previous cryptosporidiosis show a high level of T cell proliferation in response to whole crude extract of oocysts/sporozoites, fractionated sporozoite proteins, and immunodominant recombinant antigens. A recently infected donor showed a T-cell profile with a higher proportion of γ/δ T-cell receptors (γ/δ TCR) compared to donors who recovered from cryptosporidiosis years before, suggesting that the γ/δ T-cell population is involved in the early stage of the infection. The prevalent profile among proliferating T-cells was $CD4^+$ TCR α/β $CD45RO^+$ (memory phenotype) and hyper-production of IFN-γ [115]. Recent experiments conducted with recombinant *C. hominis* gp15 on PBMCs from sensitized donors showed a higher production of IFN-γ in seropositive individuals compared to seronegative ones. Both $CD4^+$ and $CD8^+$ T-cell populations from *Cryptosporidium* seropositive subjects showed a higher production of IFN-γ [176].

Humoral Response

The role of immunoglobulins in the control of cryptosporidiosis is still a matter of debate, since it seems that circulating antibodies are not determinant in blocking the course of infection [175]. As discussed in the previous section, many sporozoite antigens induce the production of specific antibodies, especially IgG and IgA, even though some of these antigens released from sporozoites on the intestinal mucosa may contribute to evasion from the host immune system. However, it is reasonable that secreted antibodies, particularly IgA, can bind invasive forms of the parasite (*e.g.* sporozoites and merozoites) and block attachment and invasion of new host cells, hence reducing the diffusion of the parasite in gut mucosa. Indeed, some investigations indicate that humoral responses induced by the entire parasite produce effective antibodies able to reduce symptoms. A pioneering study with hyper-immune bovine colostrum showed a reduction of the parasite load in infected mice; this protective effect was mostly due to presence of specific IgG1 and IgA [177]. Thus, the administration of bovine colostrum to AIDS patients with cryptosporidiosis reduced diarrhea and, in part, oocyst shedding [178]. Studies with healthy human volunteers showed that infected subjects secreted fecal IgA in response to *C. parvum* infection [179], and that those with pre-existing anti-*Cryptosporidium* IgG antibodies experienced less severe disease and had reduced diarrheal symptoms if re-infected [180, 181].

DIAGNOSIS AND DETECTION METHODS

Cryptosporidium should be suspected in all patients who present with acute gastroenteritis, most especially young children, or if the symptoms are prolonged (>3 days). Numerous techniques are used to diagnose infections in humans or animals [193]. This includes microscopic examination of feces to demonstrate the presence of oocysts, detection of *Cryptosporidium* antigens or DNA, and demonstration of life cycle stages of the parasite in biopsy by histology. This section will briefly discuss these methods, and refers interested readers to excellent publications for further details [194].

Microscopic Identification of Oocysts

Cryptosporidium oocysts are small (about 5 μm), smooth, colorless, and spherical or slightly ovoid bodies, and are therefore difficult to distinguish from other organisms or microscopic objects present in the feces when examined under a light microscope [194]. Feces can be concentrated by sedimentation using modified formol–ethyl acetate, and further by conventional flotation (*e.g.*, zinc sulphate, saturated sodium chloride or sucrose). While this results in cleaner preparations, low-density oocysts may be missed. Microscopy is an important diagnostic method because of the low cost of reagents, but good staining and visual skills are necessary. Among the staining techniques available to facilitate the recognition of oocysts, the acid-fast modified Ziehl-Nielsen technique is the most commonly employed. The use of fluorescent stains (*e.g.*, Auramine O) makes oocyst observation easier and less subjective, and slides can be examined at lower magnification than acid-fast stained slides, but a fluorescence microscope is needed. An important limitation inherent to all morphology-based diagnostics methods is that the different species and genotypes of *Cryptosporidium* show extensive overlap in the morphology of the oocysts, and thus are practically indistinguishable when observed under a microscope.

Immunological Methods

There has been a trend toward increased use of immunofluorescence microscopy (IFM), and reference laboratories in the USA and Europe often use this technique as a gold standard. Interpretation of IFM results is less subjective and increases sensitivity and specificity compared to other microscopy stains, although cross-reactions (*e.g.*, with algae) may occur when examining food, water, and environmental samples. Therefore, additional observations, *e.g.*, of internal contents by differential interference contrast and of sporozoite nuclei by staining with a nuclear fluorogen, may still be necessary [193].

Other antigen detection formats are also available, including enzyme immune assay (EIA) and immunochromatographic lateral flow (ICLF) methods. EIAs can be automated and are more sensitive than acid-fast microscopy. ICLF (also known as rapid assay) is less sensitive, but requires little skill and can be performed anywhere as a rapid test. Rapid assays are increasingly used because (1) they do not depend on trained personnel and costly equipment, and (2) they can be completed in 10 min rather than the 1 to 2 h required to perform IFM tests or EIAs.

DNA-based Diagnostic Methods

The development of the polymerase chain reaction (PCR) has permitted specific and sensitive detection of *Cryptosporidium* spp. for both clinical diagnosis and environmental monitoring. PCR-based detection methods, compared to other techniques, have higher sensitivity and specificity and allow detecting the range of pathogens present in a sample. Further, large numbers of samples can be analyzed in parellel and at relatively low cost.

The most commonly used conventional PCR assays target conserved genes, in particular the small subunit ribosomal (18S rDNA), the actin, and the heat shock protein 70 genes, to identify the different *Cryptosporidium* species and genotypes [195]. Amplification products are further analyzed by DNA sequencing or digested by endonucleases to generate species-specific restriction patterns. More recently, a variety of quantitative (real-time) PCR assays have been developed for the enumeration of *Cryptosporidium* oocysts in feces [196], sewage [197] and water [198]. qPCR is considered to have the highest sensitivity and is particularly well suited for detection of multiple pathogens in a single assay [199].

TREATMENT

Despite many years of investigation aimed at developing a satisfactory therapy for human cryptosporidiosis, it is apparent that none of the drugs used to treat infections caused by other coccidian parasites shows a robust efficacy against *Cryptosporidium* [200]. The lack of an effective chemotherapy is of particular concern, considering the benefit that an available treatment will have for risk groups such as children, the elderly and the immunocompromised (patients with HIV/AIDS, transplant recipients, patients undergoing cancer chemotherapy). Chemotherapy will also be useful in outbreaks to limit the spread of the disease.

Many different compounds have been tested, including macrolide antibiotics (spiramycin, azithromycin and clarithromycin), aminoglycoside antibiotics (paromomycin) and thiazolide antibiotics (nitazoxanide). While several compounds showed encouraging activity in *in vitro* assays, results from animal models or clinical trials have been inconsistent [200, 201]. However, Indinavir, an anti-aspartyl protease included in the antiviral therapy for HIV infected individuals, showed a certain efficacy (up to 70% reduction of the number of intracellular parasites in the ileum) in a mouse model of *C. parvum* infection [202]. Therefore, searching for specific inhibitors of *Cryptosporidium* aspartic proteases may represent a possible strategy to develop anti-cryptosporidial drugs.

Why *Cryptosporidium* is so resistant to therapies is still to be elucidated. Several factors have been considered, including: (1) the unique location of the parasite in the host cell, which may affect drug concentration; (2) the lack of specific targets or differences in targets either at the molecular or structural levels; (3) differences in biochemical pathways; and (4) existence of transport proteins or efflux pumps that transport drugs out of the parasite [201].

To date, nitazoxanide [2-acetyloloxy-N-(5-nitro-2-thiazolyl) benzamide, NTZ] is the only drug approved by the FDA for the treatment of cryptosporidiosis in immunocompetent patients. NTZ has a demonstrated activity against a broad range of parasites as well as some bacteria. Nitaxozanide is rapidly hydrolysed to its active desacetyl metabolite, named tizoxanide, which is further metabolised by glucuronidation to tizoxanide glucuronide. The mechanism of action of NTZ is still undefined. However, NTZ can cause dose-dependent toxicity in cell cultures [203], and can induce apoptosis in proliferating mammalian cells [204]. This suggests that toxicity/apoptosis induced by NTZ on host cells may account for the observed reduction in parasite numbers, even in the lack of a direct effect against *Cryptosporidium*.

Recent research on new drugs has taken advantage of the availability of genome information to focus on features that may be unique to *Cryptosporidium*. The parasite has acquired a number of genes from bacteria, or used biochemical pathways that differ substantially from those of other eukaryotes [78, 79]. This may represent interesting targets for the development of novel drugs. Notable examples are polyamine biosynthesis [205], fatty acid synthase [206] and dihydrofolate reductase [207]. Inhibitors of key enzymes in these pathways have shown strong anti-parasitic activities in *in vitro* assays [200].

PERSPECTIVES OF CONTROL

Preventive measures are by far the most effective approach to control cryptosporidiosis. As transmission commonly takes places through water or by direct contact with infected person or animals, drinking water quality and general hygiene (especially in hospitals, wards, *etc.*) are essential for the prevention of the infection. To control infection within families or at day-care centres, effective measures are those appropriate for any diarrheal disease. Specific guidelines have been issued following the occurrence of cryptosporidiosis in settings where public contact with animals is permitted, such as educational farms [207].

Hospitals involved in the care of severely immunocompromised patients should be particular vigilant in the management of cryptosporidiosis, and provision of safe water should be guaranteed, also to avoid difficulties in case a notice to boil is issued. Immunocompromised persons, including people with a primary immunodeficiency affecting T cells and children with acute leukemia, should avoid contact with animals with diarrhea, young pets, and should avoid swallowing water while swimming. It should be remembered, however, that the evidence for the effectiveness of preventive interventions is almost lacking [208].

CONCLUDING REMARKS

Parasites of the genus *Cryptosporidium* have been increasingly recognized as important pathogens of both humans and animals, worldwide. The burden of disease is higher in developing regions of the world, and infection in children, especially malnourished children, is associated with high morbidity and mortality. In industrialized countries, many waterborne outbreaks have occurred, in some cases causing infection in thousands of individuals. Despite the relevance of cryptosporidiosis, our ability to reduce the burden of disease is strongly limited by the lack of (a) practical point-of-care diagnostics, (b) effective treatment for immunocompromised patients, (c) adequate coverage with antiretroviral therapy in developing countries and (d) promising vaccine candidates. Future studied based on different "omic" approaches (genomics, transcriptomics, proteomics and metabolomics) are likely to shed light on the biology of these parasites and may help developing new therapies and control measures.

CONFLICT OF INTEREST

The authors confirm that they have no conflict of interest to declare for this publication.

ACKNOWLEDGEMENTS

Declared none.

REFERENCES

[1] Tyzzer E. A sporozoon found in the peptic glands of the common mouse. Proc Soc Exp Biol Med 1907; 5: 12-3.
[http://dx.doi.org/10.3181/00379727-5-5]

[2] Tyzzer EE. An extracelluar coccidium, *Cryptosporidium muris* (gen. & sp. nov.), of the gastric glands of the common mouse. J Med Res 1910; 23(3): 487-509, 3.
[PMID: 19971982]

[3] Tyzzer E. *Cryptosporidium parvum* (sp. nov.), a coccidium found in the small intestine of the common mouse. Arch Protistenkd 1912; 26: 394-412.

[4] Tzipori S, Widmer G. A hundred-year retrospective on cryptosporidiosis. Trends Parasitol 2008; 24(4): 184-9.
[http://dx.doi.org/10.1016/j.pt.2008.01.002] [PMID: 18329342]

[5] Slavin D. *Cryptosporidium meleagridis* (sp. nov.). J Comp Pathol 1955; 65(3): 262-6.
[http://dx.doi.org/10.1016/S0368-1742(55)80025-2] [PMID: 13242675]

[6] Panciera RJ, Thomassen RW, Garner FM. Cryptosporidial infection in a calf. Vet Pathol 1976; 8: 479-84.

[7] Nime FA, Burek JD, Page DL, Holscher MA, Yardley JH. Acute enterocolitis in a human being infected with the protozoan *Cryptosporidium*. Gastroenterology 1976; 70(4): 592-8. [PMID: 815126]

[8] Meisel JL, Perera DR, Meligro C, Rubin CE. Overwhelming watery diarrhea associated with a cryptosporidium in an immunosuppressed patient. Gastroenterology 1976; 70(6): 1156-60. [PMID: 773738]

[9] Tzipori S. Cryptosporidiosis in perspective. Adv Parasitol 1988; 27: 63-129. [http://dx.doi.org/10.1016/S0065-308X(08)60353-X] [PMID: 3289331]

[10] Mac Kenzie WR, Hoxie NJ, Proctor ME, *et al.* A massive outbreak in Milwaukee of *cryptosporidium* infection transmitted through the public water supply. N Engl J Med 1994; 331(3): 161-7. [http://dx.doi.org/10.1056/NEJM199407213310304] [PMID: 7818640]

[11] Carreno RA, Martin DS, Barta JR. *Cryptosporidium* is more closely related to the gregarines than to coccidia as shown by phylogenetic analysis of apicomplexan parasites inferred using small-subunit ribosomal RNA gene sequences. Parasitol Res 1999; 85(11): 899-904. [http://dx.doi.org/10.1007/s004360050655] [PMID: 10540950]

[12] Templeton TJ, Enomoto S, Chen WJ, *et al.* A genome-sequence survey for *Ascogregarina taiwanensis* supports evolutionary affiliation but metabolic diversity between a Gregarine and *Cryptosporidium*. Mol Biol Evol 2010; 27(2): 235-48. [http://dx.doi.org/10.1093/molbev/msp226] [PMID: 19778951]

[13] Xiao L, Fayer R, Ryan U, Upton SJ. *Cryptosporidium* taxonomy: recent advances and implications for public health. Clin Microbiol Rev 2004; 17(1): 72-97. [http://dx.doi.org/10.1128/CMR.17.1.72-97.2004] [PMID: 14726456]

[14] Ryan U, Hijjawi N. New developments in *Cryptosporidium* research. Int J Parasitol 2015; 45(6): 367-73. [http://dx.doi.org/10.1016/j.ijpara.2015.01.009] [PMID: 25769247]

[15] Ryan UM, Xiao L. Taxonomy and Molecular Taxonomy. In: Cacciò SM, Widmer G, Eds. Cryptosporidium: Parasite and Disease. New York: Springer 2014; pp. 1-22. [http://dx.doi.org/10.1007/978-3-7091-1562-6_1]

[16] Fayer R. General biology. In: Fayer R, Xiao L, Eds. Cryptosporidium and Cryptosporidiosis. Boca Raton: CRC Press 2008; pp. 1-42.

[17] Cacciò SM, Putignani L. Epidemiology of human cryptosporidiosis. In: Cacciò SM, Widmer G, Eds. Cryptosporidium: Parasite and Disease. New York: Springer 2014; pp. 43-80. [http://dx.doi.org/10.1007/978-3-7091-1562-6_2]

[18] Checkley W, White AC Jr, Jaganath D, *et al.* A review of the global burden, novel diagnostics, therapeutics, and vaccine targets for cryptosporidium. Lancet Infect Dis 2015; 15(1): 85-94. [http://dx.doi.org/10.1016/S1473-3099(14)70772-8] [PMID: 25278220]

[19] Xiao L, Ryan UM. Cryptosporidiosis: an update in molecular epidemiology. Curr Opin Infect Dis 2004; 17(5): 483-90. [http://dx.doi.org/10.1097/00001432-200410000-00014] [PMID: 15353969]

[20] Cacciò SM, Thompson RC, McLauchlin J, Smith HV. Unravelling *Cryptosporidium* and *Giardia* epidemiology. Trends Parasitol 2005; 21(9): 430-7. [http://dx.doi.org/10.1016/j.pt.2005.06.013] [PMID: 16046184]

[21] Robertson LJ, Bjorkman C, Axen C, Fayer R. Cryptosporidiosis in farmed animals. In: Cacciò SM, Widmer G, Eds. Cryptosporidium: Parasite and Disease. New York: Springer 2014; pp. 149-236. [http://dx.doi.org/10.1007/978-3-7091-1562-6_4]

[22] Xiao L. Molecular epidemiology of cryptosporidiosis: an update. Exp Parasitol 2010; 124(1): 80-9. [http://dx.doi.org/10.1016/j.exppara.2009.03.018] [PMID: 19358845]

[23] Nichols GL, Chalmers RM, Hadfield SJ. Molecular epidemiology of human cryptosporidiosis. In: Cacciò SM, Widmer G, Eds. Cryptosporidium: Parasite and Disease. New York: Springer 2014; pp. 81-148. [http://dx.doi.org/10.1007/978-3-7091-1562-6_3]

[24] Stensvold CR, Beser J, Axén C, Lebbad M. High applicability of a novel method for gp60-based subtyping of *Cryptosporidium meleagridis.* J Clin Microbiol 2014; 52(7): 2311-9. [http://dx.doi.org/10.1128/JCM.00598-14] [PMID: 24740082]

[25] Stensvold CR, Elwin K, Winiecka-Krusnell J, Chalmers RM, Xiao L, Lebbad M. Development and application of a gp60-based typing assay for *Cryptosporidium viatorum.* J Clin Microbiol 2015; 53(6): 1891-7. [http://dx.doi.org/10.1128/JCM.00313-15] [PMID: 25832304]

[26] Robinson G, Chalmers RM. Assessment of polymorphic genetic markers for multi-locus typing of *Cryptosporidium parvum* and *Cryptosporidium hominis.* Exp Parasitol 2012; 132(2): 200-15. [http://dx.doi.org/10.1016/j.exppara.2012.06.016] [PMID: 22781277]

[27] Feng Y, Torres E, Li N, Wang L, Bowman D, Xiao L. Population genetic characterisation of dominant *Cryptosporidium parvum* subtype IIaA15G2R1. Int J Parasitol 2013; 43(14): 1141-7. [http://dx.doi.org/10.1016/j.ijpara.2013.09.002] [PMID: 24126186]

[28] Li N, Xiao L, Cama VA, *et al.* Genetic recombination and *Cryptosporidium hominis* virulent subtype IbA10G2. Emerg Infect Dis 2013; 19(10): 1573-82. [http://dx.doi.org/10.3201/eid1910.121361] [PMID: 24050210]

[29] Cacciò SM, de Waele V, Widmer G. Geographical segregation of *Cryptosporidium parvum* multilocus genotypes in Europe. Infect Genet Evol 2015; 31: 245-9. [http://dx.doi.org/10.1016/j.meegid.2015.02.008] [PMID: 25687913]

[30] Widmer G, Lee Y, Hunt P, Martinelli A, Tolkoff M, Bodi K. Comparative genome analysis of two *Cryptosporidium parvum* isolates with different host range. Infect Genet Evol 2012; 12(6): 1213-21. [http://dx.doi.org/10.1016/j.meegid.2012.03.027] [PMID: 22522000]

[31] Lima AA, Samie A, Guerrant RL. Cryptosporidiosis. In: Guerrant RL, Walker DH, Weller PF, Eds. Tropical Infectious Diseases. Philadelphia, Pa: Elsevier-Churchill Livingstone 2011; pp. 640-63.

[32] Shirley DA, Moonah SN, Kotloff KL. Burden of disease from cryptosporidiosis. Curr Opin Infect Dis 2012; 25(5): 555-63. [http://dx.doi.org/10.1097/QCO.0b013e328357e569] [PMID: 22907279]

[33] Moore SR, Lima NL, Soares AM, *et al.* Prolonged episodes of acute diarrhea reduce growth and increase risk of persistent diarrhea in children. Gastroenterology 2010; 139(4): 1156-64. [http://dx.doi.org/10.1053/j.gastro.2010.05.076] [PMID: 20638937]

[34] Guerrant DI, Moore SR, Lima AA, Patrick PD, Schorling JB, Guerrant RL. Association of early childhood diarrhea and cryptosporidiosis with impaired physical fitness and cognitive function four-seven years later in a poor urban community in northeast Brazil. Am J Trop Med Hyg 1999; 61(5): 707-13. [PMID: 10586898]

[35] Kotloff KL, Nataro JP, Blackwelder WC, *et al.* Burden and aetiology of diarrhoeal disease in infants and young children in developing countries (the Global Enteric Multicenter Study, GEMS): a prospective, case-control study. Lancet 2013; 382(9888): 209-22. [http://dx.doi.org/10.1016/S0140-6736(13)60844-2] [PMID: 23680352]

[36] Hannah J, Riordan T. Case to case spread of cryptosporidiosis; evidence from a day nursery outbreak. Public Health 1988; 102(6): 539-44. [http://dx.doi.org/10.1016/S0033-3506(88)80023-4] [PMID: 3231695]

[37] Heijbel H, Slaine K, Seigel B, *et al.* Outbreak of diarrhea in a day care center with spread to household members: the role of *Cryptosporidium*. Pediatr Infect Dis J 1987; 6(6): 532-5. [http://dx.doi.org/10.1097/00006454-198706000-00009] [PMID: 3615068]

[38] Lee MB, Greig JD. A review of gastrointestinal outbreaks in schools: effective infection control interventions. J Sch Health 2010; 80(12): 588-98. [http://dx.doi.org/10.1111/j.1746-1561.2010.00546.x] [PMID: 21087255]

[39] Baxby D, Hart CA, Taylor C. Human cryptosporidiosis: a possible case of hospital cross infection. Br Med J (Clin Res Ed) 1983; 287(6407): 1760-1. [http://dx.doi.org/10.1136/bmj.287.6407.1760-a] [PMID: 6416580]

[40] Martino P, Gentile G, Caprioli A, *et al.* Hospital-acquired cryptosporidiosis in a bone marrow transplantation unit. J Infect Dis 1988; 158(3): 647-8. [http://dx.doi.org/10.1093/infdis/158.3.647] [PMID: 3045217]

[41] Bruce BB, Blass MA, Blumberg HM, Lennox JL, del Rio C, Horsburgh CR Jr. Risk of *Cryptosporidium parvum* transmission between hospital roommates. Clin Infect Dis 2000; 31(4): 947-50. [http://dx.doi.org/10.1086/318147] [PMID: 11049775]

[42] Pandak N, Zeljka K, Cvitkovic A. A family outbreak of cryptosporidiosis: Probable nosocomial infection and person-to-person transmission. Wien Klin Wochenschr 2006; 118(15-16): 485-7. [http://dx.doi.org/10.1007/s00508-006-0637-7] [PMID: 16957980]

[43] Pedersen C, Danner S, Lazzarin A, *et al.* Epidemiology of cryptosporidiosis among European AIDS patients. Genitourin Med 1996; 72(2): 128-31. [PMID: 8698361]

[44] Hellard ME, Sinclair MI, Hogg GG, Fairley CK. Prevalence of enteric pathogens among community based asymptomatic individuals. J Gastroenterol Hepatol 2000; 15(3): 290-3. [http://dx.doi.org/10.1046/j.1440-1746.2000.02089.x] [PMID: 10764030]

[45] Abeywardena H, Jex AR, Gasser RB. A perspective on *Cryptosporidium* and *Giardia*, with an emphasis on bovines and recent epidemiological findings. Adv Parasitol 2015; 88: 243-301. [http://dx.doi.org/10.1016/bs.apar.2015.02.001] [PMID: 25911369]

[46] Gait R, Soutar RH, Hanson M, Fraser C, Chalmers R. Outbreak of cryptosporidiosis among veterinary students. Vet Rec 2008; 162(26): 843-5. [http://dx.doi.org/10.1136/vr.162.26.843] [PMID: 18587060]

[47] Kinross P, Beser J, Troell K, *et al.* *Cryptosporidium parvum* infections in a cohort of veterinary students in Sweden. Epidemiol Infect 2015; 30: 1-9.

[48] Silverlås C, Mattsson JG, Insulander M, Lebbad M. Zoonotic transmission of *Cryptosporidium meleagridis* on an organic Swedish farm. Int J Parasitol 2012; 42(11): 963-7. [http://dx.doi.org/10.1016/j.ijpara.2012.08.008] [PMID: 23022616]

[49] Wang Y, Yang W, Cama V, *et al.* Population genetics of *Cryptosporidium meleagridis* in humans and birds: evidence for cross-species transmission. Int J Parasitol 2014; 44(8): 515-21. [http://dx.doi.org/10.1016/j.ijpara.2014.03.003] [PMID: 24727090]

[50] Puleston RL, Mallaghan CM, Modha DE, *et al.* The first recorded outbreak of cryptosporidiosis due to *Cryptosporidium cuniculus* (formerly rabbit genotype), following a water quality incident. J Water Health 2014; 12(1): 41-50.
[http://dx.doi.org/10.2166/wh.2013.097] [PMID: 24642431]

[51] Kváč M, Saková K, Květoňová D, *et al.* Gastroenteritis caused by the *Cryptosporidium* hedgehog genotype in an immunocompetent man. J Clin Microbiol 2014; 52(1): 347-9.
[http://dx.doi.org/10.1128/JCM.02456-13] [PMID: 24131692]

[52] Baldursson S, Karanis P. Waterborne transmission of protozoan parasites: review of worldwide outbreaks - an update 20042010. Water Res 2011; 45(20): 6603-14.
[http://dx.doi.org/10.1016/j.watres.2011.10.013] [PMID: 22048017]

[53] DAntonio RG, Winn RE, Taylor JP, *et al.* A waterborne outbreak of cryptosporidiosis in normal hosts. Ann Intern Med 1985; 103(6 (Pt 1)): 886-8.
[http://dx.doi.org/10.7326/0003-4819-103-6-886] [PMID: 4062089]

[54] Hayes EB, Matte TD, OBrien TR, *et al.* Large community outbreak of cryptosporidiosis due to contamination of a filtered public water supply. N Engl J Med 1989; 320(21): 1372-6.
[http://dx.doi.org/10.1056/NEJM198905253202103] [PMID: 2716783]

[55] Richardson AJ, Frankenberg RA, Buck AC, *et al.* An outbreak of waterborne cryptosporidiosis in Swindon and Oxfordshire. Epidemiol Infect 1991; 107(3): 485-95.
[http://dx.doi.org/10.1017/S0950268800049189] [PMID: 1752298]

[56] Chalmers RM. Waterborne outbreaks of cryptosporidiosis. Ann Ist Super Sanita 2012; 48(4): 429-46.
[http://dx.doi.org/10.4415/ANN_12_04_10] [PMID: 23247139]

[57] Widerström M, Schönning C, Lilja M, *et al.* Large outbreak of *Cryptosporidium hominis* infection transmitted through the public water supply, Sweden. Emerg Infect Dis 2014; 20(4): 581-9.
[http://dx.doi.org/10.3201/eid2004.121415] [PMID: 24655474]

[58] Beach MJ. Waterborne: recreational water. In: Fayer R, Xiao L, Eds. Cryptosporidium and Cryptosporidiosis. Boca Raton: CRC Press 2008; pp. 335-70.

[59] Robertson LJ, Chalmers RM. Foodborne cryptosporidiosis: is there really more in Nordic countries? Trends Parasitol 2013; 29(1): 3-9.
[http://dx.doi.org/10.1016/j.pt.2012.10.003] [PMID: 23146217]

[60] Ortega YR, Cama VA. Foodborne transmission. In: Fayer R, Xiao L, Eds. Cryptosporidium and Cryptosporidiosis. Boca Raton: CRC Press 2008; pp. 289-304.

[61] Smith HV, Cacciò SM, Cook N, Nichols RA, Tait A. *Cryptosporidium* and *Giardia* as foodborne zoonoses. Vet Parasitol 2007; 149(1-2): 29-40.
[http://dx.doi.org/10.1016/j.vetpar.2007.07.015] [PMID: 17728067]

[62] Dixon B, Parrington L, Cook A, Pollari F, Farber J. Detection of *Cyclospora, Cryptosporidium*, and *Giardia* in ready-to-eat packaged leafy greens in Ontario, Canada. J Food Prot 2013; 76(2): 307-13.
[http://dx.doi.org/10.4315/0362-028X.JFP-12-282] [PMID: 23433379]

[63] Cook N, Paton CA, Wilkinson N, Nichols RA, Barker K, Smith HV. Towards standard methods for the detection of *Cryptosporidium parvum* on lettuce and raspberries. Part 1: development and optimization of methods. Int J Food Microbiol 2006; 109(3): 215-21.
[http://dx.doi.org/10.1016/j.ijfoodmicro.2005.12.015] [PMID: 16529835]

[64] Cook N, Paton CA, Wilkinson N, Nichols RA, Barker K, Smith HV. Towards standard methods for the detection of *Cryptosporidium parvum* on lettuce and raspberries. Part 2: validation. Int J Food Microbiol 2006; 109(3): 222-8.
[http://dx.doi.org/10.1016/j.ijfoodmicro.2005.12.014] [PMID: 16546283]

[65] Robertson LJ. The potential for marine bivalve shellfish to act as transmission vehicles for outbreaks of protozoan infections in humans: a review. Int J Food Microbiol 2007; 120(3): 201-16. [http://dx.doi.org/10.1016/j.ijfoodmicro.2007.07.058] [PMID: 17928081]

[66] Sponseller JK, Griffiths JK, Tzipori S. The evolution of respiratory Cryptosporidiosis: evidence for transmission by inhalation. Clin Microbiol Rev 2014; 27(3): 575-86. [http://dx.doi.org/10.1128/CMR.00115-13] [PMID: 24982322]

[67] Poirot JL, Deluol AM, Antoine M, *et al.* Broncho-pulmonary cryptosporidiosis in four HIV-infected patients. J Eukaryot Microbiol 1996; 43(5): 78S-9S. [http://dx.doi.org/10.1111/j.1550-7408.1996.tb05007.x] [PMID: 8822872]

[68] Mor SM, Tumwine JK, Ndeezi G, *et al.* Respiratory cryptosporidiosis in HIV-seronegative children in Uganda: potential for respiratory transmission. Clin Infect Dis 2010; 50(10): 1366-72. [http://dx.doi.org/10.1086/652140] [PMID: 20377408]

[69] Nichols G. Epidemiology. In: Fayer R, Xiao L, Eds. Cryptosporidium and Cryptosporidiosis. Boca Raton: CRC Press 2008; pp. 79-118.

[70] Hunter PR, Hughes S, Woodhouse S, *et al.* Health sequelae of human cryptosporidiosis in immunocompetent patients. Clin Infect Dis 2004; 39(4): 504-10. [http://dx.doi.org/10.1086/422649] [PMID: 15356813]

[71] Lake IR, Harrison FC, Chalmers RM, *et al.* Case-control study of environmental and social factors influencing cryptosporidiosis. Eur J Epidemiol 2007; 22(11): 805-11. [http://dx.doi.org/10.1007/s10654-007-9179-1] [PMID: 17891460]

[72] Cama VA, Bern C, Roberts J, *et al.* *Cryptosporidium* species and subtypes and clinical manifestations in children, Peru. Emerg Infect Dis 2008; 14(10): 1567-74. [http://dx.doi.org/10.3201/eid1410.071273] [PMID: 18826821]

[73] Molloy SF, Tanner CJ, Kirwan P, *et al.* Sporadic *Cryptosporidium* infection in Nigerian children: risk factors with species identification. Epidemiol Infect 2011; 139(6): 946-54. [http://dx.doi.org/10.1017/S0950268810001998] [PMID: 20800006]

[74] Roy SL, DeLong SM, Stenzel SA, *et al.* Risk factors for sporadic cryptosporidiosis among immunocompetent persons in the United States from 1999 to 2001. J Clin Microbiol 2004; 42(7): 2944-51. [http://dx.doi.org/10.1128/JCM.42.7.2944-2951.2004] [PMID: 15243043]

[75] Robertson B, Sinclair MI, Forbes AB, *et al.* Case-control studies of sporadic cryptosporidiosis in Melbourne and Adelaide, Australia. Epidemiol Infect 2002; 128(3): 419-31. [http://dx.doi.org/10.1017/S0950268802006933] [PMID: 12113486]

[76] Goh S, Reacher M, Casemore DP, *et al.* Sporadic cryptosporidiosis, North Cumbria, England, 19962000. Emerg Infect Dis 2004; 10(6): 1007-15. [http://dx.doi.org/10.3201/10.3201/eid1006.030325] [PMID: 15207050]

[77] Putignani L, Menichella D. Global distribution, public health and clinical impact of the protozoan pathogen *Cryptosporidium*. Interdiscip Perspect Infect Dis 2010.

[78] Abrahamsen MS, Templeton TJ, Enomoto S, *et al.* Complete genome sequence of the apicomplexan, *Cryptosporidium parvum*. Science 2004; 304(5669): 441-5. [http://dx.doi.org/10.1126/science.1094786] [PMID: 15044751]

[79] Xu P, Widmer G, Wang Y, *et al.* The genome of *Cryptosporidium hominis*. Nature 2004; 431(7012): 1107-12. [http://dx.doi.org/10.1038/nature02977] [PMID: 15510150]

[80] Widmer G, Sullivan S. Genomics and population biology of *Cryptosporidium* species. Parasite Immunol 2012; 34(2-3): 61-71.
[http://dx.doi.org/10.1111/j.1365-3024.2011.01301.x] [PMID: 21595702]

[81] Guo Y, Tang K, Rowe LA, *et al.* Comparative genomic analysis reveals occurrence of genetic recombination in virulent *Cryptosporidium hominis* subtypes and telomeric gene duplications in *Cryptosporidium parvum.* BMC Genomics 2015; 16: 320.
[http://dx.doi.org/10.1186/s12864-015-1517-1] [PMID: 25903370]

[82] Peterson LW, Artis D. Intestinal epithelial cells: regulators of barrier function and immune homeostasis. Nat Rev Immunol 2014; 14(3): 141-53.
[http://dx.doi.org/10.1038/nri3608] [PMID: 24566914]

[83] Argenzio RA, Liacos JA, Levy ML, Meuten DJ, Lecce JG, Powell DW. Villous atrophy, crypt hyperplasia, cellular infiltration, and impaired glucose-Na absorption in enteric cryptosporidiosis of pigs. Gastroenterology 1990; 98(5 Pt 1): 1129-40.
[http://dx.doi.org/10.1016/0016-5085(90)90325-U] [PMID: 2323506]

[84] Farthing MJ. Clinical aspects of human cryptosporidiosis. Contrib Microbiol 2000; 6: 50-74.
[http://dx.doi.org/10.1159/000060368] [PMID: 10943507]

[85] Huang DB, White AC. An updated review on *Cryptosporidium* and *Giardia.* Gastroenterol Clin North Am 2006; 35(2): 291-314, viii.
[http://dx.doi.org/10.1016/j.gtc.2006.03.006] [PMID: 16880067]

[86] Robinson P, Okhuysen PC, Chappell CL, *et al.* Substance P expression correlates with severity of diarrhea in cryptosporidiosis. J Infect Dis 2003; 188(2): 290-6.
[http://dx.doi.org/10.1086/376836] [PMID: 12854086]

[87] Zhang Y, Lee B, Thompson M, *et al.* Lactulose-mannitol intestinal permeability test in children with diarrhea caused by rotavirus and *cryptosporidium.* Diarrhea Working Group, Peru. J Pediatr Gastroenterol Nutr 2000; 31(1): 16-21.
[http://dx.doi.org/10.1097/00005176-200007000-00006] [PMID: 10896065]

[88] Buret AG, Chin AC, Scott KG. Infection of human and bovine epithelial cells with *Cryptosporidium andersoni* induces apoptosis and disrupts tight junctional ZO-1: effects of epidermal growth factor. Int J Parasitol 2003; 33(12): 1363-71.
[http://dx.doi.org/10.1016/S0020-7519(03)00138-3] [PMID: 14527519]

[89] Mele R, Gomez Morales MA, Tosini F, Pozio E. *Cryptosporidium parvum* at different developmental stages modulates host cell apoptosis *in vitro.* Infect Immun 2004; 72(10): 6061-7.
[http://dx.doi.org/10.1128/IAI.72.10.6061-6067.2004] [PMID: 15385510]

[90] Liu J, Deng M, Lancto CA, Abrahamsen MS, Rutherford MS, Enomoto S. Biphasic modulation of apoptotic pathways in *Cryptosporidium parvum*-infected human intestinal epithelial cells. Infect Immun 2009; 77(2): 837-49.
[http://dx.doi.org/10.1128/IAI.00955-08] [PMID: 19075026]

[91] Chen XM, Gores GJ, Paya CV, LaRusso NF. *Cryptosporidium parvum* induces apoptosis in biliary epithelia by a Fas/Fas ligand-dependent mechanism. Am J Physiol 1999; 277(3 Pt 1): G599-608.
[PMID: 10484385]

[92] Current WL, Garcia LS. Cryptosporidiosis. Clin Microbiol Rev 1991; 4(3): 325-58.
[http://dx.doi.org/10.1128/CMR.4.3.325] [PMID: 1889046]

[93] Chen XM, Keithly JS, Paya CV, LaRusso NF. Cryptosporidiosis. N Engl J Med 2002; 346(22): 1723-31.
[http://dx.doi.org/10.1056/NEJMra013170] [PMID: 12037153]

[94] Manabe YC, Clark DP, Moore RD, *et al.* Cryptosporidiosis in patients with AIDS: correlates of disease and survival. Clin Infect Dis 1998; 27(3): 536-42.
[http://dx.doi.org/10.1086/514701] [PMID: 9770154]

[95] Hunter PR, Nichols G. Epidemiology and clinical features of *Cryptosporidium* infection in immunocompromised patients. Clin Microbiol Rev 2002; 15(1): 145-54.
[http://dx.doi.org/10.1128/CMR.15.1.145-154.2002] [PMID: 11781272]

[96] Certad G, Benamrouz S, Guyot K, *et al.* Fulminant cryptosporidiosis after near-drowning: a human *Cryptosporidium parvum* strain implicated in invasive gastrointestinal adenocarcinoma and cholangiocarcinoma in an experimental model. Appl Environ Microbiol 2012; 78(6): 1746-51.
[http://dx.doi.org/10.1128/AEM.06457-11] [PMID: 22247151]

[97] Benamrouz S, Guyot K, Gazzola S, *et al. Cryptosporidium parvum* infection in SCID mice infected with only one oocyst: qPCR assessment of parasite replication in tissues and development of digestive cancer. PLoS One 2012; 7(12): e51232.
[http://dx.doi.org/10.1371/journal.pone.0051232] [PMID: 23272093]

[98] Benamrouz S, Conseil V, Chabé M, *et al. Cryptosporidium parvum*-induced ileo-caecal adenocarcinoma and Wnt signaling in a mouse model. Dis Model Mech 2014; 7(6): 693-700.
[http://dx.doi.org/10.1242/dmm.013292] [PMID: 24652769]

[99] Sears CL, Kaper JB. Enteric bacterial toxins: mechanisms of action and linkage to intestinal secretion. Microbiol Rev 1996; 60(1): 167-215.
[PMID: 8852900]

[100] Guarino A, Canani RB, Pozio E, Terracciano L, Albano F, Mazzeo M. Enterotoxic effect of stool supernatant of *Cryptosporidium*-infected calves on human jejunum. Gastroenterology 1994; 106(1): 28-34.
[http://dx.doi.org/10.1016/S0016-5085(94)94093-2] [PMID: 8276202]

[101] Guarino A, Canani RB, Casola A, *et al.* Human intestinal cryptosporidiosis: secretory diarrhea and enterotoxic activity in Caco-2 cells. J Infect Dis 1995; 171(4): 976-83.
[http://dx.doi.org/10.1093/infdis/171.4.976] [PMID: 7706827]

[102] Chen XM, Splinter PL, OHara SP, LaRusso NF. A cellular micro-RNA, let-7i, regulates Toll-like receptor 4 expression and contributes to cholangiocyte immune responses against *Cryptosporidium parvum* infection. J Biol Chem 2007; 282(39): 28929-38.
[http://dx.doi.org/10.1074/jbc.M702633200] [PMID: 17660297]

[103] Bouzid M, Hunter PR, McDonald V, Elwin K, Chalmers RM, Tyler KM. A new heterogeneous family of telomerically encoded *Cryptosporidium* proteins. Evol Appl 2012; •••
[http://dx.doi.org/10.1111/j.1752-4571.2012.00277.x] [PMID: 23467513]

[104] Matsubayashi M, Teramoto-Kimata I, Uni S, *et al.* Elongation factor-1α is a novel protein associated with host cell invasion and a potential protective antigen of *Cryptosporidium parvum.* J Biol Chem 2013; 288(47): 34111-20.
[http://dx.doi.org/10.1074/jbc.M113.515544] [PMID: 24085304]

[105] Wanyiri JW, Techasintana P, OConnor RM, Blackman MJ, Kim K, Ward HD. Role of CpSUB1, a subtilisin-like protease, in *Cryptosporidium parvum* infection *in vitro.* Eukaryot Cell 2009; 8(4): 470-7.
[http://dx.doi.org/10.1128/EC.00306-08] [PMID: 19168760]

[106] OConnor RM, Wanyiri JW, Wojczyk BS, Kim K, Ward H. Stable expression of *Cryptosporidium parvum* glycoprotein gp40/15 in *Toxoplasma gondii.* Mol Biochem Parasitol 2007; 152(2): 149-58.
[http://dx.doi.org/10.1016/j.molbiopara.2007.01.003] [PMID: 17275106]

[107] Paluszynski J, Monahan Z, Williams M, *et al.* Biochemical and functional characterization of CpMuc4, a *Cryptosporidium* surface antigen that binds to host epithelial cells. Mol Biochem Parasitol 2014; 193(2): 114-21.
[http://dx.doi.org/10.1016/j.molbiopara.2014.03.005] [PMID: 24690740]

[108] Tilley M, Upton SJ, Fayer R, *et al.* Identification of a 15-kilodalton surface glycoprotein on sporozoites of *Cryptosporidium parvum.* Infect Immun 1991; 59(3): 1002-7.
[PMID: 1705238]

[109] Barnes DA, Bonnin A, Huang JX, *et al.* A novel multi-domain mucin-like glycoprotein of *Cryptosporidium parvum* mediates invasion. Mol Biochem Parasitol 1998; 96(1-2): 93-110.
[http://dx.doi.org/10.1016/S0166-6851(98)00119-4] [PMID: 9851610]

[110] Tosini F, Cacciò S, Tamburrini A, La Rosa G, Pozio E. Identification and characterisation of three antigenic proteins from Cryptosporidium parvum sporozoites using a DNA library expressing poly-histidine tagged peptides. Int J Parasitol 1999; 29(12): 1925-33.
[http://dx.doi.org/10.1016/S0020-7519(99)00158-7] [PMID: 10961848]

[111] Trasarti E, Pizzi E, Pozio E, Tosini F. The immunological selection of recombinant peptides from *Cryptosporidium parvum* reveals 14 proteins expressed at the sporozoite stage, 7 of which are conserved in other apicomplexa. Mol Biochem Parasitol 2007; 152(2): 159-69.
[http://dx.doi.org/10.1016/j.molbiopara.2006.12.010] [PMID: 17267054]

[112] Blackman MJ, Bannister LH. Apical organelles of Apicomplexa: biology and isolation by subcellular fractionation. Mol Biochem Parasitol 2001; 117(1): 11-25.
[http://dx.doi.org/10.1016/S0166-6851(01)00328-0] [PMID: 11551628]

[113] Enriquez FJ, Riggs MW. Role of immunoglobulin A monoclonal antibodies against P23 in controlling murine *Cryptosporidium parvum* infection. Infect Immun 1998; 66(9): 4469-73.
[PMID: 9712802]

[114] Tosini F, Agnoli A, Mele R, Gomez Morales MA, Pozio E. A new modular protein of *Cryptosporidium parvum*, with ricin B and LCCL domains, expressed in the sporozoite invasive stage. Mol Biochem Parasitol 2004; 134(1): 137-47.
[http://dx.doi.org/10.1016/j.molbiopara.2003.11.014] [PMID: 14747151]

[115] Bonafonte MT, Smith LM, Mead JR Jr. A 23-kDa recombinant antigen of Cryptosporidium parvum induces a cellular immune response on in vitro stimulated spleen and mesenteric lymph node cells from infected mice. Exp Parasitol 2000; 96(1): 32-41.
[http://dx.doi.org/10.1006/expr.2000.4545] [PMID: 11038318]

[116] Gomez Morales MA, Mele R, Ludovisi A, *et al. Cryptosporidium parvum*-specific CD4 Th1 cells from sensitized donors responding to both fractionated and recombinant antigenic proteins. Infect Immun 2004; 72(3): 1306-10.
[http://dx.doi.org/10.1128/IAI.72.3.1306-1310.2004] [PMID: 14977932]

[117] Wanyiri JW, OConnor R, Allison G, *et al.* Proteolytic processing of the *Cryptosporidium* glycoprotein gp40/15 by human furin and by a parasite-derived furin-like protease activity. Infect Immun 2007; 75(1): 184-92.
[http://dx.doi.org/10.1128/IAI.00944-06] [PMID: 17043102]

[118] Perkins ME, Riojas YA, Wu TW, Le Blancq SM. CpABC, a *Cryptosporidium parvum* ATP-binding cassette protein at the host-parasite boundary in intracellular stages. Proc Natl Acad Sci USA 1999; 96(10): 5734-9.
[http://dx.doi.org/10.1073/pnas.96.10.5734] [PMID: 10318953]

[119] Elliott DA, Clark DP. *Cryptosporidium parvum* induces host cell actin accumulation at the host-parasite interface. Infect Immun 2000; 68(4): 2315-22.
[http://dx.doi.org/10.1128/IAI.68.4.2315-2322.2000] [PMID: 10722635]

[120] Okhuysen PC, Chappell CL, Kettner C, Sterling CR. *Cryptosporidium parvum* metalloaminopeptidase inhibitors prevent *in vitro* excystation. Antimicrob Agents Chemother 1996; 40(12): 2781-4. [PMID: 9124840]

[121] Padda RS, Tsai A, Chappell CL, Okhuysen PC. Molecular cloning and analysis of the *Cryptosporidium parvum* aminopeptidase N gene. Int J Parasitol 2002; 32(2): 187-97. [http://dx.doi.org/10.1016/S0020-7519(01)00317-4] [PMID: 11812496]

[122] OHara SP, Yu JR, Lin JJ. A novel *Cryptosporidium parvum* antigen, CP2, preferentially associates with membranous structures. Parasitol Res 2004; 92(4): 317-27. [http://dx.doi.org/10.1007/s00436-003-1057-5] [PMID: 14727189]

[123] Xiao D, Yin C, Zhang Q, *et al.* Selection and identification of a new adhesion protein of *Cryptosporidium parvum* from a cDNA library by ribosome display. Exp Parasitol 2011; 129(2): 183-9. [http://dx.doi.org/10.1016/j.exppara.2011.06.004] [PMID: 21740903]

[124] De Graaf DC, De Coninck H, Petry F, Eeckhout IB, Peeters JE. Specific bovine antibody response against a new recombinant Cryptosporidium parvum antigen containing 4 zinc-finger motifs. Korean J Parasitol 2002; 40(1): 59-64. [http://dx.doi.org/10.3347/kjp.2002.40.1.59] [PMID: 11949214]

[125] Jenkins MC, Widmer G, OBrien C, *et al.* A highly divergent 33 kDa *Cryptosporidium parvum* antigen. J Parasitol 2014; 100(4): 527-31. [http://dx.doi.org/10.1645/13-433.1] [PMID: 24601821]

[126] LaGier MJ, Zhu G, Keithly JS. Characterization of a heavy metal ATPase from the apicomplexan *Cryptosporidium parvum.* Gene 2001; 266(1-2): 25-34. [http://dx.doi.org/10.1016/S0378-1119(01)00382-1] [PMID: 11290416]

[127] Putignani L, Possenti A, Cherchi S, Pozio E, Crisanti A, Spano F. The thrombospondin-related protein CpMIC1 (CpTSP8) belongs to the repertoire of micronemal proteins of *Cryptosporidium parvum.* Mol Biochem Parasitol 2008; 157(1): 98-101. [http://dx.doi.org/10.1016/j.molbiopara.2007.09.004] [PMID: 17981348]

[128] OConnor RM, Burns PB, Ha-Ngoc T, *et al.* Polymorphic mucin antigens CpMuc4 and CpMuc5 are integral to *Cryptosporidium parvum* infection *in vitro.* Eukaryot Cell 2009; 8(4): 461-9. [http://dx.doi.org/10.1128/EC.00305-08] [PMID: 19168754]

[129] OConnor RM, Kim K, Khan F, Ward HD. Expression of Cpgp40/15 in *Toxoplasma gondii*: a surrogate system for the study of *Cryptosporidium* glycoprotein antigens. Infect Immun 2003; 71(10): 6027-34. [http://dx.doi.org/10.1128/IAI.71.10.6027-6034.2003] [PMID: 14500524]

[130] Steele MI, Kuhls TL, Nida K, *et al.* A *Cryptosporidium parvum* genomic region encoding hemolytic activity. Infect Immun 1995; 63(10): 3840-5. [PMID: 7558289]

[131] Khramtsov NV, Tilley M, Blunt DS, Montelone BA, Upton SJ. Cloning and analysis of a *Cryptosporidium parvum* gene encoding a protein with homology to cytoplasmic form Hsp70. J Eukaryot Microbiol 1995; 42(4): 416-22. [http://dx.doi.org/10.1111/j.1550-7408.1995.tb01605.x] [PMID: 7620467]

[132] Woods KM, Tilley M, Iseli A, Upton SJ, Montelone BA, Khramtsov NV. Sequence of the gene encoding hsp90e from *Cryptosporidium parvum.* DNA Seq 1999; 10(4-5): 339-42. [http://dx.doi.org/10.3109/10425179909033961] [PMID: 10727089]

[133] Camero L, Shulaw WP, Xiao L. Characterization of a *Cryptosporidium parvum* gene encoding a protein with homology to long chain fatty acid synthetase. J Eukaryot Microbiol 2003; 50 (Suppl.): 534-8.
[http://dx.doi.org/10.1111/j.1550-7408.2003.tb00621.x] [PMID: 14736154]

[134] Enriquez FJ, Riggs MW. Role of immunoglobulin A monoclonal antibodies against P23 in controlling murine *Cryptosporidium parvum* infection. Infect Immun 1998; 66(9): 4469-73.
[PMID: 9712802]

[135] Bhat N, Joe A, PereiraPerrin M, Ward HD. Cryptosporidium p30, a galactose/N-acetylgalactosamin--specific lectin, mediates infection in vitro. J Biol Chem 2007; 282(48): 34877-87.
[http://dx.doi.org/10.1074/jbc.M706950200] [PMID: 17905738]

[136] Spano F, Putignani L, Naitza S, Puri C, Wright S, Crisanti A. Molecular cloning and expression analysis of a *Cryptosporidium parvum* gene encoding a new member of the thrombospondin family. Mol Biochem Parasitol 1998; 92(1): 147-62.
[http://dx.doi.org/10.1016/S0166-6851(97)00243-0] [PMID: 9574918]

[137] Chalmers RM, Davies AP. Minireview: clinical cryptosporidiosis. Exp Parasitol 2010; 124(1): 138-46.
[http://dx.doi.org/10.1016/j.exppara.2009.02.003] [PMID: 19545516]

[138] Desai NT, Sarkar R, Kang G. Cryptosporidiosis: An under-recognized public health problem. Trop Parasitol 2012; 2(2): 91-8.
[http://dx.doi.org/10.4103/2229-5070.105173] [PMID: 23767015]

[139] Budu-Amoako E, Greenwood SJ, Dixon BR, Barkema HW, McClure JT. Foodborne illness associated with *Cryptosporidium* and *Giardia* from livestock. J Food Prot 2011; 74(11): 1944-55.
[http://dx.doi.org/10.4315/0362-028X.JFP-11-107] [PMID: 22054199]

[140] Borad A, Ward H. Human immune responses in cryptosporidiosis. Future Microbiol 2010; 5(3): 507-19.
[http://dx.doi.org/10.2217/fmb.09.128] [PMID: 20210556]

[141] McDonald V, Korbel DS, Barakat FM, Choudhry N, Petry F. Innate immune responses against *Cryptosporidium parvum* infection. Parasite Immunol 2013; 35(2): 55-64.
[http://dx.doi.org/10.1111/pim.12020] [PMID: 23173616]

[142] Worthley DL, Bardy PG, Mullighan CG. Mannose-binding lectin: biology and clinical implications. Intern Med J 2005; 35(9): 548-55.
[http://dx.doi.org/10.1111/j.1445-5994.2005.00908.x] [PMID: 16105157]

[143] Kelly P, Jack DL, Naeem A, *et al.* Mannose-binding lectin is a component of innate mucosal defense against *Cryptosporidium parvum* in AIDS. Gastroenterology 2000; 119(5): 1236-42.
[http://dx.doi.org/10.1053/gast.2000.19573] [PMID: 11054381]

[144] Carmolli M, Duggal P, Haque R, *et al.* Deficient serum mannose-binding lectin levels and MBL2 polymorphisms increase the risk of single and recurrent *Cryptosporidium* infections in young children. J Infect Dis 2009; 200(10): 1540-7.
[http://dx.doi.org/10.1086/606013] [PMID: 19827946]

[145] Chen XM, OHara SP, Nelson JB, *et al.* Multiple TLRs are expressed in human cholangiocytes and mediate host epithelial defense responses to *Cryptosporidium parvum via* activation of NF-kappaB. J Immunol 2005; 175(11): 7447-56.
[http://dx.doi.org/10.4049/jimmunol.175.11.7447] [PMID: 16301652]

[146] OHara SP, Small AJ, Gajdos GB, Badley AD, Chen XM, Larusso NF. HIV-1 Tat protein suppresses cholangiocyte toll-like receptor 4 expression and defense against *Cryptosporidium parvum.* J Infect Dis 2009; 199(8): 1195-204.
[http://dx.doi.org/10.1086/597387] [PMID: 19265483]

[147] OHara SP, Bogert PS, Trussoni CE, Chen X, LaRusso NF. TLR4 promotes *Cryptosporidium parvum* clearance in a mouse model of biliary cryptosporidiosis. J Parasitol 2011; 97(5): 813-21. [http://dx.doi.org/10.1645/GE-2703.1] [PMID: 21506806]

[148] Chen XM, Levine SA, Splinter PL, *et al. Cryptosporidium parvum* activates nuclear factor kappaB in biliary epithelia preventing epithelial cell apoptosis. Gastroenterology 2001; 120(7): 1774-83. [http://dx.doi.org/10.1053/gast.2001.24850] [PMID: 11375958]

[149] Rogers KA, Rogers AB, Leav BA, *et al.* MyD88-dependent pathways mediate resistance to *Cryptosporidium parvum* infection in mice. Infect Immun 2006; 74(1): 549-56. [http://dx.doi.org/10.1128/IAI.74.1.549-556.2006] [PMID: 16369011]

[150] Kawai T, Akira S. The role of pattern-recognition receptors in innate immunity: update on Toll-like receptors. Nat Immunol 2010; 11(5): 373-84. [http://dx.doi.org/10.1038/ni.1863] [PMID: 20404851]

[151] Petry F, Jakobi V, Tessema TS. Host immune response to *Cryptosporidium parvum* infection. Exp Parasitol 2010; 126(3): 304-9. [http://dx.doi.org/10.1016/j.exppara.2010.05.022] [PMID: 20685209]

[152] Blikslager AT, Moeser AJ, Gookin JL, Jones SL, Odle J. Restoration of barrier function in injured intestinal mucosa. Physiol Rev 2007; 87(2): 545-64. [http://dx.doi.org/10.1152/physrev.00012.2006] [PMID: 17429041]

[153] Barakat FM, McDonald V, Foster GR, Tovey MG, Korbel DS. *Cryptosporidium parvum* infection rapidly induces a protective innate immune response involving type I interferon. J Infect Dis 2009; 200(10): 1548-55. [http://dx.doi.org/10.1086/644601] [PMID: 19821721]

[154] Gomez Morales MA, Ausiello CM, Guarino A, *et al.* Severe, protracted intestinal cryptosporidiosis associated with interferon gamma deficiency: pediatric case report. Clin Infect Dis 1996; 22(5): 848-50. [http://dx.doi.org/10.1093/clinids/22.5.848] [PMID: 8722945]

[155] White AC, Robinson P, Okhuysen PC, *et al.* Interferon-gamma expression in jejunal biopsies in experimental human cryptosporidiosis correlates with prior sensitization and control of oocyst excretion. J Infect Dis 2000; 181(2): 701-9. [http://dx.doi.org/10.1086/315261] [PMID: 10669358]

[156] You X, Mead JR. Characterization of experimental *Cryptosporidium parvum* infection in IFN-gamma knockout mice. Parasitology 1998; 117(Pt 6): 525-31. [http://dx.doi.org/10.1017/S0031182098003424] [PMID: 9881376]

[157] Jakobi V, Petry F. Humoral immune response in IL-12 and IFN-gamma deficient mice after infection with *Cryptosporidium parvum.* Parasite Immunol 2008; 30(3): 151-61. [http://dx.doi.org/10.1111/j.1365-3024.2007.01013.x] [PMID: 18179628]

[158] Dann SM, Wang HC, Gambarin KJ, *et al.* Interleukin-15 activates human natural killer cells to clear the intestinal protozoan *cryptosporidium.* J Infect Dis 2005; 192(7): 1294-302. [http://dx.doi.org/10.1086/444393] [PMID: 16136475]

[159] Bedi B, McNair NN, Förster I, Mead JR. IL-18 cytokine levels modulate innate immune responses and cryptosporidiosis in mice. J Eukaryot Microbiol 2015; 62(1): 44-50. [http://dx.doi.org/10.1111/jeu.12164] [PMID: 25155632]

[160] Guesdon W, Auray G, Pezier T, *et al.* CCL20 displays antimicrobial activity against *Cryptosporidium parvum*, but its expression is reduced during infection in the intestine of neonatal mice. J Infect Dis 2015; 212(8): 1332-40. [http://dx.doi.org/10.1093/infdis/jiv206] [PMID: 25838265]

[161] Ambros V. The functions of animal microRNAs. Nature 2004; 431(7006): 350-5. [http://dx.doi.org/10.1038/nature02871] [PMID: 15372042]

[162] Kozomara A, Griffiths-Jones S. miRBase: integrating microRNA annotation and deep-sequencing data. Nucleic Acids Res 2011; 39(Database issue): D152-7. [http://dx.doi.org/10.1093/nar/gkq1027] [PMID: 21037258]

[163] Zhou R, Feng Y, Chen XM. Non-coding RNAs in epithelial immunity to *Cryptosporidium* infection. Parasitology 2014; 141(10): 1233-43. [http://dx.doi.org/10.1017/S0031182014000614] [PMID: 24828969]

[164] OHara SP, Splinter PL, Gajdos GB, *et al.* NFkappaB p50-CCAAT/enhancer-binding protein beta (C/EBPbeta)-mediated transcriptional repression of microRNA let-7i following microbial infection. J Biol Chem 2010; 285(1): 216-25. [http://dx.doi.org/10.1074/jbc.M109.041640] [PMID: 19903813]

[165] Zhou R, Hu G, Liu J, Gong AY, Drescher KM, Chen XM. NF-kappaB p65-dependent transactivation of miRNA genes following *Cryptosporidium parvum* infection stimulates epithelial cell immune responses. PLoS Pathog 2009; 5(12): e1000681. [http://dx.doi.org/10.1371/journal.ppat.1000681] [PMID: 19997496]

[166] Hu G, Zhou R, Liu J, *et al.* MicroRNA-98 and let-7 confer cholangiocyte expression of cytokine-inducible Src homology 2-containing protein in response to microbial challenge. J Immunol 2009; 183(3): 1617-24. [http://dx.doi.org/10.4049/jimmunol.0804362] [PMID: 19592657]

[167] Hu G, Zhou R, Liu J, Gong AY, Chen XM. MicroRNA-98 and let-7 regulate expression of suppressor of cytokine signaling 4 in biliary epithelial cells in response to *Cryptosporidium parvum* infection. J Infect Dis 2010; 202(1): 125-35. [http://dx.doi.org/10.1086/653212] [PMID: 20486857]

[168] Xie H, Lei N, Gong AY, Chen XM, Hu G. *Cryptosporidium parvum* induces SIRT1 expression in host epithelial cells through downregulating let-7i. Hum Immunol 2014; 75(8): 760-5. [http://dx.doi.org/10.1016/j.humimm.2014.05.007] [PMID: 24862934]

[169] Gong AY, Zhou R, Hu G, *et al. Cryptosporidium parvum* induces B7-H1 expression in cholangiocytes by down-regulating microRNA-513. J Infect Dis 2010; 201(1): 160-9. [http://dx.doi.org/10.1086/648589] [PMID: 19916867]

[170] Gong AY, Hu G, Zhou R, *et al.* MicroRNA-221 controls expression of intercellular adhesion molecule-1 in epithelial cells in response to *Cryptosporidium parvum* infection. Int J Parasitol 2011; 41(3-4): 397-403. [http://dx.doi.org/10.1016/j.ijpara.2010.11.011] [PMID: 21236259]

[171] Zhou R, Gong AY, Eischeid AN, Chen XM. miR-27b targets KSRP to coordinate TLR4-mediated epithelial defense against *Cryptosporidium parvum* infection. PLoS Pathog 2012; 8(5): e1002702. [http://dx.doi.org/10.1371/journal.ppat.1002702] [PMID: 22615562]

[172] Zhou R, Gong AY, Chen D, Miller RE, Eischeid AN, Chen XM. Histone deacetylases and NF-kB signaling coordinate expression of CX3CL1 in epithelial cells in response to microbial challenge by suppressing miR-424 and miR-503. PLoS One 2013; 8(5): e65153. [http://dx.doi.org/10.1371/journal.pone.0065153] [PMID: 23724129]

[173] Théry C. Exosomes: secreted vesicles and intercellular communications. F1000 Biol Rep 2011; 3: 15. [http://dx.doi.org/10.3410/B3-15] [PMID: 21876726]

[174] Hu G, Gong AY, Roth AL, *et al.* Release of luminal exosomes contributes to TLR4-mediated epithelial antimicrobial defense. PLoS Pathog 2013; 9(4): e1003261. [http://dx.doi.org/10.1371/journal.ppat.1003261] [PMID: 23592986]

[175] Riggs MW. Recent advances in cryptosporidiosis: the immune response. Microbes Infect 2002; 4(10): 1067-80.
[http://dx.doi.org/10.1016/S1286-4579(02)01631-3] [PMID: 12191657]

[176] Preidis GA, Wang HC, Lewis DE, *et al.* Seropositive human subjects produce interferon gamma after stimulation with recombinant *Cryptosporidium hominis* gp15. Am J Trop Med Hyg 2007; 77(3): 583-5.
[PMID: 17827383]

[177] Fayer R, Guidry A, Blagburn BL. Immunotherapeutic efficacy of bovine colostral immunoglobulins from a hyperimmunized cow against cryptosporidiosis in neonatal mice. Infect Immun 1990; 58(9): 2962-5.
[PMID: 2387630]

[178] Nord J, Ma P, DiJohn D, Tzipori S, Tacket CO. Treatment with bovine hyperimmune colostrum of cryptosporidial diarrhea in AIDS patients. AIDS 1990; 4(6): 581-4.
[http://dx.doi.org/10.1097/00002030-199006000-00015] [PMID: 2201320]

[179] Dann SM, Okhuysen PC, Salameh BM, DuPont HL, Chappell CL. Fecal antibodies to *Cryptosporidium parvum* in healthy volunteers. Infect Immun 2000; 68(9): 5068-74.
[http://dx.doi.org/10.1128/IAI.68.9.5068-5074.2000] [PMID: 10948127]

[180] Moss DM, Chappell CL, Okhuysen PC, *et al.* The antibody response to 27-, 17-, and 15-kDa *Cryptosporidium* antigens following experimental infection in humans. J Infect Dis 1998; 178(3): 827-33.
[http://dx.doi.org/10.1086/515377] [PMID: 9728553]

[181] Chappell CL, Okhuysen PC, Sterling CR, Wang C, Jakubowski W, Dupont HL. Infectivity of *Cryptosporidium parvum* in healthy adults with pre-existing anti-*C. parvum* serum immunoglobulin G. Am J Trop Med Hyg 1999; 60(1): 157-64.
[PMID: 9988341]

[182] Ungar BL, Kao TC, Burris JA, Finkelman FD. *Cryptosporidium* infection in an adult mouse model. Independent roles for IFN-gamma and CD4+ T lymphocytes in protective immunity. J Immunol 1991; 147(3): 1014-22.
[PMID: 1677668]

[183] Hayward AR, Chmura K, Cosyns M. Interferon-gamma is required for innate immunity to *Cryptosporidium parvum* in mice. J Infect Dis 2000; 182(3): 1001-4.
[http://dx.doi.org/10.1086/315802] [PMID: 10950807]

[184] Dann SM, Wang HC, Gambarin KJ, *et al.* Interleukin-15 activates human natural killer cells to clear the intestinal protozoan *cryptosporidium.* J Infect Dis 2005; 192(7): 1294-302.
[http://dx.doi.org/10.1086/444393] [PMID: 16136475]

[185] Lean IS, Lacroix-Lamandé S, Laurent F, McDonald V. Role of tumor necrosis factor alpha in development of immunity against *Cryptosporidium parvum* infection. Infect Immun 2006; 74(7): 4379-82.
[http://dx.doi.org/10.1128/IAI.00195-06] [PMID: 16790816]

[186] Choudhry N, Petry F, van Rooijen N, McDonald V. A protective role for interleukin 18 in interferon γ-mediated innate immunity to *Cryptosporidium parvum* that is independent of natural killer cells. J Infect Dis 2012; 206(1): 117-24.
[http://dx.doi.org/10.1093/infdis/jis300] [PMID: 22517912]

[187] Laurent F, Eckmann L, Savidge TC, *et al. Cryptosporidium parvum* infection of human intestinal epithelial cells induces the polarized secretion of C-X-C chemokines. Infect Immun 1997; 65(12): 5067-73.
[PMID: 9393797]

[188] Seydel KB, Zhang T, Champion GA, *et al. Cryptosporidium parvum* infection of human intestinal xenografts in SCID mice induces production of human tumor necrosis factor alpha and interleukin-8. Infect Immun 1998; 66(5): 2379-82.
[PMID: 9573136]

[189] Wang HC, Dann SM, Okhuysen PC, *et al.* High levels of CXCL10 are produced by intestinal epithelial cells in AIDS patients with active cryptosporidiosis but not after reconstitution of immunity. Infect Immun 2007; 75(1): 481-7.
[http://dx.doi.org/10.1128/IAI.01237-06] [PMID: 17043107]

[190] Laurent F, Kagnoff MF, Savidge TC, Naciri M, Eckmann L. Human intestinal epithelial cells respond to *Cryptosporidium parvum* infection with increased prostaglandin H synthase 2 expression and prostaglandin E2 and F2alpha production. Infect Immun 1998; 66(4): 1787-90.
[PMID: 9529115]

[191] McDonald V, Pollok RC, Dhaliwal W, Naik S, Farthing MJ, Bajaj-Elliott M. A potential role for interleukin-18 in inhibition of the development of *Cryptosporidium parvum.* Clin Exp Immunol 2006; 145(3): 555-62.
[http://dx.doi.org/10.1111/j.1365-2249.2006.03159.x] [PMID: 16907926]

[192] Tarver AP, Clark DP, Diamond G, *et al.* Enteric beta-defensin: molecular cloning and characterization of a gene with inducible intestinal epithelial cell expression associated with *Cryptosporidium parvum* infection. Infect Immun 1998; 66(3): 1045-56.
[PMID: 9488394]

[193] Chalmers RM, Katzer F. Looking for *Cryptosporidium*: the application of advances in detection and diagnosis. Trends Parasitol 2013; 29(5): 237-51.
[http://dx.doi.org/10.1016/j.pt.2013.03.001] [PMID: 23566713]

[194] Smith H. Diagnostics. In: Fayer R, Xiao L, Eds. Cryptosporidium and Cryptosporidiosis. Boca Raton: CRC Press 2008; pp. 173-208.

[195] Xiao L, Ryan UM. Molecular epidemiology. In: Fayer R, Xiao L, Eds. Cryptosporidium and Cryptosporidiosis. Boca Raton: CRC Press 2008; pp. 119-72.

[196] Hadfield SJ, Robinson G, Elwin K, Chalmers RM. Detection and differentiation of *Cryptosporidium* spp. in human clinical samples by use of real-time PCR. J Clin Microbiol 2011; 49(3): 918-24.
[http://dx.doi.org/10.1128/JCM.01733-10] [PMID: 21177904]

[197] Alonso JL, Amorós I, Cañigral I. Development and evaluation of a real-time PCR assay for quantification of *Giardia* and *Cryptosporidium* in sewage samples. Appl Microbiol Biotechnol 2011; 89(4): 1203-11.
[http://dx.doi.org/10.1007/s00253-010-2984-6] [PMID: 21061006]

[198] Loganthan S, Yang R, Bath A, Gordon C, Ryan U. Prevalence of *Cryptosporidium* species in recreational *versus* non-recreational water sources. Exp Parasitol 2012; 131(4): 399-403.
[http://dx.doi.org/10.1016/j.exppara.2012.04.015] [PMID: 22609970]

[199] Maas L, Dorigo-Zetsma JW, de Groot CJ, Bouter S, Plötz FB, van Ewijk BE. Detection of intestinal protozoa in paediatric patients with gastrointestinal symptoms by multiplex real-time PCR. Clin Microbiol Infect 2014; 20(6): 545-50.
[http://dx.doi.org/10.1111/1469-0691.12386] [PMID: 24131443]

[200] Mead JT, Arrowood MJ. Treatment of cryptosporidiosis. In: Cacciò SM, Widmer G, Eds. Cryptosporidium: Parasite and Disease. New York: Springer 2014; pp. 455-88.
[http://dx.doi.org/10.1007/978-3-7091-1562-6_11]

[201] Cacciò SM, Pozio E. Advances in the epidemiology, diagnosis and treatment of cryptosporidiosis. Expert Rev Anti Infect Ther 2006; 4(3): 429-43.
[http://dx.doi.org/10.1586/14787210.4.3.429] [PMID: 16771620]

[202] Mele R, Gomez Morales MA, Tosini F, Pozio E. Indinavir reduces *Cryptosporidium parvum* infection in both *in vitro* and *in vivo* models. Int J Parasitol 2003; 33(7): 757-64.
[http://dx.doi.org/10.1016/S0020-7519(03)00093-6] [PMID: 12814654]

[203] Gargala G, Delaunay A, Li X, Brasseur P, Favennec L, Ballet JJ. Efficacy of nitazoxanide, tizoxanide and tizoxanide glucuronide against *Cryptosporidium parvum* development in sporozoite-infected HCT-8 enterocytic cells. J Antimicrob Chemother 2000; 46(1): 57-60.
[http://dx.doi.org/10.1093/jac/46.1.57] [PMID: 10882689]

[204] Müller J, Sidler D, Nachbur U, Wastling J, Brunner T, Hemphill A. Thiazolides inhibit growth and induce glutathione-S-transferase Pi (GSTP1)-dependent cell death in human colon cancer cells. Int J Cancer 2008; 123(8): 1797-806.
[http://dx.doi.org/10.1002/ijc.23755] [PMID: 18688861]

[205] Keithly JS, Zhu G, Upton SJ, Woods KM, Martinez MP, Yarlett N. Polyamine biosynthesis in *Cryptosporidium parvum* and its implications for chemotherapy. Mol Biochem Parasitol 1997; 88(1-2): 35-42.
[http://dx.doi.org/10.1016/S0166-6851(97)00063-7] [PMID: 9274865]

[206] Zhu G, Marchewka MJ, Woods KM, Upton SJ, Keithly JS. Molecular analysis of a Type I fatty acid synthase in *Cryptosporidium parvum.* Mol Biochem Parasitol 2000; 105(2): 253-60.
[http://dx.doi.org/10.1016/S0166-6851(99)00183-8] [PMID: 10693747]

[207] Compendium of measures to prevent disease associated with animals in public settings, 2011: National Association of State Public Health Veterinarians, Inc. MMWR Recomm Rep 2011; 60(RR-04): 1-24.
[PMID: 21546893]

[208] Abubakar I, Aliyu SH, Arumugam C, Hunter PR, Usman NK. Prevention and treatment of cryptosporidiosis in immunocompromised patients. Cochrane Database Syst Rev 2007; 1(1): CD004932.
[PMID: 17253532]

CHAPTER 3

Cyclospora cayetanensis

Ynes R. Ortega*

Center for Food Safety, University of Georgia, Griffin, Georgia, USA

Abstract: The coccidian parasite *Cyclospora cayetanensis* can cause gastrointestinal illness. It has been associated with consumption of contaminated foods, particularly those eaten raw or minimally processed. First considered a cyanobacteria or coccidian-like organism, in the early 1990s it was characterized as a member of the *Cyclospora* genus based on morphological properties. This parasite has been associated with travelers' diarrhea. To date, there is no animal reservoir identified for *C. cayetanensis*. Waterborne outbreaks have also been reported. In July of 1990, 21 household staff physicians developed diarrheal illness and analysis of their stool specimens showed organisms consistent with *Cyclospora* oocysts. Epidemiological investigations implicated tap water as the likely source of the outbreak. In another instance, in 1994, British soldiers and their dependents stationed in Nepal developed cyclosporiasis. Local drinking water with chlorine concentrations of 0.3-0.8 ppm was associated with this outbreak and *Cyclospora* oocysts were isolated from the water. Additionally, imported berries, salad greens, and herbs were associated with foodborne outbreaks of cyclosporiasis in the U.S., Canada, and European countries in the late 1990s. In 2013, 2014, and 2015 *C. cayetanensis* caused significant outbreaks in the U.S. and were epidemiologically associated with imported salad greens and cilantro. *Cyclospora* is resistant to many commonly used sanitizers and certain environmental conditions which are found in many locations used to process/package produce. Oocysts can be inactivated by either freezing or heating. Tools for traceback studies and *in vitro* or *in vivo* models to propagate this parasite are currently not available.

Keywords: Coccidia, Control, *Cyclospora cayetanensis*, Eimeriidae, Epidemiology, Foodborne, Transmission, Waterborne.

INTRODUCTION

Cyclospora cayetanensis is a coccidian parasite that causes gastrointestinal illness in humans, and it is a frequent cause of traveler's diarrhea in people visiting endemic areas such as Nepal, Haiti, Mexico, Guatemala, and Peru. The seasonality of this parasite makes the risk of acquiring this infection higher during

* **Corresponding author Ynes R. Ortega:** Center for Food Safety, University of Georgia, Griffin, Georgia, USA; Tel: +1-678-427-8089; E-mail: ortega@uga.edu

Fabrizio Bruschi (Ed.)

particular months of the year of each endemic location. Since 1995, *Cyclospora* outbreaks have been reported in the U.S. and Canada. Foodborne transmission has been confirmed and in most instances, linked to imported food commodities such as salad greens, herbs, and fruits. In this chapter the biology, epidemiology, treatment, diagnosis, prevention, the effectiveness of water treatment processes and recommendations for *Cyclospora cayetanensis* will be described.

History

Reports of travelers' diarrhea caused by coccidian/cyanobacterian-like bodies were perhaps the first reports of *Cyclospora.* Ashford reported three cases in Papua New Guinea which he thought were caused by *Isospora* spp. (now denominated *Cystoisospora*) [1]. Later in 1986, Soave described 4 cases of patients with organisms resembling *Cyclospora* in their stool samples. In 1989, 55 immunocompetent travellers and expatriates in Nepal presented with gastrointestinal illness, particularly diarrhea, and had algae-like organisms in their stools. At that time, it was thought that the outbreak had occurred during the months of May-June, but epidemiological studies associated with this outbreak were not performed [2]. We now know that these months correspond to the annual high season of *Cyclospora* in Nepal and that *Cyclospora* is endemic in that country. In the early 1990's, a team of scientists in Peru working on diarrheal illnesses in children noticed the presence of cyst-like structures twice the size of *Cryptosporidium.* Further investigation showed that if these structures were left in dichromate, they would differentiate from a morula-like form to two sporocysts within a round oocyst. These organisms or CLBs, at that time called "big *Cryptosporidium*", were in fact oocysts of a new coccidian parasite. This new parasite was classified into the genus *Cyclospora* based on morphologic characteristics and the parasite species was later named '*cayetanensis*' in recognition of the work performed at the Cayetano Heredia University in Lima, Peru [3, 4]. At the same time, treatments were being explored for this illness and the same Peruvian team identified trimethoprim sulfamethoxazole as the drug of choice to effectively control cyclosporiasis [5, 6]. These results were later validated by other groups of investigators [7 - 9].

MORPHOLOGY AND LIFE CYCLE

Cyclospora has three stages in its life cycle. The first stage occurs outside the host. The unsporulated oocysts are not differentiated or infectious when excreted and have a morula-like form. In about 7 to 15 days outside the host and under favorable environmental conditions, the unsporulated oocyst will differentiate to form the sporulated oocyst that consists of two sporocysts, each with two sporozoites. The second phase starts when the oocysts are ingested *via* contaminated water or foods; these oocysts will excyst and the sporozoites will infect intestinal epithelial cells. The first asexual multiplication results on the formation of a type I meront containing 8 merozoites. This is followed by the formation of a type II meront with 4 merozoites. The meronts are localized in the cytoplasm of intestinal epithelial cells in the jejunum. The third phase occurs when sexual multiplication takes place and merozoites differentiate to form gametocytes. A microgametocyte will fertilize a macrogametocyte to form a zygote which in turn will develop into an unsporulated oocyst that will be excreted in the feces of an infected individual (Fig. **1**).

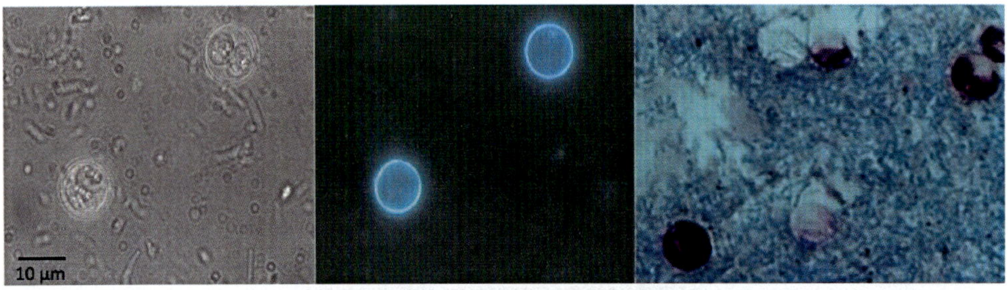

Fig. (1). *Cyclospora cayetanensis* oocysts observed using Differential Interference Contrast (DIC) microscopy, autofluorescence, and acid fast staining.

GLOBAL EPIDEMIOLOGY AND RISK FACTORS

Cyclospora is endemic in various locations worldwide and in-depth epidemiological studies have been performed in Nepal, Peru, Haiti, Guatemala, and Mexico [2, 3, 6, 7, 10 - 13]. Sporadic cases of *Cyclospora* have also been reported in Canada, Germany, Australia, and the U.S. and several outbreaks have been reported in countries considered not endemic for *C. cayetanensis* [14 - 17]. This infection can be acquired by ingestion of contaminated water and foods containing *Cyclospora* oocysts. Person to person transmission is unlikely since oocysts require at least 7 days to sporulate in the environment. Reservoirs and mechanical vectors have also been suggested to play a role in the life cycle of this parasite [18 - 20], but it is currently considered exclusively anthroponotic. *Cyclospora* oocysts have been found in feces of chickens, ducks, and dogs in several studies [18 - 20], but attempts to reproduce infection in laboratory animals have been unsuccessful and surveys in some endemic locations have come up negative in these and other animals. It has been suggested that coprophagic activities of these animals may have led to the reported identifications in those studies [21].

Cyclospora waterborne outbreaks have also been reported. Two notable outbreak were reported, one in the U.S. in Chicago and the other in Nepal. In July 1990, several house staff physicians attending a party in Chicago developed diarrhea and *Cyclospora* oocysts were identified in their stool samples. The epidemiological investigation of this outbreak consisted of a house staff cohort and an administrative employee's cohort. The house staff cohort study indicated that there was a significant association with drinking tap water (11.6 RR) and attending a house staff party. The administrative employees' cohort also indicated drinking tap water and attending a house staff party (RR of 3.8 and 5.5 respectively). Environmental assessments indicated that a tap water pump was not working the day before the first cases of diarrhea. The pump was repaired and the chlorine levels were 0.2-0.4 ppm. Coliforms and algae, particularly diatoms, were identified in the water at the bottom of the storage tanks while *Cyclospora* oocysts were not detected in any of the samples. The authors concluded that this outbreak was most likely associated with drinking or exposure to tap water. It was unclear, however, if other potential modes of exposure were also included in the questionnaires used. Since most cases of *Cyclospora* led to the development of diarrhea 3-5 days post exposure, the authors did not address the impact of the water pump failure 24 h before the outbreak [22].

In June 1994, British soldiers and their dependents developed persistent diarrhea caused by *Cyclospora*. The epidemiological association indicated that treated drinking water with 0.3-0.8 ppm of free chlorine consisting of a mix of river and municipal water was the likely source of infection. Indications were confirmed when oocysts were isolated from a two-liter water sample which was concentrated using a 5um Millipore filter [23].

Other studies reported the presence of *Cyclospora* in drinking water in Guatemala, Haiti, Egypt, Ghana, and Nepal [14] (Table **1**).

Table 1. Description of *Cyclospora cayetanensis* in water.

Location	N. Positive/ Tested Samples	Water	References
Accra, Ghana	59% of 27	500 ml sachets drinking water	[24]
Guatemala City, rural	5/12	Water fed by spring or well	
Egypt	0.24% of 840	Drinking water	[25]
Guatemala and Sacatepequez, Guatemala	2/30	River water	[12]
Haiti	1/ 55	Well and other	[26]

(Table 1) contd.....

Location	N. Positive/ Tested Samples	Water	References
Hanoi, Vietnam	63.6%	river and lake water samples	[27]
Nepal	n/d	Sewage water	[28]
Italy	21.3% of 94, 6.2% of 16	Treated water, well water	[49]
Egypt	11.9%	Pond and canal water	[50]
Arizona, US	25%, 17%, 18% of 48	Wastewater influent and effluent wastewater treatment plant	[51]
Spain	9% of 223	Wastewater treatment plant	[45]

Drinking water from well water, surface water, and water from treatment plants in Vietnam were collected over a period of twelve months. *Cyclospora* was not detected in well water or in treated water (n=18); however, 63.6% of river water (n=64) and lake water (n=64) samples tested positive. The distribution of these positive water samples varied according to geographical locations [27]. Seasonality was also identified in environmental samples in Nepal. Both sewage and drinking water were found to be contaminated with *Cyclospora* during the months of June, July, August, and November [28].

Although water is implicated as a source of transmission, it is not the only source of cyclosporiasis. It is estimated that 90% of the *Cyclospora* infections in the U.S. are attributed to foodborne transmission including salad greens, basil, sugar peas, raspberries, and blackberries. All of these items are consumed raw and are minimally processed. Survey studies have identified *C. cayetanensis* oocysts in vegetables that are usually consumed raw (Table **2**).

Table 2. Detection of *Cyclospora cayetanensis* oocysts in foods.

Location	% Positive/Total Samples	Food Type	Reference
Italy	12.2% of 49	Vegetables, fennel	[49]
Peru	1.8%, 1.6% of 110, 62	Lettuce, mint, black mint	[52]
Vietnam	8.4% of 287	herbs	[53]
Cambodia	8% of 54	Water spinach	[54]
Costa Rica	8% of 25	lettuce	[55]
Egypt	21.3% of 300	Rocket, lettuce, parsley, leek, green onion	[56]

Recent outbreaks in the U.S. have been associated with fresh foods. In 2013, between mid June and mid July, 631 cases of cyclosporiasis were reported in 23 states and New York City with a median patient age of 52 years. Epidemiological and traceback investigations showed that there were actually two outbreaks occurring at that time. Cases in Iowa and Nebraska implicated salad mix produced by Taylor Farms in Mexico while another outbreak in Texas was linked to fresh cilantro from Puebla, Mexico. No oocysts were identified in any of the implicated product items [48]. In 2014, 304 cases were reported in 29 states and New York City. Most of these cases (64%) were in Texas and the median patient age was 49 years. Epidemiological and traceback investigations indicated that some cases were likely linked to fresh cilantro from Puebla, Mexico. In 2015, 495 cases of cyclosporiasis were also reported in 30 states in 2015. The cases reported in Wisconsin and Texas were linked as well to cilantro as the likely source of the outbreak [48].

The lack of an animal reservoir for *C. cayetanensis* and the number of days required for the unsporulated oocysts to sporulate, suggest that contamination for salad greens and cilantro would more likely occur in the fields. Events such as application of pesticides prepared with contaminated water or irrigation with contaminated water containing *Cyclospora* oocysts could result in contaminated products.

PATHOPHYSIOLOGY

Cyclospora primarily affects the small intestine. Expatriates in Nepal presented with marked erythema of the distal duodenum. All had mild to moderate inflammation and an increase of plasma cells was observed in the lamina propria. Focal vacuolization of the tips of the surface epithelium, loss of brush border, and alteration of the cells from columnar to cuboid shape, partial villous atrophy, and crypt hyperplasia was observed in individuals with cyclosporiasis [29]. Similar histopathological observations were noted in studies performed in patients of other endemic areas. Diffuse edema and mixed inflammatory cell infiltration, including the presence of plasma cells and lymphocytes in the villous mucosa, has also been reported in these studies [30 - 32]. In some patients, high numbers of eosinophils has also been noted [32]. Myelin-like material between the base and the sides of the enterocytes were also present in some patients suggesting an ongoing inflammatory or immunologic process [29].

IMMUNOLOGY AND IMMUNOPATHOLOGY

In endemic locations, children usually present with acute diarrhea and repeated infections. Older children present with a less severe illness and of shorter duration, suggesting development of an immune response to the parasite after repeated exposures [11]. Attempts to describe the immune response to cyclosporiasis started in the early 1980s. To date, suitable methods to determine immune responses of individuals with acute or convalescent cyclosporiasis are lacking. This is largely due to the lack of confirmed animal models to study this infection.

SYMPTOMS

Cyclosporiasis is characterized with acute and persistent diarrhea. In endemic locations the clinical presentation is usually less severe in adults; it is more frequently identified in children older than 1 year of age. Each repeat infection results in less severe illness and shorter duration [11]. Nausea, vomiting, abdominal pain, anorexia, fatigue, low grade fever, and weight loss have also been associated with cyclosporiasis. Biliary disease has also been reported after reported *Cyclospora* infections [4]. Acalculous cholecystitis [36] was described in HIV and AIDS patients that resolved after treatment. Guillain-Barre [33] and Reiter [34] syndromes have also been reported following *Cyclospora* infection.

DIAGNOSIS AND DETECTION METHODS

In the clinical diagnosis setting, there are various alternatives to detect *Cyclospora* from fecal samples. With conventional ova and parasite examination, oocysts can be identified at 40X either by bright field or phase contrast microscopy. If the examination is done within a week of excretion or in fixed specimens, the oocysts will have the typical morphology of unsporulated oocysts with the distinctive morula. Oocysts can also be observed by epifuorescence microscopy with oocysts fluorescing a pale blue color (Fig **2**). Oocysts can be stained using the modified Ziehl-Neelsen and safranin (with and without microwave heating) [4, 37]. *Cyclospora* oocysts do not stain using iodine, iron hematoxylin, Gomori's methenamine silver (GMS) and periodic acid-shift (PAS), Giemsa, Gram, methylene blue, or trichrome [30]. Oocysts can be concentrated using the Ritchie protocol. Molecular tools are also available to detect *Cyclospora* oocysts by targeting the 18S rRNA, HSP-70, and ITS-1 genes [38 - 41]. Environmental samples require additional testing for specificity including PCR-RFLP and sequencing as amplification can be nonspecific and coccidian parasites not considered infectious to humans may be present [42, 43].

An immunofluorescence assay was evaluated using immobilized whole and sectioned oocysts; testing showed no difference among patient groups [35]. A western blot assay was also developed, however very large quantities of oocysts are needed making this assay difficult to implement in outbreak investigations (Ortega unpublished data).

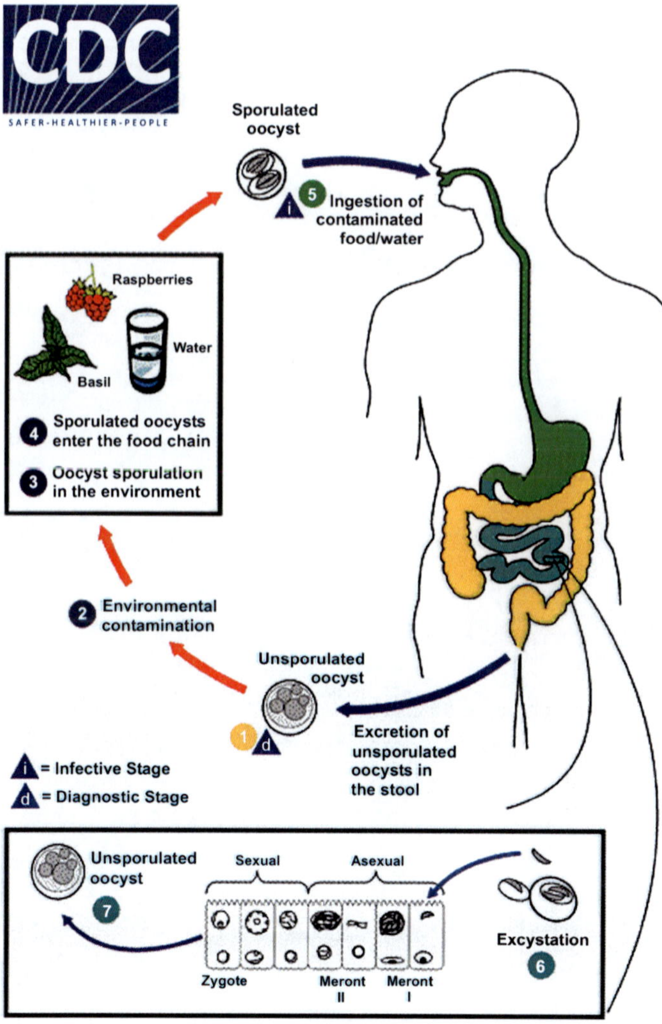

Fig. (2). Life cycle of Cyclospora cayetanensis in humans. http://www.cdc.gov/parasites/cyclosporiasis/biology.html.

TREATMENT

A seven-day treatment with trimethoprim sulfamethoxazole (TMP-SMX) successfully controls diarrhea and infection in both adults and children [5 - 7]. *Cyclospora* oocysts are no longer observed in stool samples after two days of treatment. A double blind placebo controlled study was performed in healthy adult expatriates in Nepal with 21 receiving co-trimoxazole (160 mg trimethoprim, 800 mg sulphamethoxazole) and 19 were given a placebo. Oocyst excretion ended three days after receiving co-trimoxazole twice daily for 7 days. After day 7 of treatment, one patient from the treatment group (6%) was excreting oocysts compared to 88% of the placebo group [44]. In another study in Haiti, 43 patients with AIDS and cyclosporiasis received oral treatment four times a day for 10 days. Diarrhea ceased after 2.5 days of treatment with TMP-SMX. Of those treated, 44% patients had recurrent infection 1-3 months post treatment. These patients received oral prophylaxis three times a week. A third episode of cyclosporiasis was detected in one patient 7 months later [7].

Ciprofloxacin was evaluated in AIDS patients and was found as an alternative treatment for individuals who are sulfa allergic [9].

PERSPECTIVES OF CONTROL

Cyclospora infections can be acquired by consuming contaminated foods and water. Chlorination of drinking water does not seem to inactivate *Cyclospora* oocysts; a characteristic shared by many other parasites related to *Cyclospora.*

Cyclospora cayetanensis was detected in drinking water and wastewater treatment plants in a study in Madrid, Spain. Oocysts were identified in 17 of 20 samples tested. The treatment plants met European and local regulations on water sanitary quality, suggesting that a reevaluation of controls and regulations for drinking water, wastewater, and recreational waters should be explored [45]. These results are supported by another study performed in Arizona, U.S. where *Cyclospora* oocysts were detected in the influent and effluent of two wastewater treatment plants [46].

In water and food matrices, oocysts will not sporulate when heated or frozen; therefore rendering oocysts non-infectious. Oocysts are resistant to chemical treatments and pesticides commonly used by the fresh produce industry [47].

CONCLUSION

Cyclospora was originally considered a cyanobacteria-like organism in the 1980's and later fully characterized as a coccidian parasite of the family Eimeriidae. The treatment and epidemiological associations with foods and water have been studied extensively. Molecular tools for traceback studies still remain to be fully implemented, however with the development and use of MLS tools for the identification of contaminated produce sources (unpublished data), molecular approaches for identification of *C. cayetanensis* show promise.

Our major challenge is the lack of *in vitro* cultivation and animal models to propagate *Cyclospora.* Therefore, any studies on this organism currently rely on the availability of oocysts shed by naturally infected individuals. Viability testing for evaluation of treatments or inactivation strategies in foods is limited. We still do not understand the factors that trigger the seasonality in endemic locations nor where these organisms remain dormant until the high season begins. The worldwide distribution of *Cyclospora* remains to be determined.

CONFLICT OF INTEREST

The author confirms that author has no conflict of interest to declare for this publication.

ACKNOWLEDGEMENTS

Declared none.

REFERENCES

[1] Ashford RW. Occurrence of an undescribed coccidian in man in Papua New Guinea. Ann Trop Med Parasitol 1979; 73(5): 497-500.
[http://dx.doi.org/10.1080/00034983.1979.11687291] [PMID: 534451]

[2] Shlim DR, Cohen MT, Eaton M, Rajah R, Long EG, Ungar BL. An alga-like organism associated with an outbreak of prolonged diarrhea among foreigners in Nepal. Am J Trop Med Hyg 1991; 45(3): 383-9.
[PMID: 1928575]

[3] Ortega YR, Gilman RH, Sterling CR. A new coccidian parasite (Apicomplexa: Eimeriidae) from humans. J Parasitol 1994; 80(4): 625-9.
[http://dx.doi.org/10.2307/3283201] [PMID: 8064531]

[4] Ortega YR, Sterling CR, Gilman RH, Cama VA, Díaz F. Cyclospora speciesa new protozoan pathogen of humans. N Engl J Med 1993; 328(18): 1308-12.
[http://dx.doi.org/10.1056/NEJM199305063281804] [PMID: 8469253]

[5] Madico G, Gilman RH, Miranda E, Cabrera L, Sterling CR. Treatment of Cyclospora infections with co-trimoxazole. Lancet 1993; 342(8863): 122-3.
[http://dx.doi.org/10.1016/0140-6736(93)91330-O] [PMID: 8100895]

[6] Madico G, McDonald J, Gilman RH, Cabrera L, Sterling CR. Epidemiology and treatment of Cyclospora cayetanensis infection in Peruvian children. Clin Infect Dis 1997; 24(5): 977-81.
[http://dx.doi.org/10.1093/clinids/24.5.977] [PMID: 9142805]

[7] Pape JW, Verdier RI, Boncy M, Boncy J, Johnson WD Jr. Cyclospora infection in adults infected with HIV. Clinical manifestations, treatment, and prophylaxis. Ann Intern Med 1994; 121(9): 654-7.
[http://dx.doi.org/10.7326/0003-4819-121-9-199411010-00004] [PMID: 7944073]

[8] Shlim DR, Pandey P, Rabold JG, Walch A, Rajah R. An Open Trial of Trimethoprim Alone against Cyclospora Infections. J Travel Med 1997; 4(1): 44-5.
[http://dx.doi.org/10.1111/j.1708-8305.1997.tb00773.x] [PMID: 9815477]

[9] Verdier RI, Fitzgerald DW, Johnson WD Jr, Pape JW. Trimethoprim-sulfamethoxazole compared with ciprofloxacin for treatment and prophylaxis of Isospora belli and Cyclospora cayetanensis infection in HIV-infected patients. A randomized, controlled trial. Ann Intern Med 2000; 132(11): 885-8.
[http://dx.doi.org/10.7326/0003-4819-132-11-200006060-00006] [PMID: 10836915]

[10] Sherchand JB, Cross JH. Emerging pathogen Cyclospora cayetanensis infection in Nepal. Southeast Asian J Trop Med Public Health 2001; 32 (Suppl. 2): 143-50.
[PMID: 12041579]

[11] Bern C, Ortega Y, Checkley W, *et al.* Epidemiologic differences between cyclosporiasis and cryptosporidiosis in Peruvian children. Emerg Infect Dis 2002; 8(6): 581-5.
[http://dx.doi.org/10.3201/eid0806.01-0331] [PMID: 12023913]

[12] Bern C, Hernandez B, Lopez MB, *et al.* Epidemiologic studies of Cyclospora cayetanensis in Guatemala. Emerg Infect Dis 1999; 5(6): 766-74.
[http://dx.doi.org/10.3201/eid0506.990604] [PMID: 10603209]

[13] Orozco-Mosqueda GE, Martínez-Loya OA, Ortega YR. Cyclospora cayetanensis in a pediatric hospital in Morelia, México. Am J Trop Med Hyg 2014; 91(3): 537-40.
[http://dx.doi.org/10.4269/ajtmh.13-0535] [PMID: 24957545]

[14] Ortega YR, Sanchez R. Update on Cyclospora cayetanensis, a food-borne and waterborne parasite. Clin Microbiol Rev 2010; 23(1): 218-34.
[http://dx.doi.org/10.1128/CMR.00026-09] [PMID: 20065331]

[15] Update: outbreaks of Cyclospora cayetanensis infectionUnited States and Canada, 1996. MMWR Morb Mortal Wkly Rep 1996; 45(28): 611-2.
[PMID: 8676854]

[16] Gibbs RA, Nanyonjo R, Pingault NM, *et al.* An outbreak of Cyclospora infection on a cruise ship. Epidemiol Infect 2013; 141(3): 508-16.
[http://dx.doi.org/10.1017/S0950268812001197] [PMID: 22687637]

[17] Petry F, Hofstätter J, Schulz BK, Deitrich G, Jung M, Schirmacher P. Cyclospora cayetanensis: first imported infections in Germany. Infection 1997; 25(3): 167-70.
[http://dx.doi.org/10.1007/BF02113606] [PMID: 9181385]

[18] Zerpa R, Uchima N, Huicho L. Cyclospora cayetanensis associated with watery diarrhoea in Peruvian patients. J Trop Med Hyg 1995; 98(5): 325-9.
[PMID: 7563260]

[19] García-López HL, Rodríguez-Tovar LE, Medina-De la Garza CE. Identification of Cyclospora in poultry. Emerg Infect Dis 1996; 2(4): 356-7.
[http://dx.doi.org/10.3201/eid0204.960414] [PMID: 8969254]

[20] Yai LE, Bauab AR, Hirschfeld MP, de Oliveira ML, Damaceno JT. The first two cases of Cyclospora in dogs, São Paulo, Brazil. Rev Inst Med Trop Sao Paulo 1997; 39(3): 177-9.
[http://dx.doi.org/10.1590/S0036-46651997000300011] [PMID: 9460261]

[21] Eberhard ML, Ortega YR, Hanes DE, *et al.* Attempts to establish experimental Cyclospora cayetanensis infection in laboratory animals. J Parasitol 2000; 86(3): 577-82.
[http://dx.doi.org/10.1645/0022-3395(2000)086[0577:ATEECC]2.0.CO;2] [PMID: 10864257]

[22] Huang P, Weber JT, Sosin DM, *et al.* The first reported outbreak of diarrheal illness associated with Cyclospora in the United States. Ann Intern Med 1995; 123(6): 409-14.
[http://dx.doi.org/10.7326/0003-4819-123-6-199509150-00002] [PMID: 7639439]

[23] Rabold JG, Hoge CW, Shlim DR, Kefford C, Rajah R, Echeverria P. Cyclospora outbreak associated with chlorinated drinking water. Lancet 1994; 344(8933): 1360-1.
[http://dx.doi.org/10.1016/S0140-6736(94)90716-1] [PMID: 7968039]

[24] Kwakye-Nuako G, Borketey P, Mensah-Attipoe I, Asmah R, Ayeh-Kumi P. Sachet drinking water in accra: the potential threats of transmission of enteric pathogenic protozoan organisms. Ghana Med J 2007; 41(2): 62-7.
[PMID: 17925844]

[25] Elshazly AM, Elsheikha HM, Soltan DM, Mohammad KA, Morsy TA. Protozoal pollution of surface water sources in Dakahlia Governorate, Egypt. J Egypt Soc Parasitol 2007; 37(1): 51-64.
[PMID: 17580568]

[26] Lopez AS, Bendik JM, Alliance JY, *et al.* Epidemiology of Cyclospora cayetanensis and other intestinal parasites in a community in Haiti. J Clin Microbiol 2003; 41(5): 2047-54.
[http://dx.doi.org/10.1128/JCM.41.5.2047-2054.2003] [PMID: 12734247]

[27] Miegeville M, Koubi V, Dan LC, Barbier JP, Cam PD. [Cyclospora cayetanensis presence in aquatic surroundings in Hanoi (Vietnam). Environmental study (well water, lakes and rivers)]. Bull Soc Pathol Exot 2003; 96(3): 149-52. [Cyclospora cayetanensis presence in aquatic surroundings in Hanoi (Vietnam). Environmental study (well water, lakes and rivers)].
[PMID: 14582286]

[28] Sherchand JB, Cross JH, Jimba M, Sherchand S, Shrestha MP. Study of Cyclospora cayetanensis in health care facilities, sewage water and green leafy vegetables in Nepal. Southeast Asian J Trop Med Public Health 1999; 30(1): 58-63.
[PMID: 10695790]

[29] Connor BA, Reidy J, Soave R. Cyclosporiasis: clinical and histopathologic correlates. Clin Infect Dis 1999; 28(6): 1216-22.
[http://dx.doi.org/10.1086/514780] [PMID: 10451156]

[30] Sun T, Ilardi CF, Asnis D, *et al.* Light and electron microscopic identification of Cyclospora species in the small intestine. Evidence of the presence of asexual life cycle in human host. Am J Clin Pathol 1996; 105(2): 216-20.
[http://dx.doi.org/10.1093/ajcp/105.2.216] [PMID: 8607448]

[31] Deluol AM, Teilhac MF, Poirot JL, Heyer F, Beaugerie L, Chatelet FP. Cyclospora sp:life cycle studies in patient by electron-microscopy. J Eukaryot Microbiol 1996; 43(5): 128S-9S.
[http://dx.doi.org/10.1111/j.1550-7408.1996.tb05042.x] [PMID: 8822906]

[32] Ortega YR, Nagle R, Gilman RH, *et al.* Pathologic and clinical findings in patients with cyclosporiasis and a description of intracellular parasite life-cycle stages. J Infect Dis 1997; 176(6): 1584-9.
[http://dx.doi.org/10.1086/514158] [PMID: 9395371]

[33] Richardson RF Jr, Remler BF, Katirji B, Murad MH. Guillain-Barré syndrome after Cyclospora infection. Muscle Nerve 1998; 21(5): 669-71.
[http://dx.doi.org/10.1002/(SICI)1097-4598(199805)21:5<669::AID-MUS20>3.0.CO;2-P] [PMID: 9572253]

[34] Connor BA, Johnson EJ, Soave R. Reiter syndrome following protracted symptoms of Cyclospora infection. Emerg Infect Dis 2001; 7(3): 453-4.
[http://dx.doi.org/10.3201/eid0703.017317] [PMID: 11384527]

[35] Clarke SC, McIntyre M. An attempt to demonstrate a serological immune response in patients infected with Cyclospora cayetanensis. Br J Biomed Sci 1997; 54(1): 73-4.
[PMID: 9167312]

[36] Sifuentes-Osornio J, Porras-Cortés G, Bendall RP, Morales-Villarreal F, Reyes-Terán G, Ruiz-Palacios GM. Cyclospora cayetanensis infection in patients with and without AIDS: biliary disease as another clinical manifestation. Clin Infect Dis 1995; 21(5): 1092-7.
[http://dx.doi.org/10.1093/clinids/21.5.1092] [PMID: 8589126]

[37] Visvesvara GS, Moura H, Kovacs-Nace E, Wallace S, Eberhard ML. Uniform staining of Cyclospora oocysts in fecal smears by a modified safranin technique with microwave heating. J Clin Microbiol 1997; 35(3): 730-3.
[PMID: 9041421]

[38] Relman DA, Schmidt TM, Gajadhar A, *et al.* Molecular phylogenetic analysis of Cyclospora, the human intestinal pathogen, suggests that it is closely related to Eimeria species. J Infect Dis 1996; 173(2): 440-5.
[http://dx.doi.org/10.1093/infdis/173.2.440] [PMID: 8568307]

[39] Sulaiman IM, Ortega Y, Simpson S, Kerdahi K. Genetic characterization of human-pathogenic Cyclospora cayetanensis parasites from three endemic regions at the 18S ribosomal RNA locus. Infect Genet Evol 2014; 22: 229-34.
[http://dx.doi.org/10.1016/j.meegid.2013.07.015] [PMID: 23891667]

[40] Sulaiman IM, Torres P, Simpson S, Kerdahi K, Ortega Y. Sequence characterization of heat shock protein gene of Cyclospora cayetanensis isolates from Nepal, Mexico, and Peru. J Parasitol 2013; 99(2): 379-82.
[http://dx.doi.org/10.1645/GE-3114.1] [PMID: 22924935]

[41] Adam RD, Ortega YR, Gilman RH, Sterling CR. Intervening transcribed spacer region 1 variability in Cyclospora cayetanensis. J Clin Microbiol 2000; 38(6): 2339-43.
[PMID: 10834999]

[42] Jinneman KC, Wetherington JH, Hill WE, *et al.* An oligonucleotide-ligation assay for the differentiation between Cyclospora and Eimeria spp. polymerase chain reaction amplification products. J Food Prot 1999; 62(6): 682-5.
[http://dx.doi.org/10.4315/0362-028X-62.6.682] [PMID: 10382662]

[43] Jinneman KC, Wetherington JH, Hill WE, *et al.* Template preparation for PCR and RFLP of amplification products for the detection and identification of Cyclospora sp. and Eimeria spp. Oocysts directly from raspberries. J Food Prot 1998; 61(11): 1497-503.
[http://dx.doi.org/10.4315/0362-028X-61.11.1497] [PMID: 9829192]

[44] Hoge CW, Shlim DR, Ghimire M, *et al.* Placebo-controlled trial of co-trimoxazole for Cyclospora infections among travellers and foreign residents in Nepal. Lancet 1995; 345(8951): 691-3.
[http://dx.doi.org/10.1016/S0140-6736(95)90868-4] [PMID: 7885125]

[45] Galván AL, Magnet A, Izquierdo F, *et al.* Molecular characterization of human-pathogenic microsporidia and Cyclospora cayetanensis isolated from various water sources in Spain: a year-long longitudinal study. Appl Environ Microbiol 2013; 79(2): 449-59.
[http://dx.doi.org/10.1128/AEM.02737-12] [PMID: 23124243]

[46] Kitajima M, Haramoto E, Iker BC, Gerba CP. Occurrence of Cryptosporidium, Giardia, and Cyclospora in influent and effluent water at wastewater treatment plants in Arizona. Sci Total Environ 2014; 484: 129-36.
[http://dx.doi.org/10.1016/j.scitotenv.2014.03.036] [PMID: 24695096]

[47] Erickson MC, Ortega YR. Inactivation of protozoan parasites in food, water, and environmental systems. J Food Prot 2006; 69(11): 2786-808.
[http://dx.doi.org/10.4315/0362-028X-69.11.2786] [PMID: 17133829]

[48] CDC. http://www.cdc.gov/parasites/cyclosporiasis/outbreaks/investigation-2013.html.

[49] Giangaspero A, Marangi M, Koehler AV, *et al.* Molecular detection of Cyclospora in water, soil, vegetables and humans in southern Italy signals a need for improved monitoring by health authorities. Int J Food Microbiol 2015; 211: 95-100.
[http://dx.doi.org/10.1016/j.ijfoodmicro.2015.07.002] [PMID: 26188495]

[50] Khalifa RM, Ahmad AK, Abdel-Hafeez EH, Mosllem FA. Present status of protozoan pathogens causing water-borne disease in northern part of El-Minia Governorate, Egypt. J Egypt Soc Parasitol 2014; 44(3): 559-66.
[http://dx.doi.org/10.12816/0007860] [PMID: 25643498]

[51] Kitajima M, Haramoto E, Iker BC, Gerba CP. Occurrence of Cryptosporidium, Giardia, and Cyclospora in influent and effluent water at wastewater treatment plants in Arizona. Sci Total Environ 2014; 484: 129-36.
[http://dx.doi.org/10.1016/j.scitotenv.2014.03.036] [PMID: 24695096]

[52] Ortega YR, Roxas CR, Gilman RH, *et al.* Isolation of Cryptosporidium parvum and Cyclospora cayetanensis from vegetables collected in markets of an endemic region in Peru. Am J Trop Med Hyg 1997; 57(6): 683-6.
[PMID: 9430527]

[53] Tram NT, Hoang LM, Cam PD, *et al.* Cyclospora spp. in herbs and water samples collected from markets and farms in Hanoi, Vietnam. Trop Med Int Health 2008; 13(11): 1415-20.
[http://dx.doi.org/10.1111/j.1365-3156.2008.02158.x] [PMID: 18983280]

[54] Vuong TA, Nguyen TT, Klank LT, Phung DC, Dalsgaard A. Faecal and protozoan parasite contamination of water spinach (Ipomoea aquatica) cultivated in urban wastewater in Phnom Penh, Cambodia. Trop Med Int Health 2007; 12 (Suppl. 2): 73-81.
[http://dx.doi.org/10.1111/j.1365-3156.2007.01944.x] [PMID: 18005318]

[55] Calvo M, Carazo M, Arias ML, Chaves C, Monge R, Chinchilla M. [Prevalence of Cyclospora sp., Cryptosporidium sp, microsporidia and fecal coliform determination in fresh fruit and vegetables consumed in Costa Rica]. Arch Latinoam Nutr 2004; 54(4): 428-32.
[PMID: 15969268]

[56] Said DE. Detection of parasites in commonly consumed raw vegetables. Alexandria J Medicine 2012; 48(4): 345-52.
[http://dx.doi.org/10.1016/j.ajme.2012.05.005]

CHAPTER 4

Cystoisospora belli

Massimiliano Galdiero*, **Settimia Alfano, Maria Elena Della Pepa, Gianluigi Franci, Claudia Ledda** and **Marilena Galdiero**

Department of Experimental Medicine, Division of Microbiology and Clinical Microbiology, Second University of Naples, Italy

Abstract: Cystoisosporiasis is a human intestinal disease caused by the protozoan parasite *Cystoisospora belli.* It is an obligate intracellular parasite that invades intestinal epithelial cells. It causes a self-limited or prolonged diarrhea, depending on the host's immune status. *C. belli* is cosmopolitan, but more common in the tropics and subtropics. The parasite can be spread by ingesting contaminated food or water. *C. belli* oocysts are ellipsoid and approximately 25 µm to 33 µm by 10 µm to 20 µm and contain two sporocysts, which contain 4 sporozoites each. Feces are the common way of immature form of the parasite release. The parasite, usually needs between 1 or 2 days in the environment (outside of host) to performe the adequate maturation steps to be able infect further subjects. Watery diarrhea is the most common symptom. Other symptoms can include cramps, abdominal pain, dehydration, loss of appetite, nausea, fever and vomiting. Diagnosis is achieved by oocysts observation *via* microscopic stool examination. Enhanced detection is obtained by staining stool samples *via* modified acid-fast stain, modified safranin stain or by fluorescence microscopy. Molecular diagnostic methods have the advantage of high sensibility for an early detection of these coccidian parasites in stool samples. If untreated, immunocompromised individuals may increase the risk for prolonged and severe illness. The usual treatment is with trimethoprim-sulfamethoxazole. Cystoisosporiasis can be prevented with adequate sanitation, measures to protect food and water supplies, and increased public awareness of the means of transmission.

Keywords: *C. belli* life cycle, Coccidia, *Cystoisospora belli*, Intracellular parasite, Parasitic infection, Protozoa.

INTRODUCTION

Cystoisospora belli (previously named *Isospora belli*) is a coccidian protozoan parasite that inhabits the gastrointestinal tract and causes cystoisosporiasis, a

* **Corresponding author Massimiliano Galdiero:** Department of Experimental Medicine, Division of Microbiology and Clinical Microbiology, Second University of Naples, Italy; Tel: 0815667697; Fax: 0815665663; E-mail: massimiliano.galdiero@unina2.it

human intestinal disease. It belonged to the genus Isospora until 2005; since then, it has been included in the genus Cystoisospora [1] (Table **1**). Morphological and molecular phylogenetic analysis [2] were fundamental to demonstrate the difference and divergence between the genus *Isospora* belonging to the Eimeriidae family and the genus *Cystoisospora* belonging to the Sarcocystidae. Both families belong to the suborder Eimeriorina and are included in the Coccidia class within the Apicomplexa phylum. *Isospora* are characterized by coccidia with tetrasporozoic, diplosporocystic oocyst with sporocysts containing stiedae bodies (an ultrastructural organelle located at the polar region of the sporocyst with a knoblike structure occluding a hole in the sporocyst), while *Cystoisospora* are coccidian with a similar structural organization but lacking stiedae bodies in their sporocysts.

Table 1. ***Cystoisospora belli*** **Taxonomy Classification.**

Cystoisospora belli **Taxonomy Classification from the NCBI Taxonomy Browser (2015) (http://www.ncbi.nlm.nih.gov/Taxonomy)**	
Domain	Eukaryota
Kingdom	Chromalveolata
Superphylum	Alveolata
Phylum	Apicomplexa
Class	Conoidasida
Subclass	Coccidiasina
Order	Eucoccidiorida
Suborder	Eimeriorina
Family	Sarcocystidae
Genus	*Cystoisospora*
Species	*Cystoisospora belli* *Cystoisospora felis* *Cystoisospora ohioensis* *Cystoisospora suis* *Cystoisospora timoni* *Cystoisospora ef. ohioensis* *Cystoisospora sp. 1-MM* *Cystoisospora sp. 2-MM* *Cystoisospora sp. Ex-vulpes lagor*

Studies using phylogenetic analysis of the rRNA 18S gene and of the rRNA internal transcribed spacer 1 (ITS1) have shown that *Cystoisospora* is more closely related to *Toxoplasma*, *Neospora* and *Sarcocystis* spp. than *Eimeria* [3 - 5].

Humans are the only known hosts for *C. belli*. Several mammalian species are infected by other *Cystoisopora* species, such as *C. felis* and *C. rivolta* in cats, *C. ohioensis* in dogs, *C. suis* in pigs and *C. timoni* in meerkats.

History

In 1860 Virchow described a species of Isospora isolated from human samples. In 1915 Dr. H.M. Woodcock reported again this species when he observed coccidian oocysts in the stools of European soldiers stationed in the Middle East, therefore, the parasite was named belli (from latin: bellum = war). Also during World War II it frequently diagnosed as a coccidial pathogen among troops serving in the Pacific [6]. In the Americas, isosporiasis was first reported only in the 1960s [7]. However, relatively few cases of disease were registered until the parasite was recognized as an opportunistic infection in immunocompromised patients [8]. Actually, both names (*isospora* or *cystoisospora*) are used in the medical field, often intended as synonymous by clinicians, but for the rest of the chapter we prefer to refer to the most appropriate taxonomic name as earlier defined, that is *Cystoisospora belli*. The parasite is one of the most commonly identified causes of severe and prolonged diarrhoea in immunodeficient individuals, particularly in AIDS patients. The effects of cystoiosporiasis may lead to increases in morbidity and mortality in these patients [9]. This parasite is principally known to cause the traveler's diarrhea in immunocompetent hosts and is an opportunistic pathogen in immunocompromised hosts.

Morphology and Life Cycle

Cystoisospora oocysts measure approximately between 25 µm to 30 µm in length and between 10 µm to 20 µm wide. The sporocysts measuring 12 µm x 7 to 9 µm. The *Cystoisospora* oocysts are relatively delicate and colourless, elongated and ellipsoid. The smooth cyst wall is composed of 2 layers. The outer layer is robust, impermeable to fluids, and environmentally resistant; the inner layer is membranous and the protoplasm is finely grained. Oocysts eliminated in the environment are not yet sporulated. After maturation in the external environment, the oocysts contains two sporocysts, which contain 4 sporozoites each (Fig. **1**). Sporozoites are slender and crescent-shaped. The inner structure of sporozoites and merozoites is similar to that of other coccidians and includes polar rings, rhoptries, micronemes, conoids, microtubules, and amylopectin granules. Macrogamonts contain a single large, centrally located nucleus and wall-forming bodies. Microgamonts contain multiple nuclei that migrate to the periphery, elongate and protrude from the surface, and bud into mature flagellated microgametes 5 µm to 6 µm long [10, 11]. The intracellular development of *C.*

belli was demonstrated in different mammalian cell cultures and light microscopy was used to show the entry of sporozoites into mammalian cells and how meronts are formed by endodyogeny [12 - 14]. However, the lack of clinical specimens, susceptible host models, and the restricted ability to grow *in vitro* have hampered the augmentation of more ultrastructural studies on the *in vitro* morphological development of *C. belli*. Therefore, there are limited studies to support the identification of endogenous *C. belli* stages in tissue sections. A useful study describing the ultrastructural findings of the endogenous stages of *C. belli* cultivated in continuous cell lines is available to get a more profound insight on the basic biology of this organism [15]. The study by Resende *et al.* 2014, showed that zoites of *C. belli* displayed the standard morphological characteristics of a coccidian, with the exception that crystalloid bodies were lacking. The host cell infected with *C. belli* did not show significant morphological changes, but the proximity of the endoplasmic reticulum with the membrane of the parasitophorous vacuole (formed by layers of endoplasmic reticulum around the intracellular parasite) was aligned with the vacuole membrane in order to circumvent the entire parasite.

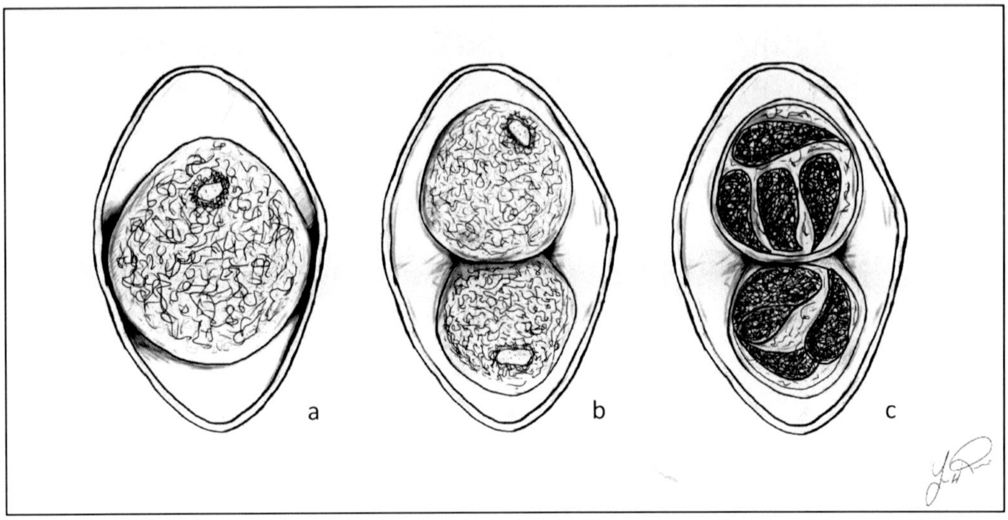

Fig. (1). **a**) immature oocyst; **b**) immature oocyst with two sporoblasts; **c**) mature oocyst contains two sporocysts, which contain 4 sporozoites each.
"kindly provided by Dr. Luigi Russo"

Overall, the various stages in *C. belli* life cycle, extra-intestinal cyst formation, and the interconnection with the host cell continue to be considerably unexplored, but in view of the opportunistic pathogenic features of *C. belli* more studies are warranted [16].

Exemplary life cycle stages of *C. belli* have, nevertheless, been analysed by light and transmission electron microscopy [9]. The life cycle of *C. belli* can be separated into two phases: asexual phase or schizogony followed by sexual phase or sporogony (Fig. **2**).

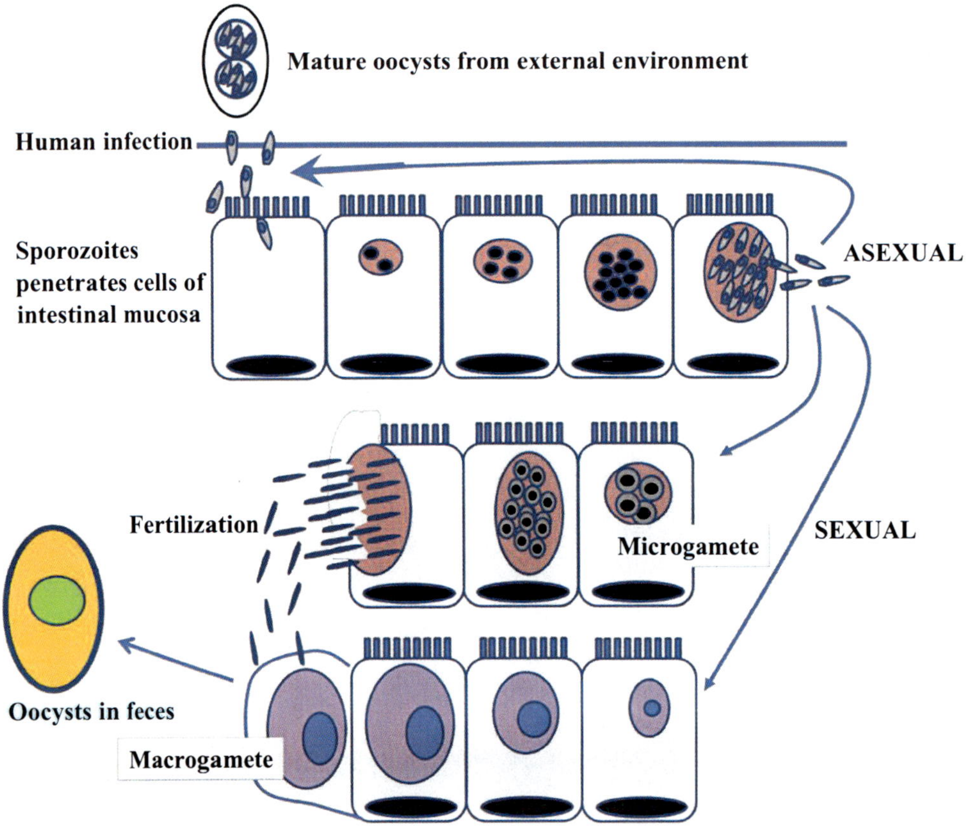

Fig. (2). Sexual and asexual replication of *C. belli.*
Human infection can be shared in several steps: i) mature oocysts from external environment are ingested; ii) the schizogony starts after that sporozoites penetrate the cells of intestinal mucosa; iii) the schizonts and their host cells burst and the merozoites are released. iv)The merozoites move into the gut lumen and penetrate into epithelial cells continuing the asexual phase; v) the sexual stage starts with the formation of male and female gametocytes; vi) fertilization lead to the development of oocysts, which are excreted in feces.

After ingestion of oocysts containing sporocysts, the sporozoites are released and they penetrate the small-intestinal epithelium where the asexual phase (schizogony) begins. Sporozoites are released in the presence of pancreatic and gastric juices and their digestion begins in the stomach and is completed in the duodenum. The sporozoites become spherical trophozoites, and with the progress of the replication and the gradually increase in volume until they fill the host cell, become schizonts. When the polynuclear schizont and its host cell burst, the merozoites are released and they move into the gut lumen and penetrate into enterocytes. The organisms either begin another round of schizogony or shift to gametogony, the sexual phase. The microgametes or male gametes appear flagellated and are attracted to macrogametes or the female gametes.

Fertilization takes place when microgametes attach to and penetrate into macrogametes. Nuclear fusion ensues and wall-forming bodies coalesce to form oocysts [11].

Oocysts are passed in feces unsporulated or partially sporulated (sporoblast stage). In the external environment oocysts can sporulate at different times depending on the environmental conditions. The mature oocysts are the infectious element and can remain viable in the environment for months (Fig. **3**). Human coccidiosis due *C. belli* is acquired by the ingestion of oocysts, which turn infective after sporulation as a consequence of the adequate moisture and moderate temperature. The presence of water has been determined to be a facilitating condition for both the development and transmission of coccidian oocysts. The parasite is transmitted by oocysts from person-to-person by the fecal-oral route or *via* contaminated water or food or with an increased probability with a poor personal hygiene.

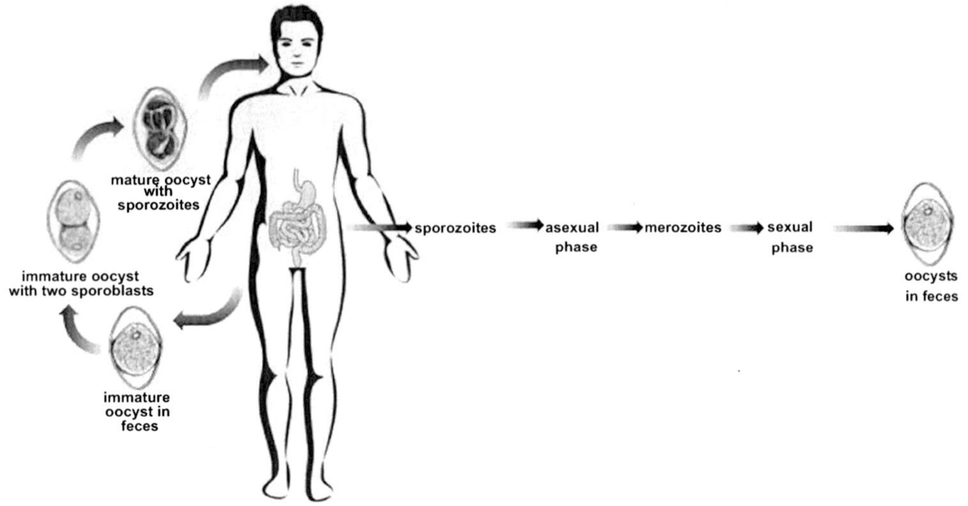

Fig. (3). The life cycle of *C. belli.*
"kindly provided by Dr. Luigi Russo"

Epidemiology and Transmission

C. belli infection has a worldwide distribution, especially in tropical and subtropical areas [17]. It is endemic in South Africa, South America, and the Southwest Pacific islands, India and south-east Asia. Reported infection rates are different in different part of the world and, for example, range from 0.1% to 1.8% in South America, from 2.5% to 31% in India, while globally (world) are considered in the range from 0.8% to 16% [17 - 20]. (Fig. **4**).

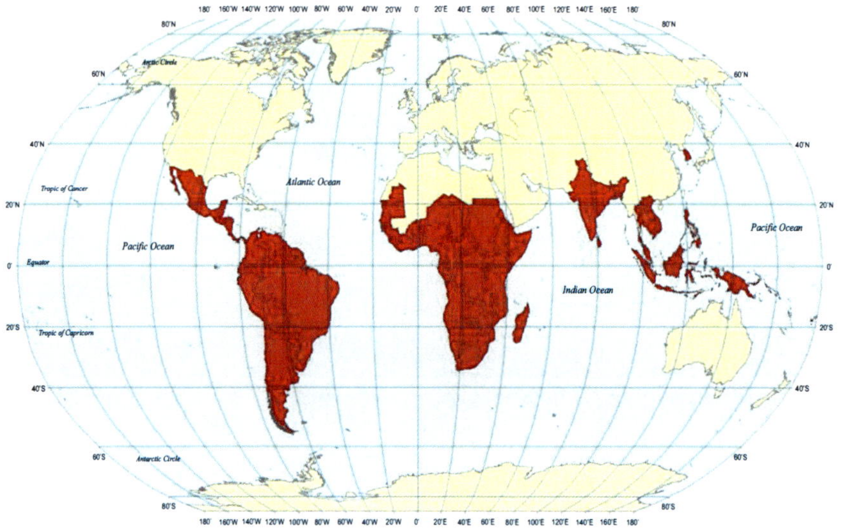

Fig. (4). Prevalence of *C. belli* infection. The countries marked in red are areas where *C. belli* is endemic. "kindly provided by Dr. Vincenzo Musella"

In the Western countries, the infection is uncommon and is largely associated to travels or immigration from tropical countries [21, 22]. However, the prevalence of this infection can be easily underestimated because only small quantities of oocysts are generally eliminated with feces, therefore, the diagnosis is often missed in spite of actual infection. Moreover, *C. belli* oocysts could be overlooked in direct fecal smears due to their commonly transparent appearance [23].

It is of note that infections with *C. belli* usually produce more aggressive and prolonged periods of clinical symptoms, even with extraintestinal dissemination of parasites in immunocompromised patients (*i.e.* AIDS, malignancies, and organ transplantations) rather than in immunocompetent individuals [24, 25].

Cystoisosporiasis is more prevalent in children, but the exact prevalence of the disease is not known. It occurs equally in both sexes. Immunodeficiency, malnutrition, and poor hygiene are known risk factors [7, 11, 23].

Pathophysiology

C. belli infection could be subdivided in two principal phases: i) acute and ii) chronic. During the acute phase of the disease the parasites are localized in the small intestine epithelium, while in the chronic phase they are found in the subepithelium of the villi. When both the host cell and the schizont burst, the relased merozoitos causes extensive destruction of the intestinal epithelium. The small bowel mucosa is clearly damaged during infections and a serious villous atrophy with shortening, blunting, fusion and clubbing of villi can be observed by light microscopy. Moreover, crypt hyperplasia and hypertrophy can be observed. The lamina propria is generally present with a higher abundance of infiltrated eosinophils, lymphocytes and plasma cells [11, 12].

It has been described that the mucosal system plays an important role in the pathogenesis of this infection and patients with a low count in CD4+ cells (less than 200/μl) are more vulnerable as observed in HIV-infected patients. The extent of the damage is directly proportional to the parasitic burden. In some patients affected by AIDS or *Cytomegalovirus* enteritis, dissemination of the organisms to mesenteric and tracheobronchial lymph nodes characterized by granulomatous foci has been described.

Immunology and Immunopathology

In immunocompetent individuals, cystoisosporiasis is manifested with a benign and self-limiting course; whereas it can be noxious in subjects with immune disorders, especially those with acquired immunodeficiency syndrome (HIV/AIDS). Immunocompromised patients and particularly those with AIDS are more severely affected and the disease is more protracted in time leading also to malfunctioning of bile acids enterohepatic circulation and of the absorption of nutrients. AIDS deaths annually are 1.8 million worldwide and opportunistic infections aggravate the condition of immunodeficient HIV patients causing increased morbidity and mortality [17]. In many Western countries [8], cistoisosporiasis is usually not present in HIV positive patients because of the routine prophylaxis against *Pneumocystis carinii* with Trimethoprim/sulfamethoxazole (TMP-SMX), which is also effective against *C. belli*, when CD4+ cell counts are low.

Cystoisosporiasis predominates among patients with CD4+ counts <200 cells/mm^3 but the incidence has considerably decreased since the introduction of high antiretroviral therapy (HAART) [26].

Cystoisisporiasis often presents a chronic and persisting nature which increases both morbidity and mortality of immunocompromised patients [27]. This chronicization competence is generally ascribed to the ability of *C. belli* to infect extraintestinal sites and form unizoic cysts that are resistant to treatment [27, 28].

When autoptic samples are carefully examined, they can reveal stages of *C. belli* in the intestine, mesenteric and mediastinal lymph nodes, liver, and spleen. Sporadically, merozoites have been found in the intestinal lumen, in the lamina propria, and within lymphatic channels. The confirmation of the ability of merozoites to conquer lymphatic channels can serve as a mechanism of *C. belli* circulation to lymph nodes and to other tissues.

Cases of cystoiosporiasis have been reported in renal transplant patients, and in patients suffering from congenital diserythropoesis and bronchoalveolar carcinoma. Therefore, cystoisosporiasis should be suspected not only in immunocompromised and HIV patients, but also in all transplant or cancer patients in which diarrhea is persistent and chronic, and there are abdominal cramps and weight loss [29]. In 2007 the first cases of a renal transplant patient [30], a liver transplant recipient [31] and cardiac transplant recipient [32] with cystoisosporiasis were described. More importantly, *C. belli* has also been observed in patients with lymphoproliferative disorders, acute lymphoblastic leukemia and human T-cell leukemia virus 1-associated adult T-cell leukemia [33].

As clearly described, clinical manifestations in *C. belli* infections are generally more severe in immunocompromised patients compared to immunocompetent patients. *C. belli* infections have been reported also in patients with concurrent Hodgkin's disease [9], non-Hodgkin's lymphoproliferative disease [34] and acute lymphoblastic leukemia [35].

Clinical Manifestations in Humans

The incubation period lasts 7-8 days and the infection may be asymptomatic or may lead to a mild self-limiting diarrhea lasting from several weeks to months without treatment.

Symptoms of *C. belli* infection include diarrhea, steatorrhea, headache, fever, malaise, abdominal pain, vomiting, dehydration, and weight loss [9, 36 - 38]. Diarrhea is typically profuse, watery, non-inflammatory and accompanied by crampy abdominal pain, while blood is not routinely detected in the feces although cases of hemorrhagic diarrhea have been sporadically reported. Evacuations occur six to twelve times a day. Generally no mucus or blood is found in the stool. After three weeks, around the time in which the parasite's cysts become apparent, the stool becomes more consistent and evacuations less numerous. Meteorism is also a constant finding. Symptoms generally attenuate until they disappear after a few weeks. The disease is usually more severe in children, but is better to remembers that generally self-limited. Fever may be present but relatively uncommon and eosinophilia is observed in some patients. The lesions caused by the protozoan replicating in the intestinal cells can be observed by microscopic examination and the main features are villous atrophy and hyperplasia of the intestinal crypts. Leukocytosis is prominent and dilation of the lymphatics can be observed [11]. If left untreated the disease can be fatal for immunocompromised individuals. Intestinal infections can also be associated with the presence of Charcot-Leyden crystals and eosinophilia. In the developing world, parasites are the most common cause of eosinophilia, therefore, a parasitic infection of nearly any bodily tissue such as infection by *C. belli* can cause eosinophilia. In developed nations, eosinophilia is generally due to concomitant allergy sometimes as adverse reaction to a pharmaceutical.

The disorder tends to be chronic with parasites present in the feces or biopsies for long lengths of time, from several months to years, resulting in malabsorption syndrome and recurrences are a common feature.

Extraintestinal sites of cystoisosporiasis are rare and more frequently occurs in immunocompromised patients, although also immunocompetent patient may present extraintestinal infection. One of the earliest reports [39] on extraintestinal infection due to *C. belli* described a 30-year-old patient, from Burkina Faso, who contracted HIV in 1990. Following her death in March 1993, mainly due to cachexia secondary to a chronic intestinal cystoisosporiasis, an autopsy was performed to reveal a massive infection of the small intestine mesenteric and mediastinal lymph nodes, liver and spleen. In all extraintestinal sites the stage in which *C. belli* was found corresponded to unizoite tissue cysts. This clinical case was of seminal importance since represented the first description of *C. belli* infection in liver and spleen. Therefore, main sites of infection and respective diseases are the small intestine with diarrhea and malabsorption, the gallbladder and biliary tree with acalculous cholecystitis and cholangitis, liver and spleen.

Diagnosis

The diagnosis of intestinal cystoisosporiasis requires identification of the characteristic parasitic oocysts in stool samples with both direct fecal smear (Fig.

5a), **5b**) and after concentration to overcome the problem of the small number of oocysts excreted with feces and to enhance the diagnostic yield. It should also be considered that fecal passage of *C. belli* oocysts may be intermittent, therefore multiple (three) stool samples are recommended before discarding the hypothesis of cystoisosporiasis infection. Either the Zinc Sulfate flotation or Sheather sugar flotation methods are appropriate to concentrate *C. belli* oocysts [11]. Furthermore *C. belli* oocysts can be detected using the innovative FLOTAC dual technique®, a method developed to detect and quantify the number of eggs, oocysts, cysts and parasite larvae found in animals and humans. The method is based on the centrifugation of stool samples in a flotation solution (FS) with a fixed specific gravity which is followed by the physical translation of the upper part of the fecal suspension which is likely to contain helminth eggs and intestinal protozoon cysts for observation under light-microscopy. For identification of oocysts two different flotation solutions are used, namaly FS4 (315 g of NaNO plus 500 ml of H2O; specific density = 1.200) and FS7 (685 g of ZnSO4 plus 685 ml of H2O; specific density = 1.350) [40, 41]. Stained fecal smears made from concentrated samples are also more sensitive than direct smears and may aid in the detection of *C. belli* oocysts.

Several technique are used for the stool analysis, in this scenario the most common is the wet mount examination where oocysts of *C. belli* are directly observable *via* light microscopy [42]. Under bright-field microscopy, the oocysts are large (25 to 30 by 10 to 19 μm) and have a typical ellipsoidal shape. Fresh stool samples are likely to only contain unsporulated oocysts, therefore a specific differential diagnosis sometime requires a limited incubation in potassium dichromate at room temperature to facilitate sporulation.

Considering that oocysts of *C. belli* are acid resistant they can be detected by acid-fast stains such as modified Ziehl-Neelsen stain (using 10% H_2SO_4) where pink oocysts containing bright red sporonts or sporoblasts can be detected (Fig. **5c**) [42]. Also Giemsa, Safranin and modified Kinyoun (another acid-fast stain using carbolfuchsin and a bright-green counterstain) staining procedures can be employed. Fluorescent staining methods using acridine orange and auramine (AuO)-rhodamine can also be employed for screening. AuO stains the nucleic acids present within the sporocysts which are readily detectable as brightly fluorescent spherical bodies, positioned inside the egg-shaped non-fluorescent oocyst wall. Sometimes a heterogeneously distributed fluorescence within the cystoisospora oocysts can be observed, representing a heterogeneous distribution of nucleic acids when the sporocysts are absent [43, 44].

An alternative method is the use of UV fluorescence microscopy, which also yields significant results in terms of sensibility and it is a simple and fast technique based on the autofluorescence of *C. belli* [45]. The oocyst wall autofluoresces and can be visualized with a bluish-violet color under UV light excitation (365 nm) (Fig. **5d**). A green fluorescence of minor intensity can be obtained with blue excitation (450 to 490 nm). Also Cyclospora cayetanensis oocysts have the property of autofluorescence, but the two parasitic oocysts can be distinguished one from the other based on their morphology: C. belli oocysts are oval and measure approximately 25-30 μm by 10-20 μm and often one or two immature sporoblasts can be observed.

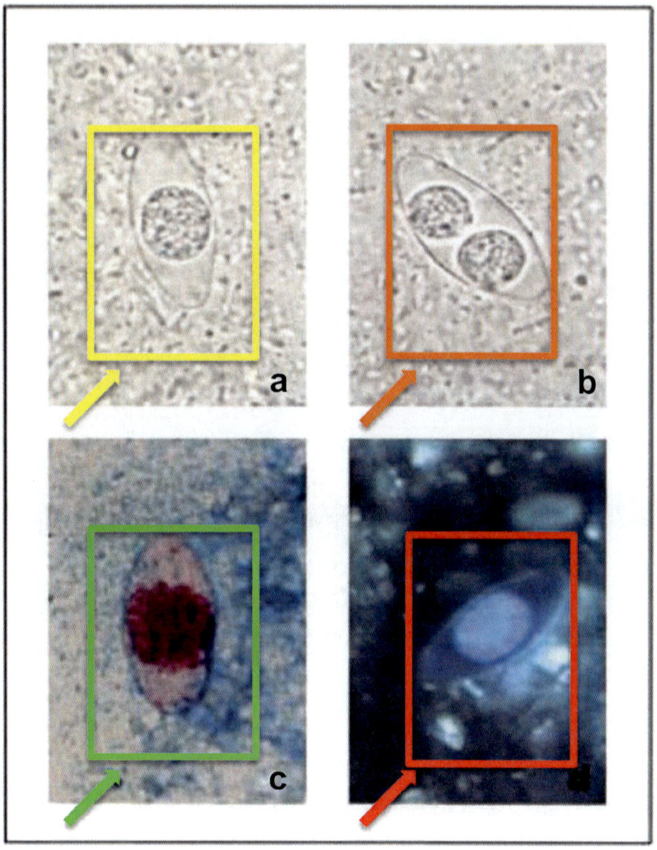

Fig. (5). *C. belli* oocysts detected by different diagnostic techniques. Direct fecal smear: unsporulated oocyst (**a**) and sporulated oocyst (**b**). Modified Ziehl-Neelsen stain (**c**) and fluorescence microscopy (**d**). Modified figure where (**a**) and (**b**) are derived by Atlas of human intestinal protozoa - Microscopic identification, www.atlas-protozoa.com; **c**) is derived by MEDICAL- LABS - Medical Laboratories Portal, www.medicals-labs.net; **d**) is derived by CDC - Centers for disease control and prevention, www.cdc.gov/parasites/cystoisospora).

If oocysts are not detected in feces, diagnosis can be attempt through intestinal biopsies [46 - 48], duodenal aspirates [38] or the duodenal string test [49], followed by routine histological staining which can demonstrate intraepithelial *C. belli* parasites. Biopsy of the villous epithelium, particularly of the small bowel, is a powerful diagnostic procedure, but has the inconvenience to be invasive. Serologic methods for detection of *C. belli* infections are not available.

Although the first form of diagnosis of parasitic infections is still by light microscopy of stools by an experienced microscopist, *C. belli* and other eukaryotic enteropathogens remain difficult to detect with conventional methods, therefore we are experiencing a strong drive forward molecular assays for such pathogens. The possibility to directly extract DNA from clinical samples and the ability to amplify it by polymerase chain reaction (PCR) have yielded increasingly feasible and affordable molecular methods with the advantage of overcome current diagnostic limitations [50]. At the present, only limited reports describe molecular methods for detection of *C. belli*, and their principal application can be either diagnosis o minor infections or follow-up of patients under under therapy.

Specific primer pairs and hybridization probe for the detection of *C. belli* in human samples by nested PCR and Southern blot hybridization have been described by Müller *et al.* [51]. The nested PCR methodology used an outer primer pair, IsoFO (5'-GTGCCTCTTCCTCTGGAAGG-3') and IsoRO (5'-GCACTCCACCCAGTTAAGTGC-3'), and an inner primer pair, IsoFI (5'-CGATGGATCATTCAAGTTTC-3') and IsoRI (5'-ACCACGTACACACCC--TAAG-3'). The first pair amplified a 559 bp DNA fragment while the inner couple of primers amplified a 396 bp DNA fragment from 10 ml of PCR product from the first PCR. After gel electrophoresis the PCR products of the first PCR were blotted on nylon membranes and hybridized with a fluorescein-labeled 24 bp oligonucleotide IsoOP [5'-GT{T/A}ATGGCTTCGGCCGGCGATGGA-3']) and visualized by using an enhanced chemiluminescence detection kit. The hybridization probe and the primer pairs are based on variable regions of the SSU-rRNA of *C.belli*. The correct DNA fragments were amplified from clinical samples of patients with *C.belli* infections and confirmation of DNA fragments amplified during the first PCR was done by nested PCR using the second primer pair IsoFI/IsoRI or by Southern blot hybridization with the probe IsoOP. The specificity of the nested PCR/Southern hybridization was proved by the absence of signal from samples not infected with *C. belli* but infected with other pathogens (*Cryptosporidium parvum, Giardia lamblia, Blastocystis hominis, Endolimax nana, Entamoeba hartmanni Blastocystis hominis, Entamoeba histolytica Encephalitozoon intestinalis* and *Toxoplasma gondii*). Only the template DNA derived from *T. gondii* produced a weak signal at the expected length in the second round PCR, but this results may be due to some conserved level of sequence similarity between *C. belli* and *T. gondii* in the analysed region and is not surprising since the two parasites have been shown to be closely related [3].

A real-time PCR assay targeting the internal transcribed spacer 2 region of the ribosomal RNA gene was developed for the detection of *C. belli* DNA in fecal samples [52]. Using a large range of control DNA and stool samples (147 bacterial, parasitic, and fecal control DNA samples), the real-time PCR methodology described by ten Hove and collegues was showed to achieve a 100% specificity. Moreover, a sensitivity of 100% was reached extracting DNA from a selected group of 21 fecal samples from patients with a previous microscopy diagnosis of *C. belli* [52].

A multiplex PCR with 4 primer sets to amplify *C. cayetanensis*, *C. belli*, *Enterocytozoon bieneusi*, and *Encephalitozoon intestinalis* was developed by Taniuchi and colleagues in 2011 where detection of amplicons is done through specific probes coupled to Luminex beads [53]. The multiplex PCR yielded very high specificity and sensitivity without any cross reactivity. Few samples produced a microscopy negative/PCR positive result, but the lower Luminex values suggested a lower burden of infection for these samples. Furthermore, a quantitative polymerase chain reaction assay with melt curve analysis (qPCR-MCA) was developed by Lalonde *et al.* for the detection of multiple species of oocysts in human feces using universal coccidia primers targeting 18S rDNA to detect oocysts followed by MCA to identify oocyst species based on amplicon melting temperature [54]. Obviously, molecular methodologies will not be a suitable diagnostic tool in routine clinical situations in endemic areas where the resources of the laboratories are often limited.

Treatment

Cystoisosporiasis in immunocompetent patients is usually self-limited and does not require therapeutic intervention but only symptomatic support, such as antimotility drugs to control diarrhea. Optimised fluid and electrolyte management is required for patients with dehydration, and nutritional supplementation should be provided to malnourished patients.

In immunocompromised individuals, the chronicity and intensity of the diarrhea warrant intervention. The most effective therapy is the use of a combination therapy with dihydrofolate reductase inhibitors and sulfonamides, specifically Trimethoprim/sulfamethoxazole (TMP-SMX) (160/800 mg, four times a day for ten days and then three times a week for three weeks as a secondary prophylaxis). Prophylaxis is useful to reduce reoccurrence in patients with AIDS that are not under HAART and for discontinuing secondary prophylaxis reconstitution of CD4+ cells (sustained CD4+ count >200cell/µl for longer than 3 months). In the case of pediatric patients the same combination TMP/SMX is administered (TMP 10mg/kg/d and SMX 50mg/kg/d) per os in two doses daily x 10d. In patients allergic to sulfonamides, the drug pyrimethamine can be a valid alternative (50/75 mg a day for two weeks and then 25 mg a day for the secondary prophylaxis) [11]. Bone-marrow toxicity exerted by pyrimethamine can be prevented by co-administration of folinic acid (5-10 mg/d) [55].

In the case of TMP-SMX intolerance, pharmacological options also include ciprofloxacin, diclazuril, macrolide antibiotics, nitazoxanide and albendazole but data to support the use are scarce. *C. belli* infections in HIV-infected patients present fewer relapses after immune reconstitution, while chronic cystoisosporiasis was only rarely reported in immunocompetent hosts [26] (Table **2**).

Table 2. Main drugs and recommendations used in *C. belli* infections.

Drug	Recommendation
Trimethoprim/sulfamethoxazole	First-line agent, represents a high effective therapy. It is also the preferred choice for secondary prophylaxis
Pyrimethamine	A dihydrofolate synthetase inhibitor. Only used in patients allergic to sulphonamides with the addition of folinic acid to overcome toxicity.
Ciprofloxacin	A fluoroquinolone inhibiting DNA gyrase with a minor effect compared to TMP/SMX against cystoisosporiasis; appropriate when sulfa drug or pyrimethamine cannot be used
Diclazuril	A triazin anticoccidial for poultry. It is used for AIDS patients hypersensitive to both TMP/SMX and Pyrimethamine. Suggested dose 300mg twice daily
Macrolide antibiotics	Marginal efficacy against *C. belli*
Nitazoxanide	Anti-giardial compound with marginal effect on *C. belli* infections
Albendazole	Rarely used, but can represent a valid alternative

Therapy needs to be monitored and the simplest way is to search for both a clinical evidence (cessation of diarrhea) and a diagnostic confirmation (disappearance of *C. belli* oocysts from stool samples).

Perspectives of Control

The disease is transmitted *via* the fecal-oral route: food and drink, which have been contaminated by feces containing the mature oocysts, but also oral-anal sexual practices could contribute to the spread of *C. belli.* Other routes of transmission are being investigated. Many animals, including dogs, pigs, mice, rabbits, can serve as paratenic hosts but not as definitive hosts and their role as reservoirs of infection is still not clear [11]. Considering that feces of infected individuals are infectious, stool handling precautions and improved sanitation are important means of prevention. Prophylactic measures are to prevent such contamination or remove parasites with common means of sterilization. When traveling to endemic areas eating cooked food and avoiding water of uncertain origin is a wise precaution. Immunoprophylaxis is available with TMP-SMX. The prophylaxis consist in the administration of TMP-SMX for 10 week (1 tablet orally, three times per week). The use of ciprofloxacin is acceptable for patients who cannot tolerate trimethoprim-sulfamethoxazole, although ciprofloxacin is not as effective [56]. There is no vaccine.

CONCLUSION

Cystoisosporiasis is a coccidian infection caused by *C. belli*, an emerging protozoan pathogen of humans.

C. belli infection is cosmopolitan in distribution, but is more common in parts of the tropics and subtropics where sanitation is poor. The parasite can be transmitted directly from human to human *via* fecal contaminated food or water.

C. belli infection is most frequent in the immunocompromised patient and can cause disease in both adults and children. The symptomatology is characterized by watery diarrhea, abdominal cramps, fever, nausea, epigastric discomfort, cholecystitis, dehydration, weakness and weight loss. These manifestations can last from days to years with chronicity being directly related to the degree of immunosuppression present in the host.

Cystoisosporiasis is not easily diagnosed without specific diagnostic investigations. *C. belli* diagnosis is performed by identifying oocysts in fecal samples. Since this enteric parasite is not included in routine stool examinations, it is possible that it is under reported. With improved detection methods, particularly the use of fluorescence microscopy, it can be easily identified due to its autofluorescence. With enhanced capability and awareness of laboratory staff, more accurate data on the prevalence of these pathogens are expected in the near future.

Trimethoprim-sufamethoxazole (sulfa drugs) are the standard therapy and usually leads to rapid improvement in symptoms. In patients allergic to sulfonamides, pyrimethamine can be used. The prognosis is good except for immunosuppressed patients. Cystoisosporiasis can be prevented with adequate sanitation, measures to protect food and water supplies, and increased public awareness of the means of transmission.

CONFLICT OF INTEREST

The authors confirm that they have no conflict of interest to declare for this publication.

ACKNOWLEDGEMENTS

Declared none.

REFERENCES

[1] Barta JR, Schrenzel MD, Carreno R, Rideout BA. The genus Atoxoplasma (Garnham 1950) as a junior objective synonym of the genus Isospora (Schneider 1881) species infecting birds and resurrection of Cystoisospora (Frenkel 1977) as the correct genus for Isospora species infecting mammals. J Parasitol 2005; 91(3): 726-7.
[http://dx.doi.org/10.1645/GE-3341.1] [PMID: 16108579]

[2] Carreno RA, Barta JR. An Eimeriid origin of isosporoid coccidia with Stieda bodies as shown by phylogenetic analysis of small subunit ribosomal RNA gene sequences. J Parasitol 1999; 85(1): 77-83.
[http://dx.doi.org/10.2307/3285704] [PMID: 10207368]

[3] Carreno RA, Schnitzler BE, Jeffries AC, Tenter AM, Johnson AM, Barta JR. Phylogenetic analysis of coccidia based on 18S rDNA sequence comparison indicates that Isospora is most closely related to Toxoplasma and Neospora. J Eukaryot Microbiol 1998; 45(2): 184-8.
[http://dx.doi.org/10.1111/j.1550-7408.1998.tb04523.x] [PMID: 9561772]

[4] Franzen C, Müller A, Bialek R, Diehl V, Salzberger B, Fätkenheuer G. Taxonomic position of the human intestinal protozoan parasite Isospora belli as based on ribosomal RNA sequences. Parasitol Res 2000; 86(8): 669-76.
[http://dx.doi.org/10.1007/PL00008550] [PMID: 10952268]

[5] Samarasinghe B, Johnson J, Ryan U. Phylogenetic analysis of Cystoisospora species at the rRNA ITS1 locus and development of a PCR-RFLP assay. Exp Parasitol 2008; 118(4): 592-5.
[http://dx.doi.org/10.1016/j.exppara.2007.10.015] [PMID: 18067892]

[6] Marshall MM, Naumovitz D, Ortega Y, Sterling CR. Waterborne protozoan pathogens. Clin Microbiol Rev 1997; 10(1): 67-85.
[PMID: 8993859]

[7] Klasse-Fischer MK, Neafie RC, Wear DJ, Meyers WM. Topics on the Pathology of Protozoan and Invasive Arthropod Diseases. Chapter 13 Cryptosporidiosis, isosporiasis, cyclosporiasis, and sarcocystosis. Fairfax, Virginia, USA: Inova Central Laboratory 2011; pp. 1-19.

[8] Faust EC, Giraldo LE, Caicedo G, Bonfante R. Human isosporiasis in the western Hemisphere. Am J

Trop Med Hyg 1961; 10: 343-9.
[PMID: 13698368]

[9] Olusegun AF, Okaka CE, Luiz Dantas Machado R. Isosporiasis in HIV/AIDS Patients in Edo State, Nigeria. Malays J Med Sci 2009; 16(3): 41-4.
[PMID: 22589664]

[10] Jongwutiwes S, Putaporntip C, Charoenkorn M, Iwasaki T, Endo T. Morphologic and molecular characterization of Isospora belli oocysts from patients in Thailand. Am J Trop Med Hyg 2007; 77(1): 107-12.
[PMID: 17620639]

[11] Lindsay DS, Dubey JP, Blagburn BL. Biology of Isospora spp. from humans, nonhuman primates, and domestic animals. Clin Microbiol Rev 1997; 10(1): 19-34.
[PMID: 8993857]

[12] Oliveira-Silva MB, Lages-Silva E, Resende DV, Prata A, Ramirez LE, Frenkel JK. Cystoisospora belli: *in vitro* multiplication in mammalian cells. Exp Parasitol 2006; 114(3): 189-92.
[http://dx.doi.org/10.1016/j.exppara.2006.03.004] [PMID: 16674944]

[13] Resende DV, Lages-Silva E, Assis DC, Prata A, Oliveira-Silva MB. Experimental infection of murine and human macrophages with Cystoisospora belli. Acta Trop 2009; 111(2): 177-80.
[http://dx.doi.org/10.1016/j.actatropica.2009.04.010] [PMID: 19427297]

[14] Siripanth C, Punpoowong B, Amarapal P, Thima N. Development of Isospora belli in Hct-8, Hep-2, human fibroblast, BEK and Vero culture cells. Southeast Asian J Trop Med Public Health 2004; 35(4): 796-800.
[PMID: 15916071]

[15] Resende DV, Assis DC, Ribeiro MF, *et al.* Ultrastructural aspects of Cystoisospora belli (syn. Isospora belli) in continuous cell lines. Microsc Res Tech 2014; 77(6): 472-8.
[http://dx.doi.org/10.1002/jemt.22372] [PMID: 24771702]

[16] Lindsay DS, Houk AE, Mitchell SM, Dubey JP. Developmental biology of Cystoisospora (Apicomplexa: Sarcocystidae) monozoic tissue cysts. J Parasitol 2014; 100(4): 392-8.
[http://dx.doi.org/10.1645/13-494.1] [PMID: 24841928]

[17] Curry A, Smith HV. Emerging pathogens: Isospora, Cyclospora and microsporidia. Parasitology 1998; 117 (Suppl.): S143-59.
[http://dx.doi.org/10.1017/S0031182099004904] [PMID: 10660937]

[18] DeHovitz JA, Pape JW, Boncy M, Johnson WD Jr. Clinical manifestations and therapy of Isospora belli infection in patients with the acquired immunodeficiency syndrome. N Engl J Med 1986; 315(2): 87-90.
[http://dx.doi.org/10.1056/NEJM198607103150203] [PMID: 3487730]

[19] Guiguet M, Furco A, Tattevin P, Costagliola D, Molina JM. HIV-associated Isospora belli infection: incidence and risk factors in the French Hospital Database on HIV. HIV Med 2007; 8(2): 124-30.
[http://dx.doi.org/10.1111/j.1468-1293.2007.00441.x] [PMID: 17352769]

[20] Robinson RD. Parasitic infections associated with HIV/AIDS in the Caribbean. Bull Pan Am Health Organ 1995; 29(2): 129-37.
[PMID: 7640691]

[21] Franzen C, Fätkenheuer G, Salzberger B, *et al.* [Isospora belli infection in an AIDS patient]. Med Klin (Munich) 1993; 88(5): 330-2. [Isospora belli infection in an AIDS patient].
[PMID: 8326922]

[22] Sorvillo FJ, Lieb LE, Seidel J, Kerndt P, Turner J, Ash LR. Epidemiology of isosporiasis among

persons with acquired immunodeficiency syndrome in Los Angeles County. Am J Trop Med Hyg 1995; 53(6): 656-9.
[PMID: 8561272]

[23] Legua P, Seas C. Cystoisospora and cyclospora. Curr Opin Infect Dis 2013; 26(5): 479-83.
[PMID: 23982239]

[24] Kim MJ, Kim WH, Jung HC, Chai JW, Chai JY. Isospora belli Infection with Chronic Diarrhea in an Alcoholic Patient. Korean J Parasitol 2013; 51(2): 207-12.
[http://dx.doi.org/10.3347/kjp.2013.51.2.207] [PMID: 23710089]

[25] Gruz F, Fuxman C, Errea A, *et al.* Isospora belli infection after isolated intestinal transplant. Transpl Infect Dis 2010; 12(1): 69-72.
[http://dx.doi.org/10.1111/j.1399-3062.2009.00439.x] [PMID: 19708892]

[26] Boyles TH, Black J, Meintjes G, Mendelson M. Failure to eradicate Isospora belli diarrhoea despite immune reconstitution in adults with HIVa case series. PLoS One 2012; 7(8): e42844.
[http://dx.doi.org/10.1371/journal.pone.0042844] [PMID: 22880120]

[27] Frenkel JK, Silva MB, Saldanha J, *et al.* Isospora belli infection: observation of unicellular cysts in mesenteric lymphoid tissues of a Brazilian patient with AIDS and animal inoculation. J Eukaryot Microbiol 2003; 50 (Suppl.): 682-4.
[http://dx.doi.org/10.1111/j.1550-7408.2003.tb00686.x] [PMID: 14736218]

[28] Lindsay DS, Dubey JP, Toivio-Kinnucan MA, Michiels JF, Blagburn BL. Examination of extraintestinal tissue cysts of Isospora belli. J Parasitol 1997; 83(4): 620-5.
[http://dx.doi.org/10.2307/3284235] [PMID: 9267401]

[29] Koru O, Araz RE, Yilmaz YA, *et al.* Case report: Isospora belli infection in a renal transplant recipent. Turkiye Parazitol Derg 2007; 31(2): 98-100.
[PMID: 17594646]

[30] Yazar S, Tokgöz B, Yaman O, Sahin I. [Isospora belli infection in a patient with a renal transplant]. Turkiye Parazitol Derg 2006; 30(1): 22-4.
[PMID: 17106849]

[31] Atambay M, Bayraktar MR, Kayabas U, Yilmaz S, Bayindir Y. A rare diarrheic parasite in a liver transplant patient: Isospora belli. Transplant Proc 2007; 39(5): 1693-5.
[http://dx.doi.org/10.1016/j.transproceed.2007.02.080] [PMID: 17580223]

[32] Sanches BF, Morgado J, Carvalho N, Anjos R. Multiple parasitic infections in a cardiac transplant recipient. BMJ Case Rep 2015; 2015: bcr2014207033.
[http://dx.doi.org/10.1136/bcr-2014-207033] [PMID: 26109619]

[33] Greenberg SJ, Davey MP, Zierdt WS, Waldmann TA. Isospora belli enteric infection in patients with human T-cell leukemia virus type I-associated adult T-cell leukemia. Am J Med 1988; 85(3): 435-8.
[http://dx.doi.org/10.1016/0002-9343(88)90603-1] [PMID: 2901228]

[34] Hallak A, Yust I, Ratan Y, Adar U. Malabsorption syndrome, coccidiosis, combined immune deficiency, and fulminant lymphoproliferative disease. Arch Intern Med 1982; 142(1): 196-7.
[http://dx.doi.org/10.1001/archinte.1982.00340140198038] [PMID: 7053725]

[35] Westerman EL, Christensen RP. Chronic Isospora belli infection treated with co-trimoxazole. Ann Intern Med 1979; 91(3): 413-4.
[http://dx.doi.org/10.7326/0003-4819-91-3-413] [PMID: 475171]

[36] Henderson HE, Gillepsie GW, Kaplan P, Steber M. The Human Isospora. Am J Hyg 1963; 78: 302-9.
[PMID: 14071353]

[37] Liebman WM, Thaler MM, DeLorimier A, Brandborg LL, Goodman J. Intractable diarrhea of infancy due to intestinal coccidiosis. Gastroenterology 1980; 78(3): 579-84.
[PMID: 7351295]

[38] Trier JS, Moxey PC, Schimmel EM, Robles E. Chronic intestinal coccidiosis in man: intestinal morphology and response to treatment. Gastroenterology 1974; 66(5): 923-35. [PMID: 4826994]

[39] Michiels JF, Hofman P, Bernard E, *et al.* Intestinal and extraintestinal Isospora belli infection in an AIDS patient. A second case report. Pathol Res Pract 1994; 190(11): 1089-93. [http://dx.doi.org/10.1016/S0344-0338(11)80908-8] [PMID: 7746744]

[40] Cringoli G, Rinaldi L, Maurelli MP, Utzinger J. FLOTAC: new multivalent techniques for qualitative and quantitative copromicroscopic diagnosis of parasites in animals and humans. Nat Protoc 2010; 5(3): 503-15. [http://dx.doi.org/10.1038/nprot.2009.235] [PMID: 20203667]

[41] Becker SL, Lohourignon LK, Speich B, *et al.* Comparison of the Flotac-400 dual technique and the formalin-ether concentration technique for diagnosis of human intestinal protozoon infection. J Clin Microbiol 2011; 49(6): 2183-90. [http://dx.doi.org/10.1128/JCM.01035-10] [PMID: 21525226]

[42] Ng E, Markell EK, Fleming RL, Fried M. Demonstration of Isospora belli by acid-fast stain in a patient with acquired immune deficiency syndrome. J Clin Microbiol 1984; 20(3): 384-6. [PMID: 6208216]

[43] Hanscheid T, Cristino JM, Salgado MJ. Screening of auramine-stained smears of all fecal samples is a rapid and inexpensive way to increase the detection of coccidial infections. Int J Infect Dis 2008; 12(1): 47-50. [http://dx.doi.org/10.1016/j.ijid.2007.04.008] [PMID: 17600749]

[44] Pacheco FT, Silva RK, Martins AS, *et al.* Differences in the detection of Cryptosporidium and Isospora (Cystoisospora) oocysts according to the fecal concentration or staining method used in a clinical laboratory. J Parasitol 2013; 99(6): 1002-8. [http://dx.doi.org/10.1645/12-33.1] [PMID: 23829170]

[45] Bialek R, Binder N, Dietz K, Knobloch J, Zelck UE. Comparison of autofluorescence and iodine staining for detection of Isospora belli in feces. Am J Trop Med Hyg 2002; 67(3): 304-5. [PMID: 12408672]

[46] Bhaijee F, Subramony C, Tang SJ, Pepper DJ. Human immunodeficiency virus-associated gastrointestinal disease: common endoscopic biopsy diagnoses. Patholog Res Int 2011. 2011:247923. [http://dx.doi.org/10.4061/2011/247923]

[47] Boldorini R, Tosoni A, Mazzucco G, *et al.* Intracellular protozoan infection in small intestinal biopsies of patients with AIDS. Light and electron microscopic evaluation. Pathol Res Pract 1996; 192(3): 249-59. [http://dx.doi.org/10.1016/S0344-0338(96)80228-7] [PMID: 8739472]

[48] Navaneethan U, Venkatesh PG, Downs-Kelly E, Shen B. Isospora belli superinfection in a patient with eosinophilic gastroenteritisa diagnostic challenge. J Crohns Colitis 2012; 6(2): 236-9. [http://dx.doi.org/10.1016/j.crohns.2011.08.010] [PMID: 22325178]

[49] Whiteside ME, Barkin JS, May RG, Weiss SD, Fischl MA, MacLeod CL. Enteric coccidiosis among patients with the acquired immunodeficiency syndrome. Am J Trop Med Hyg 1984; 33(6): 1065-72. [PMID: 6334448]

[50] Verweij JJ, Stensvold CR. Molecular testing for clinical diagnosis and epidemiological investigations of intestinal parasitic infections. Clin Microbiol Rev 2014; 27(2): 371-418. [http://dx.doi.org/10.1128/CMR.00122-13] [PMID: 24696439]

[51] Müller A, Bialek R, Fätkenheuer G, Salzberger B, Diehl V, Franzen C. Detection of Isospora belli by polymerase chain reaction using primers based on small-subunit ribosomal RNA sequences. Eur J Clin Microbiol Infect Dis 2000; 19(8): 631-4. [http://dx.doi.org/10.1007/s100960000322] [PMID: 11014628]

[52] ten Hove RJ, van Lieshout L, Brienen EA, Perez MA, Verweij JJ. Real-time polymerase chain reaction for detection of Isospora belli in stool samples. Diagn Microbiol Infect Dis 2008; 61(3): 280-3. [http://dx.doi.org/10.1016/j.diagmicrobio.2008.03.003] [PMID: 18424043]

[53] Taniuchi M, Verweij JJ, Sethabutr O, *et al.* Multiplex polymerase chain reaction method to detect Cyclospora, Cystoisospora, and Microsporidia in stool samples. Diagn Microbiol Infect Dis 2011; 71(4): 386-90. [http://dx.doi.org/10.1016/j.diagmicrobio.2011.08.012] [PMID: 21982218]

[54] Lalonde LF, Reyes J, Gajadhar AA. Application of a qPCR assay with melting curve analysis for detection and differentiation of protozoan oocysts in human fecal samples from Dominican Republic. Am J Trop Med Hyg 2013; 89(5): 892-8. [http://dx.doi.org/10.4269/ajtmh.13-0106] [PMID: 24019437]

[55] Weiss LM, Perlman DC, Sherman J, Tanowitz H, Wittner M. Isospora belli infection: treatment with pyrimethamine. Ann Intern Med 1988; 109(6): 474-5. [http://dx.doi.org/10.7326/0003-4819-109-6-474] [PMID: 3261956]

[56] Verdier RI, Fitzgerald DW, Johnson WD Jr, Pape JW. Trimethoprim-sulfamethoxazole compared with ciprofloxacin for treatment and prophylaxis of Isospora belli and Cyclospora cayetanensis infection in HIV-infected patients. A randomized, controlled trial. Ann Intern Med 2000; 132(11): 885-8. [http://dx.doi.org/10.7326/0003-4819-132-11-200006060-00006] [PMID: 10836915]

CHAPTER 5

Entamoeba histolytica and other Pathogenic Intestinal Amoebea

Eduardo R. Cobo[1] and **Kris Chadee**[2,*]

[1] *Department of Production Animal Health, Faculty of Veterinary Medicine*

[2] *Department of Microbiology, Immunology and Infectious Diseases, Gastrointestinal Research Group, Snyder Institute for Chronic Diseases, Cumming School of Medicine, University of Calgary, Calgary, Canada*

Abstract: *Entamoeba histolytica* is a parasitic enteric protozoan and the etiologic agent of amebic colitis and liver abscess in humans. Amebiasis is commonly acquired through water- and food-borne transmissions and highly prevalent in tropical areas of the world where sanitation is poor. Symptoms can be mild, including loose stools and abdominal cramping but if the parasite invades the colon it can advance to amebic dysentery characterized by frequent watery and bloody stools. Following colonic invasion, trophozoites can enter the bloodstream and lymphatics and disseminate to the liver, heart, lungs, brain, or other organs, to cause abscesses, which if left untreated can lead to death. As treatment for amebiasis with metronidazole has adverse side effects research into novel therapeutics and vaccines is still actively being pursued. The present chapter summarizes the characteristics of amebiasis, their treatments and control and the importance of water transmission.

Keywords: Amebiasis, Colitis, Cysteine proteinases, *Entamoeba dispar*, *Entamoeba histolytica*, Liver abscess, Protozoa.

INTRODUCTION

Entamoeba histolytica is an extracellular anaerobic parasitic enteric protozoan and the etiologic agent of amebic colitis and liver abscesses through water- and food-borne transmissions. Infection with *E. histolytica*, named amebiasis, affects up to 10% of the world's population (approximately 500 million people worldwide) and is a major cause of symptomatic illnesses (50 million of persons/year) and death (100,000 persons/year). Although the incidence of amebiasis may be over-

* **Corresponding author Kris Chadee:** Cumming School of Medicine, Department of Microbiology, Immunology and Infectious Diseases, University of Calgary Health Sciences Centre, 3330 Hospital Dr. NW, Calgary, Alberta T2N 4N1, Canada; Tel: (403) 210-3975; Fax: (403) 270-2772; E-mail:kchadee@ucalgary.ca

Fabrizio Bruschi (Ed.)

estimated by misdiagnosing pathogenic *E. histolytica* with non-pathogenic amoebas (*e.g.*, *E. dispar* and *E. moshkovskii*), amebiasis is considered one of the major causes of death due to parasitic infections in developing countries, after malaria and schistosomiasis [1].

History

Reports of people with bloody and mucous diarrhoea (probably amebiasis) were written 1000-600BC when Assyrian and Babylonian texts mentioned cases of dysentery in the Tigris-Euphrates basin [2]. By the 16th century, amebiasis was commonly described in developed Europe, mostly due to the growth of European colonies and increased world trade. At that time, there was a distinction of hepatic and intestinal forms of amebiasis. The first accurate description of both forms of the disease came from the book "*Researches into the Causes, Nature and Treatment of the More Prevalent Diseases of India and of Warm Climates Generally*" written by James Annersley in the 19th century [2]. With advances in microscopy, Friedrich Losch was the first to describe *E. histolytica* in Russia in 1873 from a young farmer who had been suffering from chronic dysentery [2]. In 1925, Emile Brumpt suggested that there must be two species: one that is invasive while the other is not (later named *E. dispar*) to explain why some people who were infected with *E. histolytica* (*i.e.* cysts in their stool) could never develop symptoms of the disease and spontaneously clear the infection [3]. However, it was not until the 1990's, due to biochemical, immunological and genetic tools that *E. histolytica* was reclassified as an invasive species named *E. histolytica* and the non-invasive species named *E. dispar* [3].

Morphology and Life Cycle

The life cycle of *E. histolytica* has 2 stages: the infectious cyst and the disease-inducing motile trophozoite stage. This life cycle begins when viable quadrinucleate cysts of *E. histolytica* are ingested from fecally contaminated food or water [4]. This amebic cyst has a protective wall formed of proteins and polysaccharides that make it highly resistant to most disinfectants and antiseptics commonly used in water treatments. Among those proteins, a proteomic study indicated the *E. histolytica* cysts contain specific glycoproteins Jacob, Jessie and chitinase and protein kinase, small GTPase signaling molecules, DNA repair proteins, epigenetic regulators and surface associated proteins [5]. Moreover, transcriptomic data showed that *E. histolytica* possesses at least 672 genes specific to cysteine proteases, putative DNA-binding and transcription factor-related proteins of the cyst wall [6]. The resistance of *E. histolytica* cysts is likely due to the cell walls are made of chitin fibrins, formed in small vesicles and regulated by chitin synthase and chitinase [7]. Therefore, ingested cysts transit intact through the acid stomach until the terminal ileum of the small intestine. In the intestine, cysts excyst and mature into trophozoites that migrate and colonize the colon. *E. histolytica* trophozoites in the colonic lumen feed bacteria, cellular debris and mucin and begin the process of aggregation and eventually encystation forming cysts that are then excreted in stool [4].

The presence of other *Entamoeba* spp. may not be associated with amebiasis but represents an epidemiological problem in public health. The other species that belong to the genus *Entamoeba* isolated from humans, including *E. dispar*, *E. moshkovskii*, *E. poleki* (also called *E. chattoni*), *E. coli* and *E. hartmanii*. Among the non-pathogenic amoebas, *E. dispar* is usually isolated from most individuals infected with *Entamoeba* spp. (about 90%) and patients remain as asymptomatic carriers (*i.e.*, healthy individuals without diarrhea [8], and the parasite is considered non-pathogenic (Table **1**).

Table 1. Biological characterization between pathogenic *Entamoeba histolytica* and non-pathogenic *E. dispar*.

Characteristics	*Entamoeba Histolytica*	*Entamoeba Dispar*
Trophozoite (μM)	12-30	12-40
Cyst (μM)	10-30	10-30
Localization	Colon and systemic organs (liver, pericardium, brain)	Colon
Transmission	Fecal-oral *via* contaminated food and water	Fecal-oral *via* contaminated food and water
Life Cycle-Cyst Stage	Cysts pass through the stomach and small intestine and excyst to motile trophozoites in the large intestine	Cysts pass through the stomach and small intestine and excyst to motile trophozoites in the large intestine
Life Cycle- Trophozoite Stage	Trophozoites inhabit the large intestine and encyst to cysts that are excreted with the feces. Trophozoites penetrate the intestinal mucosa and disseminate systemically to the liver and brain	Trophozoites inhabit the large intestine and encyst to cysts that are excreted with the feces
Epidemiology	Worldwide (predominantly in developing areas), human-human transmission	Worldwide, human-human transmission
Clinical Features	Amebic diarrheic colitis and extra-intestinal signs, including amebic liver abscess	None
Clinical Specimen	Intestinal stool	Intestinal stool
Treatment	Metronidazole and luminal microbicidals	None

Global Epidemiology and Risk Factors

Morbidity and mortality in amebiasis are particularly associated with children with undernourished under two years old in areas with poor sanitation and no access to potable water [9, 10]. Malnutrition causes immune deficiency, which increases susceptibility to *E. histolytica*, evidenced by more cases of infection and severe diarrhea [9]. Mechanistically, it has been showed that deficiency in adipocytokine leptin, common in malnourished children, increases ameba-induced apoptosis in host epithelial cells, addressing the relevant link between nutrition and immune defenses against amebiasis [11]. Clinical manifestation of amebiasis in children varies based on the immune and nutritional status and includes colonization with no *E. histolytica*–associated diarrhea; diarrhea with prior asymptomatic colonization and/or subsequent asymptomatic persistence; or *E. histolytica*–associated diarrhea with no evidence of colonization [10, 12, 13]. The association between amebiasis and poor living conditions is shown by the endemic prevalence of amebiasis in developing countries from all continents, including Asia (*e.g.*, Bangladesh, Malaysia), Africa (*e.g.*, Gambia, Egypt), and America (*e.g.* Mexico, Ecuador) [14 - 19]. In developed countries, *E. histolytica* can be identified by operational surveillance systems as an outbreak of food-borne and waterborne parasitic protozoa disease in the Australian continent, North America and Europe [20]. Moreover, sporadic cases of *E. histolytica* represent an important cause of chronic diarrhea in industrialized countries, affecting mostly returning travelers globally or recent immigrants [8, 10]. Of note, *E. histolytica* can be transmitted in men who have sex with men through oral-anal and oral-genital sexual contact [4, 21].

Symptoms

Infections with *E. histolytica* manifest clinically in only 4 to 10% of the cases while in the remaining 90% of infected individuals the disease courses asymptomatically [22]. Of note, some of the asymptomatic carriers could have been infected by *E. dispar* or *E. moshkovskii* that are indistinguishable from *E. histolytica* by light microscopy and routine parasitological techniques. In asymptomatic carriers, *E. histolytica* trophozoites in the colonic form cysts that pass through stool to perpetuate the life cycle. Approximately 4 to 10% of asymptomatic individuals may later developed colitis or extra-intestinal diseases [23]. Amoebic colitis is characterized by gradual onset of bloody or profuse watery diarrhea and abdominal pain and tenderness. Patients with clinical amebiasis suffer dysentery and acute diarrhea characterized by mucoid stools without fecal leukocytes [22]. Patients with amoebic colitis may develop ulcerations of the colonic mucosa with typical flask-shaped ulcers and necrotizing colitis that often lead to death. Other clinical signs of amebiasis include the formation of a fibrotic mass in the intestinal wall (amoeboma) and systemic invasion of *E. histolytica* in up to 1% of the cases provoking extra-intestinal disease characterized by abscesses in the liver, brain or lung [22]. Liver invasion by *E. histolytica* after entering the host through the digestive tract leads to the development of amoebic liver abscess (ALA). In this process, amoebas use the portal route and break through the sinusoidal endothelial barrier to reach the hepatic parenchyma [24]. The typical ALA includes necrotic lysis of the liver tissue with acute manifestation in the patients (*e.g.*, fever, pain and liver tenderness) and it is more often in young adult males [24]. The development of ALAs is the fatal feature of amebiasis and the most common extraintestinal form with high rate of mortality on untreated populations [1].

Pathophysiology

The colon is covered by gel-forming glycosylated MUC2 mucins secreted by goblet cells that protect the colonic epithelium from colonization by intestinal pathogens. Colonic mucin O-linked oligosaccharides secreted by goblet cells has the capacity to prevent the attachment of *E. histolytica* to the epithelium [25]. However, *E. histolytica* has mechanisms to overcome the protective mucin barrier, invade the intestinal mucosa and to systemically disseminate to the liver. The invasion of colonic tissue by *E. histolytica* is summarized in three consecutive steps: adherence, contact-dependent cytolysis of epithelial cells and phagocytosis. An extensive arsenal of virulence factors produced by *E. histolytica* (Table **2**) including lectins, amebapores and proteolytic enzymes (*i.e.* glycosidases, proteases, cysteine proteinases) assist trophozoites in locomotion activities and tissues penetration in the colonic mucosa through the proteolytic degradation of mucin polymer network, components of the innate defense and extracellular matrix components, such as elastin, collagen and fibrinogen proteases [4, 26].

Table 2. Characteristics of the major virulent factors in *Entamoeba histolytica*.

Virulent Factors	Biological Functions
Gal-GalNAc adherence lectin	Binding to Gal/GalNAc residues of colonic MUC2 mucin
	Induction of parasite adherence/colonization in the gut
	Induction of pro-inflammatory cytokine responses
	Activation of macrophages

(Table 2) contd.....

Virulent Factors	Biological Functions
Cysteine proteinases	Disruption of colonic MUC2 mucin
	Degradation of IgG, IgA, complement, cathelicidins
	Alteration of mucosal tight junction proteins
	Induction of cytokine response
Prostaglandins (PGE_2)	Induction of neutrophil chemokine IL-8
	Alterations in mucosal tight junction proteins
Amoebapore	Formation of cellular membrane pores
	Induction of cytolysis

Immunology and Immunopathology

Three distinct steps summarize the pathogenesis of amebiasis: first trophozoites of *E. histolytica* attach to mucosal epithelial cells followed by invasion into the intestinal mucosa forming flask-shaped ulcers that can then disseminate to the liver and other soft organs. The adherence of *E. histolytica* to the colonic epithelium is characterized by binding of the amebic adherent Gal-lectin adhesin to galactose (Gal) and N-acetyl-D-galactosamine (GalNAc) oligosaccharide mioties on colonic MUC2 mucin and other glycoproteins expressed on the surface of intestinal epithelial cells [27 - 29] (Table **2**). Recently, it has been described that *E. histolytica* invades the intestinal tissue due to an unique capacity to ingest distinct pieces of living human cells (named "trogocytosis"), which results in intracellular calcium elevation and eventual cell death [30]. This phagocytic capacity of *E. histolytica* is also evidenced by rapid internalization of colonic mucin and eventual release after high-affinity binding by the Gal-lectin [27]. In addition, *E. histolytica* enhances the secretion of preformed and newly synthesized mucin glycoprotein stores in goblet cells mediated by protein kinase C (PKC) and Ca^{2+}-dependent signaling pathways. By this process, *E. histolytica* produces mucin secretagogues that deplete the colonic mucin by exhausting goblet cells [10, 31, 32]. Amoeba Gal-lectin coupling to host cell receptors triggers a potent pro-inflammatory response. Besides binding to MUC2 mucin, the Gal-lectin interacts with host Toll like receptors 4 to stimulate the production of inflammatory cytokines IL-8, IL-6, IL-12, IL-1β, IFN-γ, and TNF-α by activating NF-κB [33]. Gal-lectin epitopes may likewise stimulate TNF-α gene and IL-1α/β gene expression partially through PKC in bone marrow macrophages [34, 35]. The roles of these cytokines in amebiasis outcome are variable. TNF-α produced by inflammatory monocytes and Kupffer cells in the liver accounted for the damaging inflammatory response [36]. However, TNF-α mRNA is rapidly degraded and the protein is not secreted in bone marrow-derived macrophages exposed to *E. histolytica*, indicating a parasite mechanisms to modulate host defense [37]. On the other hand, IFN-γ showed to be beneficial in protection against amebiasis and IFN-γ produced by natural killer T cells reduces the severity of amebic liver abscess in mice [38]. IFN-γ produced by peripheral blood mononuclear cells was shown to ameliorate diarrhea/dysentery in children infected with *E. histolytica* [39].

Cysteine proteases of *E. histolytica* (cathepsin-like enzymes) either secreted or localized on the surface of the parasite, are additional relevant virulent factors that modulate adherence, cytotoxicity, motility, disruption of the colonic epithelial barrier and the innate immune response during invasion [40 - 42] (Table **2**). Among the amoeba proteases, *E. histolytica* cysteine protease A5 (EhCP-A5), abundantly secreted and/or present on the surface of the parasite, is a major virulent factor in the pathogenesis of amebiasis. EhCP-A5 specifically disrupts the colonic mucin gel barrier [43] by depolymerizing the MUC2 polymers at the C-terminal domain, whereas the N-terminal domain remains resistant to proteolysis [41, 44]. Secreted EhCP-A5 also degrades mucin oligosaccharides independently of proteolytic activity [45] and causes selective damage to enteric neurons [46]. In addition, EhCP-A5 may initiate amoebic colitis by binding αVβ3 integrin on colonic cells *via* its RGD motif and stimulate NF-κB-mediated pro-inflammatory responses [47]. Cysteine proteases exhibit IL-1β converting enzyme activity that activates IL-1β released by damaged intestinal cells [48]. Interestingly, soluble proteins from *E. histolytica* prepared by collecting the supernatant after the amoeba were lysed, inhibit NF-κB activation and inflammation in epithelial cells and promote cell survival through activation of heat shock proteins (Hsp) 27 [49]. However, this Hsp response was induced only in epithelial cells that were primed by macrophage secretions. This suggests that amebic colitis could results from the lack of such protective responses to suppress pro-inflammatory cytokine induction in a minority of susceptible individuals [49]. Amoeba cysteine proteases are important effectors degrading serum and secretory immunoglobulin IgA and IgG and complement components C3 and C5 [50, 51]. Concomitantly, amoeba activates but resists killing by complement C5b-9 complexes deposited on the membrane surface by the Gal-lectin adhesins that bound C8 and C9 [52]. This degradation of immune factors (Igs and complement) modulates the evasion of the host immunity as mucosal local IgA antibodies have the capacity of inhibiting *E. histolytica* adherence [53] and IgA anti-lectin antibody response in stools of children was associated with acquired immune protection against *E. histolytica* colonization [23, 54]. Likewise, innate factors can be deprived by *E. histolytica* since trophozoites up regulate the synthesis of host defense cathelicidin in human intestinal epithelial cells and mice but are resistant to be killed by these peptides.

Even thought cysteine proteinases (rEhCP1) degrade cathelicidins the fragments still retain their antimicrobial activity against bacteria [55]. Additionally, an important role of cysteine proteases is in alternating intestinal barrier function. EhCP-A5 severely disrupts paracellular epithelial tight junctions proteins (claudin-2, occludin, and ZO-1) and increases pro-inflammatory cytokine IL-1β, TNF-α, IFN-γ [56]. Moreover, *E. histolytica* produces mimic proteins, such as occludin-like protein that displaces mucosal epithelial occludin-occludin tight junction interactions resulting in epithelial disruption [57].

Following *E. histolytica* degradation and dissolution of the mucus barrier, the parasite binds the colonic epithelium to evoke a rapid pro-inflammatory response in disease pathogenesis. Ameba-induced colonic inflammation leads to erosion and ulceration of the superficial epithelium and micro invasion with mild to moderate neutrophil infiltration in the lamina propria around trophozoites [58]. Chemo attraction of neutrophils by *E. histolytica* is regulated by membrane bound proteins particularly present in highly virulent amoebic strains [59]. In particular, *E. histolytica* produces prostaglandin E_2 (PGE_2) by a cyclooxygenase (COX)-like enzyme responsible for the biosynthesis of PGE_2, that induces IL-8 production in colonic epithelial cells, a potent neutrophil chemokine [60, 61] (Table **2**). *E. histolytica* also induces the production of PGE_2 in macrophages that suppresses IFN-γ production in an autocrine fashion for immune-regulation that could facilitate amoeba invasion [62]. *E. histolytica*-derived PGE_2 influences the onset of amoebic diarrhea as it alters ion permeability of tight junctions in colonic epithelia through EP4 receptors [63]. This suggests that either live *E. histolytica* or its secretory/soluble components are capable of stimulating the production IL-8 by colonic cells which in turn elicits a neutrophil response that does not require parasite-cell contact or cell injury [64]. Neutrophils are important immune effectors against amebiasis; when activated by IFN-γ and TNF-α, they exert microbicidal activity against *E. histolytica* trophozoites and resistance to amebic contact-dependent killing [65]. Macrophages are important effector cells in amebiasis as they acquire cytotoxic activity against *E. histolytica* by the synthesis of nitric oxide (NO) accompanied by O_2^- and H_2O_2 molecules [66]. Macrophages are attracted to the site of infection by the chemoattractant, monocyte chemotactic protein (MCP) secreted by colonic epithelial cells after stimulation with soluble proteins of *E. histolytica* [67]. Macrophages primed by the Gal-lectin signals *via* surface TLR4 receptors to activate NF-κB thorough MAP kinase signaling [68]. However, full activation of macrophages neccesary to kill *E. histolytica* requires IFN-γ, TNF-α and macrophage colony-stimulating factor 1 [69]. In addition, transforming growth factor β primed macrophages enhance expression of the NO synthase gene [70] and TNF-α augments NO-dependent macrophage cytotoxicity

[71]. As an evasive mechanism, *E. histolytica* may selectively overcome macrophage attack by inducing dysfunction of macrophage cytotoxicity modulating iNOS mRNA expression and NO production independent from TNF-α and PGE_2 [72].

Contact of intestinal epithelial and inflammatory cells with *E. histolytica* results in rapid cytotoxic effects including increased intracellular Ca^{2+}, reactive oxygen species production, loss of membrane integrity, DNA fragmentation, phosphatidylserine (PS) exposure, and caspase-3 activation [73]. *E. histolytica* Gal-lectin specifically elicits a contact-dependent response in macrophages that triggers the transcriptional priming of inflammasome and non-inflammasom--dependent pro-inflammatory cytokines [74]. Important in the cytolytic effect, *E. histolytica* secretes saposin-like polypeptides (77 amino acid residues arranged in four alpha-helical domains) denominated amoebapores [75] (Table **2**). Three main amoebapores (A, B and C) are synthesized by *E. histolytica*, which are phylogenetically related to mammalian membrane-permeabilizing proteins NK-lysin and granulysin as membrane-permeabilizing and pore-forming peptides [76, 77]. The function of amoebapores is essentially to induce pore formation in acidic vesicles synthetic liposomes although amoebapores do not degrade host cell DNA degradation and may require other factors to induce killing *in vivo* [77]. Thus, amoebapores are mainly cytotoxic, lysing host cells after adherence and engulfed bacteria into the phagosome [76, 78]. Invasion of colonic mucosa by *E. histolytica* trophozoites provokes necrosis of the intestinal epithelium and migration to the lamina propria where they gain access to the bloodstream and disseminate to the liver [10, 22]. In the liver, *E. histolytica* causes acute hepatic inflammation with transient T cell immunosuppression and impaired lymphocyte function [79, 80]. After a transient immunosuppressed stage, IL-2 production and low IL-4 and TNF-α (Th-1-like response) have been shown to control infection and resist reinfection in an amoebic liver abscess model in gerbils [79, 81].

E. histolytica has an infective stage where amoebas possess an encysted form to infect other person. This encysted form usually exits the host *via* feces and enters the next host *via* the oral route. The encystation involves Gal-lectin expressed on the surface of trophozoites that interact with host ligands [82]. These parasite carbohydrate lectins have been shown to initiate multicellular clumping of amoebas and form mature detergent resistant cysts. Presumably, the lectin allows binding of chitin by the secreted chitin-binding proteins to form a matrix with the secreted cyst wall components [83]. Perhaps, trophozoites use Gal-lectin binding to “sense” multivalent galactose-terminal ligands in the host that can trigger an encystation signal [82]. Lastly, it is important to note that the ability of *E.*

histolytica to provoke ALA is defined by a number of factors required for virulence and ALA development, including those already described for complement resistance, lysis (CPs and amoebapores), and cell adherence (Gal-lectin) [24].

Diagnosis and Detection Methods

A challenge in the diagnosis of amebiasis is to distinguish and differentiate the three species of amoebas: *E. dispar*, *E. moshkovskii* and *E. histolytica*. Standard diagnostic clinical methods for *E. histolytica* include microscopic detection on wet preparation, concentration and permanently stained smears [84]. However, microscopy is not a reliable method for the differentiation among *E. histolytica/E. dispar/E. moshkovskii,* and has a poor sensitivity and confounded with false-positive results due to misidentification of macrophages as trophozoites, PMN and cysts [84]. Thus, microscopic examination requires consecutive examinations for ova and parasites of stool samples to improve the detection. Nowadays, microscopy is mostly confined to tropical and subtropical developing countries that lack of facilities to conduct other most sophisticated methods [84]. Culture methods of *Entamoeba* species have been extensively described in the literature but culturing amoebas is not practical in a clinical diagnostic laboratory because it is difficult, expensive, laboratory intensive and yields low sensitivity [84]. Serological tests, including indirect hemagglutination, latex agglutination, immunoelectrophoresis, the amebic gel diffusion test, immunodiffusion, complement fixation, indirect immunofluorescence assay (IFA), and enzyme-linked immunosorbent assay (ELISA) are available but they are not frequently used because they may not distinguish past from current infection [84]. The DNA-based methods may be a good choice for diagnosing *Entamoeba* species in developed countries. PCR of the small-subunit rRNA gene (18S rDNA) is more sensitive than the best ELISA kit currently available and effective for differentiation of the three *Entamoeba* species. One disadvantage is that PCR-based methods may be still impractical in routine diagnosis due to difficulties in DNA extraction from fecal samples and that the amplification and detection of DNA are time-consuming and expensive. In this regard, real-time PCR is a quantitative method that may offer enhanced sensitivity compared to conventional PCR although there are still a limited number of probes that can be applied. Certainly, the development of molecular tools, including PCR and real-time PCR, to detect *E. histolytica, E. dispar*, and *E. moshkovskii* DNA in stool or liver abscess samples has leading position in making an accurate diagnosis of amebiasis. Antigen detection using fecal ELISAs for the detection of antigens in fecal samples, serum and liver abscess pus has demonstrated good sensitivity and specificity for detection of *E. histolytica* and are rapid to perform [84]. Moreover, these ELISAs can be used in laboratories in the developing world that do not have molecular facilities [84]. Alternatively, a combination of microscopy of a fecal specimen, serological testing, and, where indicated, by colonoscopy and biopsy of intestinal amebic lesions or by drainage of a liver abscess is most commonly attempted for the diagnosis of invasive amebiasis [84].

E. histolytica is a relevant pathogen in the majority of the aquatic ecosystems of the developing world characterized by systematic lack of healthy water and contamination of drinking water [4]. Recent studies have established *E. histolytica* as the etiological agent in approximately 3% of 325 water-associated outbreaks of parasitic protozoan disease in North America and Europe [85]. Foodborne transmission of *E. histolytica* mostly occurs by drinking fecally contaminated water or ingestion of food (*i.e.* vegetables, crops) previously irrigated with contaminated water [4]. However, foodborne transmission may be indirect and the mere consumption of drinks with ice, ice cream and raw fruit in ice contaminated with *E. histolytica* may cause infection [4]. Contamination of food by infected food handlers and person-to-person contact are other relevant means of transmission. While availability of potable water is more common in larger public water supplies, small water supplies in rural communities in developed countries remain highly vulnerable to contamination and resulting outbreaks of amebiasis and other infections [86]. In water supplies, *E. histolytica* cysts may contaminate the surface waters as cysts with chitin-rich wall allows the parasite to remain viable in suitable aquatic environments for several months at low temperature and they resist chlorination and ozone better than most enteric bacteria and viruses [4]. However, trophozoites are susceptible to most disinfectants used [87]. Thus, it is necessary to minimize the risk factors associated with waterborne and foodborne transmission. Furthermore, standardized diagnostic methods of *E. histolytica* cysts are required to maintain public health surveillance. In this regard, novel technology has been developed to limit the presence of waterborne pathogens, including *E. histolytica*. Poly(dimethylsiloxane) microfluidic devices for high-trapping and sorting of micron-size particles were shown to be effectively to contain protozoa cysts usually found in the water, including *Entamoeba*, *Giardia*, and *Cryptosporidium* (<30 μm) [88]. Likewise, chip-scale microscope system, based on automated and high-throughput imaging, showed promissory results identifying cysts of *Giardia*, *Cryptosporidium*, and *Entamoeba* [89]. Most developed countries have legislation to monitor microbial contamination, although waterborne pathogens can still prevail screening to provide water and food safety. What is needed is a worldwide implementation of water purification systems including treatment of drinking water with regimens of coagulation/flocculation, sedimentation, filtration and disinfection to remove protozoan parasites effectively [87].

Treatment

Clinical treatment relies on oral treatments with metronidazole and nitroimidazole derived compounds (*e.g.* tinidazole, ornidazole), highly effective tissue amoebicide, in combination with other luminal agents (*e.g.* paromomycin, iodoquinol) to eliminate intestinal colonization [8]. These drugs have adverse effects and there is an increasing concern of potential resistance of *E. histolytica.* Moreover, drug therapy can be unaffordable and not easily available in certain developing countries and areas. Asymptomatic carriers, usually detected in epidemiological studies or in travelers by fecal antigen detection assays with fresh stool samples, should also be treated with a luminal agent to minimize the spread of infection [90]. In the development of novel therapeutics, strategies to inhibit the activity of amoeba cysteine proteases could contribute significantly to host protection against *E. histolytica.* In this regard, auranofin, a US Food and Drug Administration (FDA)-approved drug used therapeutically for rheumatoid arthritis, was found to be a potent drug against *E. histolytica* [91]. Auranofin killed *E. histolytica* trophozoites much more effectively than metronidazole and alleviated amebic colitis in a mouse model and liver abscess in a hamster model by targeting amoeba thioredoxin reductase, preventing the reduction of thioredoxin and enhancing sensitivity of trophozoites to reactive oxygen-mediated killing [91].

Perspectives of Control

The elaboration of an effective amebiasis vaccine could improve child health in the developing world, reducing cases of amebic colitis and liver abscess. While the use of vaccines could be a cost-effective, safe, and massive alternative for mitigating amebiasis, the development of effective antigens of *E. histolytica* is still under research and no vaccine has been approved for human clinical trials to date. Experimentally, vaccinations using native and recombinant forms of the parasite Gal-lectin, serine-rich protein and the 29-kDa-reductase antigen have shown to protect animals against intestinal amebiasis and liver abscess [92]. Extensive work has been dedicated to the Gal-lectin from *E. histolytica* to identify specific parasite adherence motifs that can induce antibodies effective in preventing amebic attachment and cytolysis of host epithelial cells. In this search, a subunit vaccine candidate containing the cysteine-rich region of the 170-kDa subunit Gal-lectin of *E. histolytica* containing the carbohydrate recognition domain (CRD) has showed important advantages. The Gal-lectin can inhibited the binding of *E. histolytica* to human mucin [29] and to activate dendritic cells to produce IL-12 [93]. *In vivo*, the Gal-lectin induced a protective Th1 response in macrophages and spleen cells and reduced the severity of amebiasis through IL-12 induction in a gerbil model of liver abscess [94]. Furthermore, the Gal-lectin induced Th1-type specific cellular immune response and serum antibodies in mice that recognized a recombinant portion of the heavy subunit that inhibited the adherence of trophozoites to target cells *in vitro* [95, 96]. Mucosal intranasal administration of the Gal-lectin, combined with cytosine guanine oligodeoxynucleotide (CpG-ODN), induced lectin-specific Th1 responses, serum anti-Gal-lectin IgG and stool anti-Gal-lectin IgA [97, 98]. This immune response was capable of blocking parasite adherence to target cells *in vitro*, induced IFN-γ, IL-12 and IL-2 and protected gerbils challenged intra-hepatically with live trophozoites [97, 98]. The use of CpG-ODN as an adjuvant to augment Th1 responses is fundamental as *E. histolytica* DNA signaled through TLR9 and induced TNF-α in macrophages in a manner similar to that seen with CpG-ODN [99]. While the Gal-lectin is the most promising molecule shown to be effective in vaccine development in animal models of disease it is yet to be tested clinically.

CONCLUSION

In summary, the ability of *E. histolytica* trophozoites to invade intestinal mucosa and produce proteolytic enzymes and resist complement-mediated lyses will determine the grade of trophozoites invasion and colitis [58]. This presence of virulent factors in *E. histolytica* are specifically determinant in parasite survival, replication, and transmission, as the related non-pathogenic *E. dispar* lack these essential factors (*e.g.*, EhCP-A5) and does not cause diseases [100]. There are currently no vaccines available for enteric parasites and novel treatments have been investigated to replace or reduce conventional metronidazole or its derivatives. Ongoing research efforts in the mechanisms of pathogenesis are necessary for treatment and vaccine design against *E. histolytica* and its eradication.

CONFLICT OF INTEREST

The authors confirm that they have no conflict of interest to declare for this publication.

ACKNOWLEDGEMENTS

This work was supported by a Discovery Grant from the Natural Sciences and

Engineering Research Council of Canada (NSERC) to KC. EC was supported by an Alberta Innovates-Health Solutions (AIHS) Fellowship Award.

REFERENCES

[1] Amoebiasis WH. Amoebiasis. Wkly Epidemiol Rec 1997; 72(14): 97-9.
[PMID: 9100475]

[2] Cox FE. History of human parasitology. Clin Microbiol Rev 2002; 15(4): 595-612.
[http://dx.doi.org/10.1128/CMR.15.4.595-612.2002] [PMID: 12364371]

[3] Ackers J. The diagnostic implications of the separation of Entamoeba histolytica and Entamoeba dispar. J Biosci 2002; 3(27): 573-8.

[4] Tengku SA, Norhayati M. Public health and clinical importance of amoebiasis in Malaysia: a review. Trop Biomed 2011; 28(2): 194-222.
[PMID: 22041740]

[5] Ali IK, Haque R, Siddique A. Proteomic analysis of the cyst stage of Entamoeba histolytica. PLoS neglected tropical diseases 2012; 6(5): e1643.

[6] Ehrenkaufer GM, Hackney JA, Singh U. A developmentally regulated Myb domain protein regulates expression of a subset of stage-specific genes in Entamoeba histolytica. Cell Microbiol 2009; 11(6): 898-910.
[http://dx.doi.org/10.1111/j.1462-5822.2009.01300.x] [PMID: 19239479]

[7] Makioka A, Kumagai M, Hiranuka K, Kobayashi S, Takeuchi T. Different structure and mRNA expression of Entamoeba invadens chitinases in the encystation and excystation. Parasitol Res 2011; 109(2): 417-23.
[http://dx.doi.org/10.1007/s00436-011-2270-2] [PMID: 21286750]

[8] Fletcher SM, Stark D, Harkness J, Ellis J. Enteric protozoa in the developed world: a public health perspective. Clin Microbiol Rev 2012; 25(3): 420-49.
[http://dx.doi.org/10.1128/CMR.05038-11] [PMID: 22763633]

[9] Mondal D, Haque R, Sack RB, Kirkpatrick BD, Petri WA Jr. Attribution of malnutrition to cause-specific diarrheal illness: evidence from a prospective study of preschool children in Mirpur, Dhaka, Bangladesh. Am J Trop Med Hyg 2009; 80(5): 824-6.
[PMID: 19407131]

[10] Marie C, Petri WA Jr. Regulation of virulence of Entamoeba histolytica. Annu Rev Microbiol 2014; 68: 493-520.
[http://dx.doi.org/10.1146/annurev-micro-091313-103550] [PMID: 25002094]

[11] Marie CS, Verkerke HP, Paul SN, Mackey AJ, Petri WA Jr. Leptin protects host cells from Entamoeba histolytica cytotoxicity by a STAT3-dependent mechanism. Infect Immun 2012; 80(5): 1934-43.
[http://dx.doi.org/10.1128/IAI.06140-11] [PMID: 22331430]

[12] Mondal D, Minak J, Alam M, *et al.* Contribution of enteric infection, altered intestinal barrier function, and maternal malnutrition to infant malnutrition in Bangladesh. Clin Infect Dis 2012; 54(2): 185-92.
[http://dx.doi.org/10.1093/cid/cir807] [PMID: 22109945]

[13] Mondal D, Petri WA Jr, Sack RB, Kirkpatrick BD, Haque R. Entamoeba histolytica-associated diarrheal illness is negatively associated with the growth of preschool children: evidence from a prospective study. Trans R Soc Trop Med Hyg 2006; 100(11): 1032-8.
[http://dx.doi.org/10.1016/j.trstmh.2005.12.012] [PMID: 16730764]

[14] Taniuchi M, Sobuz SU, Begum S, *et al.* Etiology of diarrhea in Bangladeshi infants in the first year of life analyzed using molecular methods. J Infect Dis 2013; 208(11): 1794-802. [http://dx.doi.org/10.1093/infdis/jit507] [PMID: 24041797]

[15] Shahrul Anuar T, M Al-Mekhlafi H, Abdul Ghani MK, *et al.* Prevalence and risk factors associated with Entamoeba histolytica/dispar/moshkovskii infection among three Orang Asli ethnic groups in Malaysia. PLoS One 2012; 7(10): e48165. [http://dx.doi.org/10.1371/journal.pone.0048165] [PMID: 23133561]

[16] Bradbury RS, Barbé B, Jacobs J, *et al.* Enteric pathogens of food sellers in rural Gambia with incidental finding of Myxobolus species (Protozoa: Myxozoa). Trans R Soc Trop Med Hyg 2015; 109(5): 334-9. [http://dx.doi.org/10.1093/trstmh/trv020] [PMID: 25758855]

[17] Khalifa RM, Ahmad AK, Abdel-Hafeez EH, Mosllem FA. Present status of protozoan pathogens causing water-borne disease in northern part of El-Minia Governorate, Egypt. J Egypt Soc Parasitol 2014; 44(3): 559-66. [http://dx.doi.org/10.12816/0007860] [PMID: 25643498]

[18] Quihui-Cota L, Morales-Figueroa GG. Persistence of intestinal parasitic infections during the national de-worming campaign in schoolchildren of northwestern Mexico: a cross-sectional study. Ann Gastroenterol 2012; 25(1): 57-60. [PMID: 24714136]

[19] Vasco G, Trueba G, Atherton R, *et al.* Identifying etiological agents causing diarrhea in low income Ecuadorian communities. Am J Trop Med Hyg 2014; 91(3): 563-9. [http://dx.doi.org/10.4269/ajtmh.13-0744] [PMID: 25048373]

[20] Baldursson S, Karanis P. Waterborne transmission of protozoan parasites: review of worldwide outbreaks - an update 20042010. Water Res 2011; 45(20): 6603-14. [http://dx.doi.org/10.1016/j.watres.2011.10.013] [PMID: 22048017]

[21] Hung CC, Chang SY, Ji DD. Entamoeba histolytica infection in men who have sex with men. Lancet Infect Dis 2012; 12(9): 729-36. [http://dx.doi.org/10.1016/S1473-3099(12)70147-0] [PMID: 22917103]

[22] Haque R, Huston CD, Hughes M, Houpt E, Petri WA Jr. Amebiasis. N Engl J Med 2003; 348(16): 1565-73. [http://dx.doi.org/10.1056/NEJMra022710] [PMID: 12700377]

[23] Haque R, Ali IM, Sack RB, Farr BM, Ramakrishnan G, Petri WA Jr. Amebiasis and mucosal IgA antibody against the Entamoeba histolytica adherence lectin in Bangladeshi children. J Infect Dis 2001; 183(12): 1787-93. [http://dx.doi.org/10.1086/320740] [PMID: 11372032]

[24] Santi-Rocca J, Rigothier MC, Guillén N. Host-microbe interactions and defense mechanisms in the development of amoebic liver abscesses. Clin Microbiol Rev 2009; 22(1): 65-75. [Table of Contents.]. [http://dx.doi.org/10.1128/CMR.00029-08] [PMID: 19136434]

[25] Belley A, Keller K, Grove J, Chadee K. Interaction of LS174T human colon cancer cell mucins with Entamoeba histolytica: an *in vitro* model for colonic disease. Gastroenterology 1996; 111(6): 1484-92. [http://dx.doi.org/10.1016/S0016-5085(96)70009-4] [PMID: 8942726]

[26] He C, Nora GP, Schneider EL, *et al.* A novel Entamoeba histolytica cysteine proteinase, EhCP4, is key for invasive amebiasis and a therapeutic target. J Biol Chem 2010; 285(24): 18516-27. [http://dx.doi.org/10.1074/jbc.M109.086181] [PMID: 20378535]

[27] Chadee K, Johnson ML, Orozco E, Petri WA Jr, Ravdin JI. Binding and internalization of rat colonic mucins by the galactose/N-acetyl-D-galactosamine adherence lectin of Entamoeba histolytica. J Infect Dis 1988; 158(2): 398-406.
[http://dx.doi.org/10.1093/infdis/158.2.398] [PMID: 2900266]

[28] Chadee K, Ndarathi C, Keller K. Binding of proteolytically-degraded human colonic mucin glycoproteins to the Gal/GalNAc adherence lectin of Entamoeba histolytica. Gut 1990; 31(8): 890-5.
[http://dx.doi.org/10.1136/gut.31.8.890] [PMID: 2201583]

[29] Chadee K, Petri WA Jr, Innes DJ, Ravdin JI. Rat and human colonic mucins bind to and inhibit adherence lectin of Entamoeba histolytica. J Clin Invest 1987; 80(5): 1245-54.
[http://dx.doi.org/10.1172/JCI113199] [PMID: 2890655]

[30] Ralston KS, Solga MD, Mackey-Lawrence NM, Somlata , Bhattacharya A, Petri WA Jr. Trogocytosis by Entamoeba histolytica contributes to cell killing and tissue invasion. Nature 2014; 508(7497): 526-30.
[http://dx.doi.org/10.1038/nature13242] [PMID: 24717428]

[31] Chadee K, Keller K, Forstner J, Innes DJ, Ravdin JI. Mucin and nonmucin secretagogue activity of Entamoeba histolytica and cholera toxin in rat colon. Gastroenterology 1991; 100(4): 986-97.
[PMID: 2001836]

[32] Keller K, Olivier M, Chadee K. The fast release of mucin secretion from human colonic cells induced by Entamoeba histolytica is dependent on contact and protein kinase C activation. Arch Med Res 1992; 23(2): 217-21.
[PMID: 1340298]

[33] Galván-Moroyoqui JM, Del Carmen Domínguez-Robles M, Meza I. Pathogenic bacteria prime the induction of Toll-like receptor signalling in human colonic cells by the Gal/GalNAc lectin Carbohydrate Recognition Domain of Entamoeba histolytica. Int J Parasitol 2011; 41(10): 1101-12.
[http://dx.doi.org/10.1016/j.ijpara.2011.06.003] [PMID: 21787776]

[34] Seguin R, Chadee K. Entamoeba histolytica modulates TNF-alpha, IL-1 alpha/beta and c-fos gene expression in macrophages. Arch Med Res 1992; 23(2): 147-9.
[PMID: 1340279]

[35] Séguin R, Mann BJ, Keller K, Chadee K. Identification of the galactose-adherence lectin epitopes of Entamoeba histolytica that stimulate tumor necrosis factor-alpha production by macrophages. Proc Natl Acad Sci USA 1995; 92(26): 12175-9.
[http://dx.doi.org/10.1073/pnas.92.26.12175] [PMID: 8618866]

[36] Helk E, Bernin H, Ernst T, *et al.* TNFα-mediated liver destruction by Kupffer cells and Ly6Chi monocytes during Entamoeba histolytica infection. PLoS Pathog 2013; 9(1): e1003096.
[http://dx.doi.org/10.1371/journal.ppat.1003096] [PMID: 23300453]

[37] Seguin R, Keller K, Chadee K. Entamoeba histolytica stimulates the unstable transcription of c-fos and tumour necrosis factor-alpha mRNA by protein kinase C signal transduction in macrophages. Immunology 1995; 86(1): 49-57.
[PMID: 7590881]

[38] Lotter H, González-Roldán N, Lindner B, *et al.* Natural killer T cells activated by a lipopeptidophosphoglycan from Entamoeba histolytica are critically important to control amebic liver abscess. PLoS Pathog 2009; 5(5): e1000434.
[http://dx.doi.org/10.1371/journal.ppat.1000434] [PMID: 19436711]

[39] Haque R, Mondal D, Shu J, *et al.* Correlation of interferon-gamma production by peripheral blood mononuclear cells with childhood malnutrition and susceptibility to amebiasis. Am J Trop Med Hyg 2007; 76(2): 340-4.
[PMID: 17297046]

[40] Serrano-Luna J, Pina-Vazquez C, Reyes-Lopez M, Ortiz-Estrada G, de la Garza M. Proteases from Entamoeba spp. and Pathogenic Free-Living Amoebae as Virulence Factors. J Trop Med 2013. 2013: 890603.

[41] Moncada D, Keller K, Chadee K. Entamoeba histolytica cysteine proteinases disrupt the polymeric structure of colonic mucin and alter its protective function. Infect Immun 2003; 71(2): 838-44. [http://dx.doi.org/10.1128/IAI.71.2.838-844.2003] [PMID: 12540564]

[42] Moncada D, Yu Y, Keller K, Chadee K. Entamoeba histolytica cysteine proteinases degrade human colonic mucin and alter its function. Arch Med Res 2000; 31(4) (Suppl.): S224-5. [http://dx.doi.org/10.1016/S0188-4409(00)00227-7] [PMID: 11070293]

[43] Moncada D, Keller K, Ankri S, Mirelman D, Chadee K. Antisense inhibition of Entamoeba histolytica cysteine proteases inhibits colonic mucus degradation. Gastroenterology 2006; 130(3): 721-30. [http://dx.doi.org/10.1053/j.gastro.2005.11.012] [PMID: 16530514]

[44] Lidell ME, Moncada DM, Chadee K, Hansson GC. Entamoeba histolytica cysteine proteases cleave the MUC2 mucin in its C-terminal domain and dissolve the protective colonic mucus gel. Proc Natl Acad Sci USA 2006; 103(24): 9298-303. [http://dx.doi.org/10.1073/pnas.0600623103] [PMID: 16754877]

[45] Moncada D, Keller K, Chadee K. Entamoeba histolytica-secreted products degrade colonic mucin oligosaccharides. Infect Immun 2005; 73(6): 3790-3. [http://dx.doi.org/10.1128/IAI.73.6.3790-3793.2005] [PMID: 15908414]

[46] Lourenssen S, Houpt ER, Chadee K, Blennerhassett MG. Entamoeba histolytica infection and secreted proteins proteolytically damage enteric neurons. Infect Immun 2010; 78(12): 5332-40. [http://dx.doi.org/10.1128/IAI.00699-10] [PMID: 20855514]

[47] Hou Y, Mortimer L, Chadee K. Entamoeba histolytica cysteine proteinase 5 binds integrin on colonic cells and stimulates NFkappaB-mediated pro-inflammatory responses. J Biol Chem 2010; 285(46): 35497-504. [http://dx.doi.org/10.1074/jbc.M109.066035] [PMID: 20837477]

[48] Zhang Z, Wang L, Seydel KB, *et al.* Entamoeba histolytica cysteine proteinases with interleukin-1 beta converting enzyme (ICE) activity cause intestinal inflammation and tissue damage in amoebiasis. Mol Microbiol 2000; 37(3): 542-8. [http://dx.doi.org/10.1046/j.1365-2958.2000.02037.x] [PMID: 10931347]

[49] Kammanadiminti SJ, Chadee K. Suppression of NF-kappaB activation by Entamoeba histolytica in intestinal epithelial cells is mediated by heat shock protein 27. J Biol Chem 2006; 281(36): 26112-20. [http://dx.doi.org/10.1074/jbc.M601988200] [PMID: 16840786]

[50] Kelsall BL, Ravdin JI. Degradation of human IgA by Entamoeba histolytica. J Infect Dis 1993; 168(5): 1319-22. [http://dx.doi.org/10.1093/infdis/168.5.1319] [PMID: 8228372]

[51] Reed SL, Ember JA, Herdman DS, DiScipio RG, Hugli TE, Gigli I. The extracellular neutral cysteine proteinase of Entamoeba histolytica degrades anaphylatoxins C3a and C5a. J Immunol 1995; 155(1): 266-74. [PMID: 7602103]

[52] Braga LL, Ninomiya H, McCoy JJ, *et al.* Inhibition of the complement membrane attack complex by the galactose-specific adhesion of Entamoeba histolytica. J Clin Invest 1992; 90(3): 1131-7. [http://dx.doi.org/10.1172/JCI115931] [PMID: 1381719]

[53] Leyva O, Rico G, Ramos F, Morán P, Melendro EI, Ximénez C. Inhibition of adhesion process mediated by anti-Entamoeba histolytica specific monoclonal IgA antibodies. Arch Med Res 1992; 23(2): 227-9.
[PMID: 1340300]

[54] Haque R, Duggal P, Ali IM, *et al.* Innate and acquired resistance to amebiasis in bangladeshi children. J Infect Dis 2002; 186(4): 547-52.
[http://dx.doi.org/10.1086/341566] [PMID: 12195383]

[55] Cobo ER, He C, Hirata K, *et al.* Entamoeba histolytica induces intestinal cathelicidins but is resistant to cathelicidin-mediated killing. Infect Immun 2012; 80(1): 143-9.
[http://dx.doi.org/10.1128/IAI.05029-11] [PMID: 22083705]

[56] Kissoon-Singh V, Moreau F, Trusevych E, Chadee K. Entamoeba histolytica exacerbates epithelial tight junction permeability and proinflammatory responses in Muc2(-/-) mice. Am J Pathol 2013; 182(3): 852-65.
[http://dx.doi.org/10.1016/j.ajpath.2012.11.035] [PMID: 23357502]

[57] Goplen M, Lejeune M, Cornick S, Moreau F, Chadee K. Entamoeba histolytica contains an occludin-like protein that can alter colonic epithelial barrier function. PLoS One 2013; 8(9): e73339.
[http://dx.doi.org/10.1371/journal.pone.0073339] [PMID: 24058468]

[58] Eckmann L, Reed SL, Smith JR, Kagnoff MF. Entamoeba histolytica trophozoites induce an inflammatory cytokine response by cultured human cells through the paracrine action of cytolytically released interleukin-1 alpha. J Clin Invest 1995; 96(3): 1269-79.
[http://dx.doi.org/10.1172/JCI118161] [PMID: 7657801]

[59] Chadee K, Moreau F, Meerovitch E. Entamoeba histolytica: chemoattractant activity for gerbil neutrophils *in vivo* and *in vitro.* Exp Parasitol 1987; 64(1): 12-23.
[http://dx.doi.org/10.1016/0014-4894(87)90003-8] [PMID: 2886359]

[60] Dey I, Chadee K. Prostaglandin E2 produced by Entamoeba histolytica binds to EP4 receptors and stimulates interleukin-8 production in human colonic cells. Infect Immun 2008; 76(11): 5158-63.
[http://dx.doi.org/10.1128/IAI.00645-08] [PMID: 18710858]

[61] Dey I, Keller K, Belley A, Chadee K. Identification and characterization of a cyclooxygenase-like enzyme from Entamoeba histolytica. Proc Natl Acad Sci USA 2003; 100(23): 13561-6.
[http://dx.doi.org/10.1073/pnas.1835863100] [PMID: 14585927]

[62] Wang W, Chadee K. Entamoeba histolytica suppresses gamma interferon-induced macrophage class II major histocompatibility complex Ia molecule and I-A beta mRNA expression by a prostaglandin E2-dependent mechanism. Infect Immun 1995; 63(3): 1089-94.
[PMID: 7868232]

[63] Lejeune M, Moreau F, Chadee K. Prostaglandin E2 produced by Entamoeba histolytica signals *via* EP4 receptor and alters claudin-4 to increase ion permeability of tight junctions. Am J Pathol 2011; 179(2): 807-18.
[http://dx.doi.org/10.1016/j.ajpath.2011.05.001] [PMID: 21683675]

[64] Yu Y, Chadee K. Entamoeba histolytica stimulates interleukin 8 from human colonic epithelial cells without parasite-enterocyte contact. Gastroenterology 1997; 112(5): 1536-47.
[http://dx.doi.org/10.1016/S0016-5085(97)70035-0] [PMID: 9136832]

[65] Denis M, Chadee K. Human neutrophils activated by interferon-gamma and tumour necrosis factor-alpha kill Entamoeba histolytica trophozoites *in vitro.* J Leukoc Biol 1989; 46(3): 270-4.
[PMID: 2547889]

[66] Lin JY, Chadee K. Macrophage cytotoxicity against Entamoeba histolytica trophozoites is mediated by nitric oxide from L-arginine. J Immunol 1992; 148(12): 3999-4005.
[PMID: 1318338]

[67] Kammanadiminti SJ, Dey I, Chadee K. Induction of monocyte chemotactic protein 1 in colonic epithelial cells by Entamoeba histolytica is mediated *via* the phosphatidylinositol 3-kinase/p65 pathway. Infect Immun 2007; 75(4): 1765-70.
[http://dx.doi.org/10.1128/IAI.01442-06] [PMID: 17283105]

[68] Kammanadiminti SJ, Mann BJ, Dutil L, Chadee K. Regulation of Toll-like receptor-2 expression by the Gal-lectin of Entamoeba histolytica. FASEB j 2004; 18(1): 155-7.

[69] Denis M, Chadee K. Cytokine activation of murine macrophages for *in vitro* killing of Entamoeba histolytica trophozoites. Infect Immun 1989; 57(6): 1750-6.
[PMID: 2542164]

[70] Lin JY, Seguin R, Keller K, Chadee K. Transforming growth factor-beta 1 primes macrophages for enhanced expression of the nitric oxide synthase gene for nitric oxide-dependent cytotoxicity against Entamoeba histolytica. Immunology 1995; 85(3): 400-7.
[PMID: 7558128]

[71] Lin JY, Seguin R, Keller K, Chadee K. Tumor necrosis factor alpha augments nitric oxide-dependent macrophage cytotoxicity against Entamoeba histolytica by enhanced expression of the nitric oxide synthase gene. Infect Immun 1994; 62(5): 1534-41.
[PMID: 7513301]

[72] Wang W, Keller K, Chadee K. Entamoeba histolytica modulates the nitric oxide synthase gene and nitric oxide production by macrophages for cytotoxicity against amoebae and tumour cells. Immunology 1994; 83(4): 601-10.
[PMID: 7533135]

[73] Ralston KS, Petri WA Jr. Tissue destruction and invasion by Entamoeba histolytica. Trends Parasitol 2011; 27(6): 254-63.
[http://dx.doi.org/10.1016/j.pt.2011.02.006] [PMID: 21440507]

[74] Mortimer L, Moreau F, Cornick S, Chadee K. Gal-lectin-dependent contact activates the inflammasome by invasive Entamoeba histolytica. Mucosal Immunol 2014; 7(4): 829-41.
[http://dx.doi.org/10.1038/mi.2013.100] [PMID: 24253103]

[75] Leippe M, Bruhn H, Hecht O, Grötzinger J. Ancient weapons: the three-dimensional structure of amoebapore A. Trends Parasitol 2005; 21(1): 5-7.
[http://dx.doi.org/10.1016/j.pt.2004.10.009] [PMID: 15639732]

[76] Hecht O, Van Nuland NA, Schleinkofer K, *et al.* Solution structure of the pore-forming protein of Entamoeba histolytica. J Biol Chem 2004; 279(17): 17834-41.
[http://dx.doi.org/10.1074/jbc.M312978200] [PMID: 14970207]

[77] Leippe M, Andrä J, Müller-Eberhard HJ. Cytolytic and antibacterial activity of synthetic peptides derived from amoebapore, the pore-forming peptide of Entamoeba histolytica. Proc Natl Acad Sci USA 1994; 91(7): 2602-6.
[http://dx.doi.org/10.1073/pnas.91.7.2602] [PMID: 8146160]

[78] Andrä J, Herbst R, Leippe M. Amoebapores, archaic effector peptides of protozoan origin, are discharged into phagosomes and kill bacteria by permeabilizing their membranes. Dev Comp Immunol 2003; 27(4): 291-304.
[http://dx.doi.org/10.1016/S0145-305X(02)00106-4] [PMID: 12590963]

[79] Campbell D, Gaucher D, Chadee K. Serum from Entamoeba histolytica-infected gerbils selectively suppresses T cell proliferation by inhibiting interleukin-2 production. J Infect Dis 1999; 179(6): 1495-501.
[http://dx.doi.org/10.1086/314781] [PMID: 10228072]

[80] Chadee K, Denis M, Keller K. Down-regulation of murine lymphocyte responsiveness to mitogens after treatment with antigens of Entamoeba histolytica. Parasitol Res 1991; 77(7): 572-6.
[http://dx.doi.org/10.1007/BF00931015] [PMID: 1792226]

[81] Campbell D, Chadee K. Interleukin (IL)-2, IL-4, and tumor necrosis factor-alpha responses during Entamoeba histolytica liver abscess development in gerbils. J Infect Dis 1997; 175(5): 1176-83.
[http://dx.doi.org/10.1086/520355] [PMID: 9129082]

[82] Frederick JR, Petri WA Jr. Roles for the galactose-/N-acetylgalactosamine-binding lectin of Entamoeba in parasite virulence and differentiation. Glycobiology 2005; 15(12): 53R-9R.
[http://dx.doi.org/10.1093/glycob/cwj007] [PMID: 16037494]

[83] Coppi A, Eichinger D. Regulation of Entamoeba invadens encystation and gene expression with galactose and N-acetylglucosamine. Mol Biochem Parasitol 1999; 102(1): 67-77.
[http://dx.doi.org/10.1016/S0166-6851(99)00085-7] [PMID: 10477177]

[84] Fotedar R, Stark D, Beebe N, Marriott D, Ellis J, Harkness J. Laboratory diagnostic techniques for Entamoeba species. Clin Microbiol Rev 2007; 20(3): 511-32. [table of contents.].
[http://dx.doi.org/10.1128/CMR.00004-07] [PMID: 17630338]

[85] Karanis P, Kourenti C, Smith H. Waterborne transmission of protozoan parasites: a worldwide review of outbreaks and lessons learnt. J Water Health 2007; 5(1): 1-38.
[http://dx.doi.org/10.2166/wh.2006.002] [PMID: 17402277]

[86] Pitkänen T, Karinen P, Miettinen IT, *et al.* Microbial contamination of groundwater at small community water supplies in Finland. Ambio 2011; 40(4): 377-90.
[http://dx.doi.org/10.1007/s13280-010-0102-8] [PMID: 21809781]

[87] Bouzid M, Steverding D, Tyler KM. Detection and surveillance of waterborne protozoan parasites. Curr Opin Biotechnol 2008; 19(3): 302-6.
[http://dx.doi.org/10.1016/j.copbio.2008.05.002] [PMID: 18524569]

[88] Kim J, Erath J, Rodriguez A, Yang C. A high-efficiency microfluidic device for size-selective trapping and sorting. Lab Chip 2014; 14(14): 2480-90.
[http://dx.doi.org/10.1039/C4LC00219A] [PMID: 24850190]

[89] Lee SA, Erath J, Zheng G, *et al.* Imaging and identification of waterborne parasites using a chip-scale microscope. PLoS One 2014; 9(2): e89712.
[http://dx.doi.org/10.1371/journal.pone.0089712] [PMID: 24586978]

[90] Tanyuksel M, Petri WA Jr. Laboratory diagnosis of amebiasis. Clin Microbiol Rev 2003; 16(4): 713-29.
[http://dx.doi.org/10.1128/CMR.16.4.713-729.2003] [PMID: 14557296]

[91] Debnath A, Parsonage D, Andrade RM, *et al.* A high-throughput drug screen for Entamoeba histolytica identifies a new lead and target. Nat Med 2012; 18(6): 956-60.
[http://dx.doi.org/10.1038/nm.2758] [PMID: 22610278]

[92] Quach J, St-Pierre J, Chadee K. The future for vaccine development against Entamoeba histolytica. Hum Vaccin Immunother 2014; 10(6): 1514-21.
[http://dx.doi.org/10.4161/hv.27796] [PMID: 24504133]

[93] Ivory CP, Chadee K. Activation of dendritic cells by the Gal-lectin of Entamoeba histolytica drives Th1 responses *in vitro* and *in vivo*. Eur J Immunol 2007; 37(2): 385-94.
[http://dx.doi.org/10.1002/eji.200636476] [PMID: 17219364]

[94] Campbell D, Mann BJ, Chadee K. A subunit vaccine candidate region of the Entamoeba histolytica galactose-adherence lectin promotes interleukin-12 gene transcription and protein production in human macrophages. Eur J Immunol 2000; 30(2): 423-30. [http://dx.doi.org/10.1002/1521-4141(200002)30:2<423::AID-IMMU423>3.0.CO;2-0] [PMID: 10671197]

[95] Gaucher D, Chadee K. Construction and immunogenicity of a codon-optimized Entamoeba histolytica Gal-lectin-based DNA vaccine. Vaccine 2002; 20(27-28): 3244-53. [http://dx.doi.org/10.1016/S0264-410X(02)00303-1] [PMID: 12213393]

[96] Gaucher D, Chadee K. Prospect for an Entamoeba histolytica Gal-lectin-based vaccine. Parasite Immunol 2003; 25(2): 55-8. [http://dx.doi.org/10.1046/j.1365-3024.2003.00607.x] [PMID: 12791100]

[97] Ivory CP, Chadee K. Intranasal immunization with Gal-inhibitable lectin plus an adjuvant of CpG oligodeoxynucleotides protects against Entamoeba histolytica challenge. Infect Immun 2007; 75(10): 4917-22. [http://dx.doi.org/10.1128/IAI.00725-07] [PMID: 17620349]

[98] Ivory CP, Keller K, Chadee K. CpG-oligodeoxynucleotide is a potent adjuvant with an Entamoeba histolytica Gal-inhibitable lectin vaccine against amoebic liver abscess in gerbils. Infect Immun 2006; 74(1): 528-36. [http://dx.doi.org/10.1128/IAI.74.1.528-536.2006] [PMID: 16369009]

[99] Ivory CP, Prystajecky M, Jobin C, Chadee K. Toll-like receptor 9-dependent macrophage activation by Entamoeba histolytica DNA. Infect Immun 2008; 76(1): 289-97. [http://dx.doi.org/10.1128/IAI.01217-07] [PMID: 17984204]

[100] Wilson IW, Weedall GD, Hall N. Host-Parasite interactions in Entamoeba histolytica and Entamoeba dispar: what have we learned from their genomes? Parasite Immunol 2012; 34(2-3): 90-9. [http://dx.doi.org/10.1111/j.1365-3024.2011.01325.x] [PMID: 21810102]

CHAPTER 6

Free-living Amoebae

Vinay Khanna*

Department of Microbiology, Kasturba Medical College, Manipal University, Karnataka, India

Abstract: *Acanthamoeba Species* is the most common free living amoeba present in environment. It is isolated from soil, water, contact lens solutions, transplant units and various other hospital environment. There are many species of Acanthamoeba such as *A.astronyxis, A.castellani, A.culbertsoni, A.hatchetti, A.keratitis etc.* which are known to cause opportunistic infection in both immunocompetent as well as immunocompromised host. Transmission mainly occurs through direct contact. There are two described morphological forms; a trophozoite form and a cyst form. The trophozoites have characteristics pointed thorn like acanthapodias, containing one nucleus with central dense large nucleolus. The cytoplasm measures between 15-50μm, granular and contains various organelles. Clinically patients usually presents with granulomatous amoebic encephalitis (GAE) which is characterized by focal neurological deficit, headache, visual disturbances, seizures and behavioral abnormalities which develops over months to years. Laboratory diagnosis of Acanthamoeba spp. is done by examining CSF which generally shows predominant lymphocytes, elevated proteins and low glucose levels. Histopathological samples of the brain generally reveals cerebral edema, multiple necrotic and hemorrhage lesions. Acanthamoeba can easily be cultivated on non-nutrient agar with overlay of *Escherichia coli* or *Entrobacter spp.* Amoeba feeds on bacteria's and confluent growth is seen in 4-5 days of culturing. The combination therapy is advisable in proven cases of Acanthamoeba infection. Combining Amphotericin B plus Trimethoprim-Sulphamethoxazole plus rifampicin has successfully used in few cases.

Naegleria Fowleri: The organism was first reported in Australia in 1965. It is an environmental ameboflagellate parasite found in variety of water bodies such as ponds, swimming pools; aquarium *etc.* prefers temperature of 30-45°C. There are three stages seen in Naegleria life cycle: the infective trophozoites, transient flagellated and the resistant cystic stage. The portal of entity of trophozoites is *via* olfactory neuroepithelial cell lining covering the cribriform plate to reach olfactory bulb. Demyelination and myelinoclasis are observed in gray matter due to vascular blockage. These pathological changes are attributed to release of phospholipolytic enzymes which causes breaks in the lipid membrane of neuronal cells. Clinically, patients of primary amoebic meningoencephalitis (PAM) usually presented with high grade fever,

* **Corresponding author Vinay Khanna:** Department of Microbiology, Kasturba Medical College, Manipal University, Karnataka, India; Tel/Fax: 0820-2571809; E-mail:drvinay2004@yahoo.com

Fabrizio Bruschi (Ed.)

headache, photophobia and features of raised intracranial pressure. Laboratory diagnosis is done using peripheral smear, CSF examination, culture, histopathological examination and imaging modalities. Hematological findings are leukocytosis with predominant neutrophils. The CSF shows low glucose and high protein levels. Centrifugation of fresh CSF sample up to 500 RPM may reveal motile trophozoites. Morphologically, the size of trophozoites ranges between 12-25μm, with a single nucleus and centrally placed nucleolus in the absence of peripheral chromatin. Liquid culture media such as such as Nelson's medium containing ox liver digest and glucose are used with serum for growing amoebae. Mammalian cell lines can be employed to demonstrate cytopathic effect. Multiplex PCR detects free living amoeba within 6 hours but routine use in diagnostic laboratory is limited due to rarity of finding these organisms and having high cost of PCR. Brain imaging is easy to perform but restricted by nonspecific findings such as cerebral edema. Specifically, infraction involving frontal, orbital and cerebellum area can be observed in few cases of PAM. There is no optimal treatment regime for *Naegleria fowleri*. Literature suggests combination therapy works best with amphotericin, rifampicin and azithromycin.

Balamuthia Mandrillaris: Over 200 cases were reported from South America and United States. The true prevalence of disease is unknown in south East Asia. Organism is commonly isolated from soil contact with activities related to soil such as gardening, agriculture pose risk of acquiring the organism. It was first isolated in 1986 from baboon brain that died of meningoencephalitis. The portal of entry of the organism is *via* cutaneous lesions, nasal mucosa and then subsequent spread to brain. CNS lesions mimic acanthamoeba encephalitis and have chronic slowly progressive course over many years. The life cycle of *Balamuthia* involves two stages: trophozoites and the cyst. The morphologically variable trophozoites are 12- 60 microns in size containing single nucleus with large centrally placed nucleolus. Cysts are spherical in shape measuring 12-30μm and contain a single nucleus with double wall having outer ectocyst, middle fibrillar layer and inner amorphous endocyst.The trophozoites, cysts and inflammatory cells are observed in perivascular regions of the infected tissue. In CSF, elevated protiens, reduced glucose are common findings. *Balamuthia* spp. can be grown in tissue cultures such as Monkey Kidney cell lines, Human Lung fibroblast and Human Brain Microvascular Endothelial cell lines. ELISA test is very specific to detect high antibodies titers. The antibodies do not cross react with other free living amoebae. PCR is also highly specific and sensitive test in which primers are developed against mitochondrial rRNA genes. Recently, real time PCR are developed targeting RNAase P gene of *B.mandrillaris*.

Sappinia Species: Two species of Sappinia are well-known cause of CNS infections in humans, *Sappinia diplodea* and *Sappinia pedata. S.diploidea* was first isolated from lizard faeces. As this parasite is found in animal faeces, persons handing livestock are at higher risk. Only one known case of Sappinia encephalitis infection reported in literature. The diagnosis was confirmed on histopathological sample, which showed necrotizing haemorrhagic inflammation of infected tissue, containing trophozoites. The trophozoite of Sappinia is characterized by two opposing nucleus with central flattening. Diagnosis can also be done by amplifying rDNA of both *Sappinia diploidea* and *Sappinia pedata* using SSU primers. The real time PCR can also be used based on 18rRNA gene sequences. Sappinia spp. is cultivated on non- nutrient agar with overlay of *Enterobacter* or *Escherichia coli. Vahlkamphia spp.* and *Paravahlkamfia francinae* are other emerging free living amoebas that were first isolated from CSF of young patient who presented with typical symptoms of primary amoebic meningoencephalitis.

Keywords: *Acanthamoeba* spp., *Balamuthia mandrillaris, Negaleria fowleri, Paravahlkamfia francinae, Sappinia* species, *Vahlkamphia* spp.

Free-living amoebae (FLA) are aerobic, mitochondriate, eukaryotic protists and have the capability to exist as free- living organisms in nature and invading living host so they are also called as amphizoic amoebae [1]. The present classification schemes of protists, based largely on their genetic relatedness while the traditional classification scheme which is based on morphological characteristics, is no longer a valid scheme. According to the new classification scheme, the Eukaryotes have been classified into six clusters or 'Super Groups', namely Amoebozoa, Opisthokonta, Rhizaria, Archaeplastida, Chromalveolata and Excavata. The four amoebaehave been classified into two Super Groups, Amoebozoa and Excavata, *Acanthamoeba* and *Balamuthia* are under cluster Amoebozoa: Acanthamoebidae; while *Naegleria fowleri* comes under excavata: Heterolobosia: Vahlkampfiidae; and *Sappinia* under the cluster amoebozoa: Flabellinea: Thecamoebidae [2]. There is an increasing scientific interest in the field of free-living amoebae as determined by published articles over the last five decade. A pubmed search using "Acanthamoeba", "Balamuthia", Naegleria" or "Sappinia" was carried out from 1960 to 2010 which showed constant increase in number of cases from zero to nearly two hundred. Now, the question arises, why they are on the rise? There are many explanations given. Number of immunocompromised patients and contact lens wearers are on the rise [2]. Ability to act as a host or reservoir for other microbial pathogens [3]. These are model organism for motility studies which has led to a significant interest in this organism over the years. Ecologically, infections with free living amoebae have been linked to warm water such as above-ground pipelines, tropical lakes, geothermal water, heated swimming pools or discharges of industrial cooling water and soil [2]. Free living amoeba plays two major ecological roles: Influencing the structure of the microbial community and enhancing nutrient recycling. Both these activities are associated with free living amoebae (FLA). They feed on bacteria thus regulating bacterial populations in the soil. These FLAs cause 60% of the total reduction in bacterial population. Bacteria are primary decomposers of organic material but fail to release minerals from their mass, FLAs act as secondary decomposers thus releasing minerals that are tied up with bacteria. In this way, FLAs make the nutrients available that would otherwise remain inaccessible. The soil containing *Acanthamoeba* and bacteria showed significantly greater mineralization of carbon, nitrogen, and phosphorous compared to the soil containing bacteria without FLAs. Overall, these FLAs regulate the population of bacteria in ecosystem and this was demonstrated by the findings that nitrogen mineralization by *Acanthamoeba* permitted continued growth of bacteria such as *Pseudomonas paucimobilis* resulting in a greater bacterial biomass [3, 4].

ACANTHAMOEBA SPECIES

Introduction

Though many genera of free living amoeba are present in nature, humans are adversely affected by only four of them. *Acanthamoeba* species being the most common and the most widely studied amongst all of the free living amoebae.The term "acanth" (Greek: meaning spikes) denotes presence of spine-like structures, known as acanthopodia, on the surface of these organisms [5, 6]. Since, griffin purposed the name *Acanthamoeba*, more than twenty four species of *Acanthamoeba* have been discovered [6]. There are many species of *Acanthamoeba* such as *A. astronyxis, A.castellani, A.culbertsoni, A.hatchetti, A.keratitis* which are known to cause opportunistic infection in both immunocompetent as well as immunocompromised host. It mostly affects central nervous system, lungs, sinuses and skin. It is transmitted through direct contact or by inhalation of air containing cysts. The cysts are resistant to biocides, chlorine and antibiotics and often remain dormant in adverse environmental conditions. *Acanthamoeba* was discovered as a contaminant of yeast culture, *Cryptococcus pararoseus* and described to cause water borne illness first affecting lungs then spreading haematogenously to brain where it causes granulomatous acanthamoeba encephalitis (GAE) which was described in early 60s and Acanthamoeba keratitis was first reported in 1970s [7, 8].

Epidemiology

It was first isolated on 1913 by Puschkarew from dust and named Amoeba polyphagus *Acanthamoeba castellani* was named after Castellania, who discovered it in 1930 as a yeast culture plate contaminant [9]. Twenty four species of *Acanthamoeba* are presently identified based on morphological and molecular based criteria. *Acanthamoeba* species have been isolated from soil, water, cooling towers, swimming pools, air conditioners, ventilation systems, gastrointestinal washings, dialysis units, dust, contaminant of bacterial, hot tubs, contact lens solutions, IUD and contaminated fungal and mammalian cell culture. It has also been isolated from faecal samples of patients with chronic diarrhea [10]. Worldwide incidence rates of infection are not available as it is not a reportable disease.

In the United States, amoebic keratitis cases have been substantially increased in the 1980's with the introduction of disposable soft lenses. Some evidence showed that with usage of multipurpose cleaning solutions the rates have declined. The estimated rate of Acanthamoeba keratitis is 1 per 250,000 people in the US, although rates vary among studies from 1.65-2.01 per million population of 1 per 10,000 people who wears contact lenses [11, 12]. Three multicentric questionnaire based survey of Acanthamoeba keratitis took place in England. The first survey done between1992-96 found an incidence estimate of 1 case in 39370 or 0.25 per 10000 [13]. Subsequent surveys of 1997-98 and 1998-99 estimates have been corrected to one case per 32260 or 0.31 per 10000 and one case in 37090 or 0.27 per 10000 respectively [14]. Up rise in the incidence of Acanthamoeba keratitis in recent years has been widely reported in literature. In developed countries, contact lens wearers constitute more than 90% of patients. Where as keratitis in emerging nations is more common in non-contact lens users. Studies from southern parts of India reported contact lens wearers to be at risk in only 0-2% of cases of corneal ulcer. A study was done to see the longitudinal trends of incidence among predominately non-contact lens wearers in a high volume referral center in South India. A total of 38,529 unique cases of infectious keratitis were evaluated over a time period (1988-2009) of which 372 were culture positive for *Acanthamoeba* species. Only three cases (0.9%) of Acanthamoeba keratitis occurred among contact lens wearers. This study identified discreet unique Acanthamoeba keratitis epidemics among a rural, non-contact lens wearing population in South India [15]. Granulomatous amoebic encephalitis (GAE) and disseminated Acanthamoeba diseases are not very common, but rates are continuously on rise due to increasing population of immunocompromised patients. More than 200 cases of GAE have been reported till date [5].

Life Cycle

*Acanthamoeba*passes through two types of stages; one is known as vegetative state which is morphologically refers to as a trophozoites stage while other stage is known as dormant cystic stage.

The Trophozoite Stage

In Greek "trophozoites" means to nourish. As per it's name, the trophozoites voraciously feed on bacteria and organic materials. Under optimal temperature conditions (30°C) along with neutral pH, ample food supply and osmolality of 50-80 mosmol. It reproduces by binary fission.

Presence of fine, tapering thorn like acanthopodia arising from surfaces are the characteristic features of Acanthamoeba species. The size of trophozoites varies from 15-50μm in different species. These are uninucleate organisms with large densely stained nucleolus. The cytoplasm is of fine granular texture containing numerous mitochondria, ribosomes, food vacuoles and a contractile vacuole. There are seventeen genotypes of trophozoites are identified from T1 to T17 but majority of the infections in humans are due to T4 genotype. The T4 genotype of *A.castellani* have a cyst wall which is composed of 33% proteins, 4-6% lipids, 35% cellulose, 80% ash and 20% unidentified material. Trophozoites with other microorganisms such as *Legionella, Vibrio, Mycobacteria and Burkholderia* spp. are also been reported in literature [10].

The Cystic Stage

When food is limited or there is impending drying due to environmental stress, the amoeba round up and form cysts. In these conditions, the trophozoites stage undergoes cellular differentiation to emerge as double walled cyst. The outer wall of cyst is made up of proteins and polysaccharide while the inner wall composed of PAS positive cellulose. A space separates both the walls except at certain points where there is formation of opercula in the centre of ostioles, which acts as an exit point for excysting trophozoites. The endocyst exhibits various shapes such as stellate, polygonal, oval or spherical. When a favorable condition returns, the dormant amoeba gets activated and leaves the cyst by dislocating the operculum and reverting to a trophic form [16]. Gas chromatography along with spectrometric examination has revealed the rich carbohydrate composition of cyst walls with a higher percentage of glucose and galactose and small percentage of xylose and mannose, lineage analysis of cellulose indicating 1,4 linked glucosyl confirmation alongside other glycosides linkages [16, 10]. The prodromal period of acanthamoeba infection is unknown, it has been observed that several weeks or months pass before any manifestation of symptoms. There are many possible ways of transmission of *Acanthamoeba* to humans. Cysts and trophozoites of *Acanthamoeba* have been recovered from the nasal mucosa of healthy individuals suggestive of nasopharyngeal route of transmission. Amoebae may also enter through breach in the skin resulting in hematogenous spread to the lungs and brain or through inhalation of amoebic cysts. Granulomatous skin lesions demonstrate presence of both both trophozoites and cystic forms of this amoeba [10].

Immunology and Pathogenesis

Due to ubiquitous presence of these amoeba in environment, low titers of antibodies are seen in humans and animals sera but whether these antibodies are protective in nature is still unknown [5]. High titers of Anti-acanthamoeba antibodies have also been demonstrated in patients due to various species of *Acanthamoeba* causing GAE and skin infections. The studies done by Vishveshwara S & Stehr-Green showed Complement-fixing antibody to *A.culbertsoni* have been found in 2 out of 1000 serum samples collected randomly from patients 6 with old infarct and encephalitis. On autopsy of these cases, amoebae were detected in brain tissues of using immuneperoxidase staining with rabbit anti-Acanthamoeba serum. Chappell *et al.*, 2001, demonstrated that Hispanics are less likely to develop anti-acanthamoeba antibodies as compared with Caucasians [17]. The alternative complement pathway is activated in cases of acanthamoeba infection which is independent of antibody activation mechanism. In healthy hosts, immunoglobulins and complement helps in recognition of these amoebae by inflammatory cells such as neutrophils, macrophages and probably lymphocytes. The alternative complement system and antibody-activation stimulate production of neutrophils that releases lysosomal enzymes and reactive oxygen intermediates, hypochlorite and hydrogen peroxide, which leads to destruction of amoebae. According to Marciano-Cabral , 2003 immunomodulating macrophages directly engulfed the amoebae in tissue. But, in few cases of GAE, granulomatous reaction against acanthamoebae is absent [7]. *Acanthamoeba* may induce macrophage which in turns produces granulomatous type of reaction in patients suffering from GAE [6]. Based on *in vitro* studies, brain microglial cells and macrophages play important role in the killing of acanthamoebae in brain tissue. When these microglial cells are cultured with acanthamoebae, they produce a variety of interleukins such as IL-1a and b, and tumor necrosis factor. This macrophage-mediated killing of amoebae is mainly contact- dependant. Trophozoites of *Acanthamoeba* spp. destroy the host tissue by direct phagocytosis. In few studies, it was shown that several species of acanthamoeba produce food cups during ingestion of rat B103 neuroblastoma cells, and channelizing the ingested material into intra cytoplasmic food vacuoles [7]. Other studies have also confirmed the findings of food cups(amoebastome) production when amoeba is inoculated into hamster cornea that ingests detached epithelial cells thus aids in phagocytosis [8]. It has been observed that enzymes secreted by acanthamoebae, facilitate the spread and invasion of amoebae and causes lysis of host cells thus provides nutrients to amoebae. Few of the studies demonstrated that *A.culbersoni* produces more phospholipases as compared to *A.rhydops*. While, production of serine protease activity was found to be higher in pathogenic rather than nonpathogenic isolates of acanthamoebae. Subsequently, it has been shown that various species and strains of *Acanthamoebae*, produces serine and cysteine proteinases, plasminogen activators, metalloproteinase [7, 8]. *In vitro*, proteinases produced by amoebae cause cytopathic effect on cell culture. In cases of acanthamoeba keratitis, likelihood of attachment of acanthamoeba glycoproteins is twice in damaged cornea as compared with intact corneas which suggests that corneal injury is an important contributing factor of acanthamoebae induced keratitis. *In vitro*, it has been proven that rabbit corneal epithelium mannose residue inhibits the binding of acanthamoebae trophozoites [18]. It is suggested that trophozoites should be present in cornea than cystic forms as cysts are unable to attach to corneal epithelium. Host defense also plays important role in the prevention of acanthamoeba keratitis. Tear enzymes such as Lactoferrins and immunoglobulins IgA are known to have inhibitory effect on binding of acanthamoebae trophozoites to corneal epithelium and decreased level of sIgA are seen in patients with acanthamoeba keratitis [19].

Clinical Features

Granulomatous amoebic encephalitis (GAE) is a rare, chronic disease and is invariably fatal. It is one of the major public health concerns as there is an increase in the number of immunocompromised patients over the last few decades, who act as susceptible hosts.Patients under immunosuppressive therapy and excessive use of steroids or with lymphoproliferative, haemolytic malignancies, diabetes mellitus, pneumonitis, renal failure, liver cirrhosis or other hepatic diseases, agammaglobulinemia and patient undergoing organ/tissue transplantation are at high risk of acquiring this infection. The symptoms of granulomatous encephalitis are similar to symptoms produced by viral, bacterial and fungal meningitis. It is very important to consider GAE if other diagnoses have been ruled out. The features of meningoencephalitis include headache, stiff neck, nausea, vomiting, low grade fever, cerebellar ataxia, mental state abnormalities, lethargy, visual disturbance, hemiparesis, seizures and coma. Facial asymmetry due to facial nerve palsy with numbness is also common. The most heavily affected CNS tissues such as cerebrum which visually appears edematous with extensive necrosis mainly involving the parietal and occipital lobes [10, 16]. Acanthamoeba keratitis (AK) is painful vision threatening illness. If not diagnosed and treated on time, it may lead to corneal ulceration, loss of visual acuity and finally leads to blindness and unnecessary enucleation [9, 20]. Keratitis is usually preceded by trauma to the cornea or usage of contaminated contact lenses due to the usage of contaminated lens solution which is the biggest risk factor of developing acanthamoeba keratitis. Corneal epithelium may show minor erosions while wearing hard or soft contact lenses. AK is characterized by corneal inflammation, photophobia and severe ocular pain with paracentral stromal infiltrates, recurrent breakdown of the corneal epithelium and refractory corneal lesions [11,13]. Acanthamoebae induced keratitis mainly affects one eye but bilateral cases have also been reported in literature [12]. Acanthameoba keratitis occurs in individuals following corneal trauma and contaminated contact lens usage. No case of granulomatous amoebic encephalitis has been reported to have developed following acanthamoeba keratitis although there have been cases of uveitis which were associated with fatal GAE [18]. Acanthamoeba keratitis may confused with viral or fungal keratitis and is often misdiagnosed as dendritic keratitis of HSV which leads to a delay in starting the appropriate therapy. In developed countries,contact lens wearers are at high risk for getting amoebic keratitis while in India, corneal trauma is the most common etiology of acanthamoebae induced corneal infection [21]. Once Introduced into the eye, amoeba gets adhered to the corneal surfaces with the help of mannose receptors and by presence of calcium ions. The amoeba then infiltrates the stroma and causes tissue necrosis and ulceration [19].

Laboratory Diagnosis

Imaging Techniques

CT scan and MRI have been extensively used to diagnose brain lesions granulomatous amoebic encephalitis, but findings are usually nonspecific. Multifocal low density lesions in cortical and subcortical regions are also noted. Progressive hydrocephalus with meningeal thickening and pseudo tumour like lesions are identifiable features. Enhanced CT scan can identify large isolated lesions [16].

Microscopic Methods

Definitive diagnosis of GAE is done using formalin fixed, paraffin embedded and H&E stained biopsy and autopsy samples of encephalitic lesions and subsequent isolation of amaoebae in culture. Biopsy samples stained with H&E helps to demarcate brain granulomas from inflammatory infiltrates and amoebae. These amoebic trophozoites are detected occasionally in CSF wet mount or using giemsa stained smears of CSF sediments.

Electron microscopy allows visual demonstration of trophozoites and cysts in brain tissue, skin lesions and cerebrospinal fluid. These trophozoites can be differentiated from inflammatory host cells mainly due to their characteristic nuclear structure as acanthamoeba has large central nucleolus forming halo inside the rounded nucleus. However, it is impossible to tell whether trophozoites belong to *Acanthamoeba spp.* or *Balamuthia mandrillaris* just by light microscopy, as both organisms have similar type of nuclear arrangements. The cyst and trophozoites of acanthamoeba can be differentiated from host cells and another amoeba by transmission electron microscopy. However, it is expensive and time consuming technique and needs expertise. Histological stains such as periodic acid Schiff and Gomori's methenamine silver stain specifically identify the presence of cysts in tissue specimen. Other stains such as acridine orange and calcofluor white are used for fluorescent observation of the cysts in various tissues. H&E stained biopsy samples demonstrate presence of granulomas, necrosis, inflammatory infiltrate and vasculitic lesions in the affected blood vessels. Fluorescence microscopic and immunohistochemical studies of the brain and skin tissue sections are very specific methods to rule out GAE [16, 10]. Anti-acanthamoebae antibodies generated in rabbits can identify amoeba in tissue. Patient specimens are incubated, followed by a secondary anti-rabbit IgG associated with fluorescent markers to detect both trophozoites and cysts. More recently, species specific monoclonal antibodies are developed as an important diagnostic and epidemiological tool [22]. Only few false negative antibodies found to be reactive with formalin fixed paraffin embedded infected tissue [23].

Culture Methods

Brain or skin biopsy and autopsy samples are placed on 1.5% non-nutrient agar plates covered with layer of *E.coli* or with *Enterobacter aerogenes* to isolate and prepare cultures of acanthamoebae. Depending on the density of amoebae, acanthamoebae can be observed after 24 hours of inoculation. Amoeba feeds on bacteria's and confluent growth are seen with 4-5 days of culture. Synthetic media such as proteose peptone yeast glucose extract has also been used to grow these amoebae. Samples placed on tissue culture cells with antibiotics like penicillin, streptomycin and gentamycin in which case the amoebae destroys the cell monolayer in 1-2 days depending on the number of amoebae [10].

Serology

An increasing antibody titer can be an indication of acute infection. Indirect immunofluorescence is performed using serial dilution of serum from an individual suspected of having acanthamoebiasis followed by incubation of the sera with slides containing amoebic extract. Fluorescent labeled Anti IgG antibodies such as FITC can detect and titrated using fluorescent microscopy. Infected individuals possess high levels of antibody titers ranging between 1:256

to 1:1024 in the serum while healthy individuals who had environmental exposure to acanthamoeba have low antibody titer usually not more than 1:80, due to these reasons, it can be used as a confirmatory test [16]. Western immunoblot analysis specifically demonstrates antibodies to acanthamoeba in human serum. An ELISA method using whole fixed deposits rather than distributed extracts as an antigen source was developed and shown to be effective tools for identifying antibodies to Acanthamoeba in the clinical laboratory setting [17].

Molecular Techniques

Acanthamoeba detection can be rapidly achieved by using molecular methods. The detection of acanthamoeba at genus level is sufficient to recognize whether an individual is infected or not. PCR assay used in molecular identification of acanthamoeba has an advantage of detecting all known acanthamoebae subgroups. This PCR assay has successfully used for environment as well as in clinical specimen. DNA amplification from cyst is a cumbersome procedure which requires prior treatment of cyst with proteinases K for DNA extraction. Diagnosis of acanthamoeba Keratitis is made clinically by appearance of the corneal ulceration and by demonstration of amoebae in the epithelium of cornea.

Acanthamoeba keratitis diagnosed in a similar way of diagnosing GAE through microscopic and culture analysis of corneal biopsy samples. Corneal scraping is a non-invasive and efficient method for culture and isolation of amoebae. Wet mount preparation treated with 10% KOH is used for light microscopic examination. Other methods such as impression cytology and H&E and Giemsa stains have been successfully been used in diagnosis of acathamoeba keratitis [21]. Immunofluorescence assays and microscopy immunoperoxidase techniques are helpful in detecting presence of acanthamoebae in host tissue. Ultra structure of cysts and trophozoites are observed using transmission electron microscope. Use of *in vivo* confocal microscopy has been used to identify characteristics of eye lesions caused by organism.

Culture Methods

Culture of corneal biopsy or scraping and washes from contact lenses or lens cases are the most common, cost effective and efficient method for detecting Acanthamoeba spp. PCR helps in rapid and confirmatory diagnosis of amoebae in biopsy samples, contact lens solution, contact lenses and also from environment. As compared with microscopy PCR is highly specific method of diagnosing acanthamoeba keratitis [22].

Treatment

Granulomatous amoebic encephalitis (GAE) is difficult to treat due to delayed presentation and nonspecific generalized meningoencephalitis mimicking many other diseases, lack of early diagnosis and in most of the cases diagnosis is established postmortem. For diagnosed cases, few successful attempts have been made to treat acanthamoebae encephalitis with combination of pentamidine isethionate, sulfadiazine, flucytosine and fluconazole or itraconazole. For the cutaneous granulomas, topical application of chlorhexidine gluconate and ketoconazole cream in addition to the above noted antimicrobial therapy have resulted in therapeutic success but many a times side effects of these medications lead to discontinuation, noncompliance and partial treatment [24, 25]. Poor prognosis for GAE patients can be attributed to delayed diagnosis, suboptimal efficacy of antimicrobials therapy and problems inherent to the immunocompromised host.

The combination therapy is advisable in proven cases of acanthamoeba keratitis. Combining Amphotericin B plus Trimethoprim-Sulphamethoxazole plus rifampicin has successfully used in few cases. A number of drugs have invitro action against Acanthamoeba spp. such as fluconazole, azithromycin, sulfadiazine, pentamidine, miltefosine that have been used successfully in treating patients using various combinations. Miltefosine which is a hexadecylphosphorylcholine has been shown to be effective anti amoebicidal potential. Significantly, medical cure has been achieved with either polyhexamethylene biguanide (PHMB) or chlorhexidine gluconate with or without brolene. When medical treatment failed, a combination of debridement and penetrating keratoplasty has successfully used. Currently the drugs of choice are chlorhexidine gluconate, PHMB and brolene as these have found to greatly improve the prognosis for AK sufferers [9].

Prevention

Granulomatous amoebic encephalitis occurs in immunocompromised host, so there are no clear methods for prevention of infection. However, in cases of acanthamoebae induced keratitis, contact lenses and lens care solution are well known risk factors. An important means of prevention is educating lens wearers how to take proper care of contact lenses. Also they are warned not to use contact lenses during swimming or performing water sports and relaxing in water tubs *etc.* [6, 21].

NAEGLERIA FOWLERI

Introduction

Naegleria fowleri is an environmental ameboflagellate parasite found in variety of water bodies such as ponds, swimming pools; aquarium *etc.* prefers temperature of 30-45 degree Celsius for growth. History of water recreational activities poses risk for infection. It causes primary amoebic meningoencephalitis, a rare but fatal condition. Case reports of PAM have been published worldwide but not many cases have been diagnosed and verified. Of the thirty species of Naegleria, only *Naegalria fowleri* has been demonstrated to be pathogenic to humans. Another species, *N.australiensis* has been proven to pathogenic in mice and is useful to study pathogenesis of Naegleria [26].

Epidemiology

The number of reported cases of *Naegalria fowleri* infection is small due to high fatality rates seen in primary amoebic meningoencephalitis (98% death rate). In 1965, Fowler and Cutler was first to report human diseases caused by amoeboflagellates in Australia. Their work has shown, how free living amoebae can cause CNS infections and as well survive without living host. Since then, more than 300 cases have been published from more than 16 countries and there might be more unpublished data of this disease. In 1966, Dr. Butt termed the infection as primary amoebic meningoencepahalitis in order to differentiate CNS infections by other secondary invasions such as *Entamoeba histolytica* [27]. An interesting case of primary amoebic meningoencephalitis has been described by St. Summers in Ireland in 1909.

It is quite well established that there are several types of *N. fowleri*. These are speciated based on length of internal transcribed spacer I and a one base pair transition in the 5. 85S rDNA.

In a review on origin and evolution of pathogenic amoeboflagellate *Naegalria fowleri, Dr. Johan F. Jonckheere* have described seven among the eight known types of Naegleria spp. have been found in Europe while three varieties were seen in the USA with one regional unique type and only one each of the eight types found in Oceania (Australia and New Zealand) and Japan. He described that the two most common types prevalent in Asia, were, similar to one described in Europe and the USA.

With respect to virulence, there was no difference found in these two types [28, 29]. Till 2011, approximately nine cases of PAM due to *Naegalria fowleri* had been reported from India. The first case report of PAM was from Karnataka, India, in a five month old infant, it was hypothesized that he contacted the disease from underground well water in the absence of history of swimming [30]. It is important to consider PAM should be considered as differential diagnosis in cases of unexplained meningoencephalitis.

Morphology and Life cycle

There are three stages to Naegleria life cycle: the infective trophozoites, transient flagellate stage and the cystic stage. The portal of entity of trophozoites is *via* olfactory neuro epithelial cell lining covering the cribriform plate to reach olfactory bulb. The amoeba affects multiple parts of the brain. In white matter, it causes demyelination while myelinoclasis is observed in gray matter due to vascular blockage. The pathological changes in CNS are attributed due to release of phospholipolytic enzymes which causes breaks in the lipid membrane of neuronal cells.

The Trophozoites Stage

The motility of the trophozoites is aided by hemispherical lobopodia bulging at the anterior end. It exhibits active sinusoidal locomotion. It measures 10-25 µm, can be maintained under laboratory conditions on a diet of gram negative bacteria, Reproduction is by binary fission. There is a single vesicular nucleus with a prominent central nucleolus which stains densely with chromatic dyes. The cytoplasm is distinguished into an ectoplasm and endoplasm; the latter containing numerous mitochondria, ribosomes, food vacuoles, contractile vacuoles and trailing protoplasmic filaments.

The Flagellar Stage

Trophozoites transform into a pear shaped flagellated stage containing two flagella. Morphology of amoebae changes with change in ionic concentration. This change can be artificially induced in the laboratory by washing trophic amoeba in distilled water or diluted saline. This stage is also known as feeding stage in its life cycle. The size of flagellated *Naegleria spp*. ranges in length from 10-16 µm. With removal of external ionic changes, it transforms back into the amoeboid form in less than an hour.

The Cystic Stage

In adverse conditions, the trophozoites transform into resistant cystic form. The cysts are spherical in shape and have double wall with thick endocyst which closely opposed by ectocyst. There are pores in the wall which may not be well appreciated on light microscopy. Both the flagellated and cystic forms also possess a single nucleus with a prominent nucleolus. Over 30 species of Naegleria have been described as per ssRNA gene sequence [31]. However, only *Naegalria fowleri* is known to infect mice after intracerebral inoculation.

Being thermophilic, they can tolerate temperatures up to 45° C. Therefore, survives summers when environmental temperature is high. Infection occurs in the individuals engaged in aquatic activities and accidently comes in contact with amoeboflagellates. However, temperature tolerance by itself does not account for pathogenicity, as there are thermophilic Naegleria such as *Naegleria lovaniensis* that are nonpathogenic.

Pathophysiology and Immunology

Cerebral hemispheres are found to be grossly edematous and congested. The leptomeninges (arachnoid and pia mater) are also congested and show hyperemia with scanty purulent exudate within sulci, at the base of the brain, brainstem and cerebellum. Hemorrhagic necrosis with purulent exudates of olfactory bulbs are the characteristics lesions seen in PAM [32]. The cortex may show numerous superficial hemorrhagic areas. On CT scan, lesions can be seen in and around the base of the orbital, frontal and temporal lobes, base of the brain, hypothalamus, midbrain, pons, medulla oblongata and the upper portion of the spinal cord. Other radiological findings are obliteration of the cisterns around the midbrain and the subarachnoid space over the cerebral hemispheres. Marked diffuse enhancement in these regions may be seen under contrast CT [16].

Microscopically, cerebral hemispheres, brain stem, cerebellum and upper portion of the spinal cord contain fibrino-purulent leptomeningeal exudates constituting predominantly neutrophils. Few eosinophils, macrophages and lymphocytes may be found. Often, large pockets of amoebic trophozoites, without PMNs may be seen within edematous and necrotic neural tissue. Trophic amoebae are also present deep in Virchow–Robin spaces, usually surrounding blood vessels without inflammatory response. Amoebic cysts are absent in these areas. Specific antibody response to *N.fowleri* is well documented with titres as high as 1:4096 being demonstrated in serum samples by IFA of recovered patient [24]. Immunoblot based assays performed by Centers for Disease Control and

Prevention (CDC) demonstrated IgM to be the principal antibody class generated by patients who had contracted PAM. Antibody response proved to be particularly well developed against *N. fowleri* antigens 190, 66, 30 and 14 kDa. Delayed type of response to *Naegalria fowleri* was studied using animal models. Guinea pigs sensitized with intradermal injections of *N. fowleri* and other Naegleria species developed a cell mediated immune reaction when challenged with homologous or heterologous Naegleria antigens [34]. In contrast, experimental studies done using mice suggested that CMI does not play a role in protection against Naegleria. Congenitally athymic mice (T cell deficient) and euthymic mice were equally susceptible to infection by *N. fowleri* [35]. In addition, failure to impair host resistance to *N. fowleri* infection with diethylstilbestrol, which depresses delayed typed hypersensitivity. In conclusion, CMI alone does not constitute a major line of defense against *N. fowleri*. It has been demonstrated that Naegleria are resistant to lysis by cytolytic molecules from the host such as tumor necrosis factor (TNF)-α, IL-1 and the membrane attack complex (MAC) C5b-C9 of complement [36].

Clinical Features

Acute fulminating hemorrhagic meningoencephalitis mainly occurs among healthy children and young adults with a typical history of recent exposure to warm fresh water. The typical cause of PAM occurs in the hot summers when people engage in aquatic activities. The characteristic features of PAM are the rapidly progressive symptoms following exposure. In absence of prompt diagnosis and intervention, the disease progresses fast and proves to be fatal within a week or less. The first case of Primary amoebic meningoencephalitis caused by *Naegalria fowleri* was reported in 1965 by Fowler and Carter from Australia. Butt coined the term PAM [10]. Primary amoebic meningoencephalitis should be suspected in young adults and children with acute neurological symptoms following recent exposure to fresh warm water. The time of aquatic activities to the onset of clinical illness varies between 5-7 days and may sometimes be as short as 24 hours. No distinctive clinical features differentiate PAM from acute pyogenic or bacterial meningoencephalitis. It is imperative that the attending physicians should obtain a detailed history regarding fresh water exposure prior to onset of symptoms. Initial symptoms include sudden onset of bifrontal or bitemporal headaches, high fever, nuchal rigidity followed by nausea, vomiting, irritability and restlessness. Nuchal rigidity is usually associated with postive Kernig and Brudzinski signs. Photophobia occurs late in the clinical course. Neurological abnormalities includes lethargy, seizures, confusion, coma, diplopia and behavior abnormalities may lead to death within a week of onset of symptoms. Third, fourth and sixth nerve palsy may indicate brain edema and herniation. Intracranial tension is raised up to or more than 600mm of mercury. Cardiac rhythm abnormalities and myocardial necrosis are also seen sometimes.

Laboratory Diagnosis

Diagnosis of free living amoeba is often missed due to low level of suspicion. After ruling out causes of bacterial and viral meningitis, it is important to consider amoebic pathology of meningoencephalitis albeit history of exposure is rarely found. Radiological techniques such as computed tomography scanner magnetic resonance imaging may show lesions but these are usually nonspecific. Sometimes, CT scan may show obliteration of cistern surrounding the midbrain and sub arachnoid space.

Microscopic Methods

In cases of meningoencephalitis, after ruling out common causative organisms such as bacteria, fungus and viruses, diagnosis of meningoencephalitis due to Naegalria spp. should be considered. Pre mortem diagnosis of PAM is rare. CSF findings suggest low pressure, cloudy, purulent and slightly hemorrhagic with increase in cellularity, predominantly neutrophilic. CSF is characterized by low glucose and elevated proteins levels. Direct wet mount examination of CSF is the method of choice in the diagnosis of PAM as CSF contain more motile amoebae which can easily be identified under the light microscope [27, 29]. In CSF, amoeba can also be diagnosed in stained fixed preparations Wright's, Giemsa or hematoxylin and eosin (H & E) stain. Although Gram stain can also be used for the diagnosis of Naegleria but the characteristic nuclear morphology of amoeba may not be appreciated. Amoeba can sometimes be mistaken for macrophages but the nucleus of *Naegalria fowleri* contains a large central nucleolus which distinguishes it from the inflammatory cells.

Naegalria fowleri can be easily distinguishable from other free living amoebae due to its transformation into motile flagellatated forms when amoeba are placed in distilled water or saline [16, 10]. Serology Monoclonal antibodies produced in rabbits can be used to identify amoeba in tissue sections and CSF. Amoebas in the CSF are identified by specific indirect immunofluorescence antibody assays using a polyclonal or monoclonal antibody to the amoeba in conjunction with fluorescinated secondary antibodies. For histological examinations, biopsy material should be fixed in 10% KOH. Amoeba are observed in biopsied brain tissue with H & E staining or by immunoperoxidase staining using Anti-Naegleria antibodies to the amoebic antigens /organism [37].

A commercially available enzyme linked immunosorbent assay based on the use of a monoclonal antibody (5D12) that recognizes a glycosylated epitope on *Naegalria fowleri* can be used for diagnosis.

Dr. Olivier A E Sparagano (1993) and collegues have shown that monoclonal antibody 3A4 reactivity by ELISA in all *N. fowleri* tested strains. These MAbs showed reactivity with all the three morphological forms of *N. fowleri* (trophozoites, cysts, and flagellates) [37]. These monoclonal antibodies can also be used to differentiate *Naegalria fowleri* from other species of Naegleria and from other FLA in the tissue and environmental samples. However, no Ag-Ab reactivity seen with the other five species; *N. lovaniensis, N. gruberi, N. australiensis, N. jadini,* and *N. andersoni* [31].

Culture Methods

CSF or biopsied brain tissue should be kept and transported at room temperature to the laboratory. Centrifugation of fresh CSF sample upto 500 RPM may reveal motile trophozoites and can be cultured on suitable media. Morphologically, the size of trophozoites ranged from 12-25microns with nucleus and centrally placed nucleolus and absence of peripheral chromatin. Liquid culture media are used with serum such as Nelson's medium which contains ox liver digest and glucose. Mammalian cell lines can be employed to demonstrate cytopathic effect [10, 16].

Molecular Methods

Multiplex PCR detects free living amoeba within 6 hours but is not useful in routine diagnostic laboratory due to rarity of finding these organisms and high test cost.Isoenzyme analysis has been developed for the identification of *N. fowleri*. Sequencing of the 5.8S rRNA gene and the internal transcribed spacers 1 and 2 (ITS1 and ITS2) of *N. fowleri* has distinguished specific genotypes. A new sensitive, rapid and discriminating technique that uses a single primer set and the DNA intercalating dye SYT09 for real-time PCR and melting-curve analysis of the 5.8S rRNA gene flanking the ITS has been developed that distinguishes several Naegleria species in environmental samples [38]. MALDI-TOF- MS are some of the advanced technique used specifically to identify Naegleria trophozoites [39].

Treatment

There is no optimal therapeutic regime for *Naegalria fowleri*. Reviewers suggested combination therapy with Amphotericin (i.v. at 1.5 and i.t. at 1.5 mg/kg of body weight/day), Rifampicin (10 mg/kg of body weight/day) and miconazole (i.v. at 350 and i.t. at10 mg/ m2 of body surface area/ day) given for 3-4 weeks [25]. Survival rates of patients with PAM are very less even after appropriate treatment regimen.

Based on *in vitro* testing and *in vivo* mouse studies, amphotericin B was reported to be more effective than amphotericin B methyl ester, a water soluble form of the drug. *In vitro* studies of phenothiazine compounds (chlorpromazine and trifluperazine) which can accumulate in the CNS were found to have inhibitory effects on *Naegalria fowleri* [40]. Azithromycin, a macrolide antimicrobial, has been shown to be effective against Naegleria both *in vitro* and *in vivo* conditions. *Naegalria fowleri* was sensitive to the triazole compound voriconazole; low concentrations (≤10 micrograms per ml) were amoebostatic, while concentrations ≥10 micrograms per ml were amoebicidal [23].

Prevention

Naegalria fowleri is a heat loving amoeba thus proliferates in water when the ambient environmental temperature increases above 30°C. With further possibility of increased temperature due to global warming, it is quite possible that cases of PAM may be seen even in countries where it previously has not been recorded [33]. It is susceptible to chlorine in water.

Sunlight and presence of organic matter in the swimming pools can reduce the efficacy of chlorine.In high risk areas, monitoring of recreational waters is considered by local Public Health authorities especially during summer.

Domestic water supply carried over land warmed by sun was also a suspected source of this infection. Drinking water containing amoeba has never been known to cause PAM. Control of total coliform counts in water and public education are effective means in preventing primary amoebic meningoencephalitis cases.

BALAMUTHIA MANDRILLARIS

Introduction

It was first isolated from baboon's brain that died of meningoencephalitis in 1986. Over 200 cases were reported from South America and United States. The true prevalence of disease is unknown in south East Asia. It is most commonly isolated from soil and activities related with soil contact such as gardening, biking, agriculture *etc.* that poses high risk of acquiring it. It causes infections in both immunocompromised and immunocompetent individuals; the CNS lesions mimic the one being produced by Acanthamoeba spp. The disease has slow, chronic but progressive course over 2 years. The portal of entry of the organism is *via* cutaneous lesions, nasal mucosa and then inhaled cysts subsequently spread to brain through haematogenous route. This amoeba was initially described as leptomyxid amoeba, due to its resemblance with soil amoeba those included under the family leptomyxidae, but in 1993 it was recognized as being different from the leptomyxids and was designated as *Balamuthia mandrillaris* [16]. Epidemiology Till date, *B.mandrillaris* is the only known species belonging to genus Balamuthia and it causes granulomatous amoebic encephalitis, both in humans and other animals. It has been reported both in immunocompetent and immunocompromised individual's especially young children and older individuals [18]. The disease caused by Balamuthia is not very common, so it is quite possible that many cases were being misdiagnosed and probably not reported. In the US, the infection is more common among Hispanic Americans but the reason is not known. More research is needed to know the factors associated with increased trend of disease in hispanic ethnicity. The increased risk of infection can occur in immunocompromised, alcoholics and pregnancy. It is not transmitted from person to person except organ transplantation [16, 10]. The first case of Balamuthia GAE was reported from South East Asia, in Thailand in a young male who had given 6 months past history of road traffic accident prior to development of meningoencephalitis. After the accident, the patient developed a star shaped ulcer on his nose which did not heal; later on patient was found to be infected with *B.mandrillaris*, which he acquires during the accident when he was thrown in swamp. This led to speculations that *B.mandrillaris* might be present in muddy, moist and humid environment of tropical countries such as Thailand and India [41]. However, till date no case has been reported from India. Life cycle The trophozoites is pleomorphic and measures 12- 60 µm with characteristic irregular branching structure. Trophozoites contain single nucleus but occasionally binucleated forms are seen. The nucleus contains large centrally placed dense nucleolus. However, in infected tissues, amoeba with two or three nucleolus bodies have been found. Other organelles include mitochondria and endoplasmic retinaculum containing ribosomes.

Asexual reproduction by binary fission occurs in trophozoites to form daughter cells. Under extreme conditions, such as overcrowding of cells, lack of nutrients, pH or temperature changes, encystment occurs [16]. The cyst stage; Encystment ensures the survival of the organism under adverse circumstances. Cysts are smaller in size ranging from 12-30 µm, with mean size of 15 µm. Transmission electron microscope studies have shown that the cyst contains three walls, with a thin irregular outer wall, afibrillar middle, also known as mesocyst and inner most wall called endocyst. While under light microscope, it appears to contain only two walls in which outer wall appears to be wavy whereas inner wall is smooth and round. Under favorable conditions *i.e.*, temperature between 30-37° C, neutral pH and availability of nutrients, cysts differentiates to form trophozoites *via* a process known as encystment.

Clinical Presentation

The patterns of clinical presentation have been described in some patients, who initially developed skin lesions followed by neurological manifestations over weeks or months while other presented with early neurological involvement [19]. Early recognition of these conditions may help in diagnosis and treatment prior to development of CNS symptoms. In some of the cases, skin lesions may disappear in the absence of treatment. In these circumstances, Subsequent CNS involvement still develops. Patient with cutaneous involvement commonly presents with a non-ulcerated asymptomatic plaque, either single or with occasional satellite lesions. History of previous trauma is present in most of the reported cases. The most common location is the central face over the nose. Other lesions are located at chest and elbow. Occasionally the border is raised giving the appearance of an annular lesion. The lesions may vary from 1-7 cm and the colour varies from skin tone to slightly red with violaceous tone. Sometimes borders are ill- defined suggesting soft tissue infiltration. In untreated cases, tissue infiltration may progress to involve the central face. Sensation is often preserved, a sign that permits differentiation from tuberculoid leprosy. Central nervous system symptoms are due to thrombotic angiitis, leading to hemorrhage, infarction and necrosis. Patients present with neurological involvement with symptoms such as fever and malaise. Initial symptoms may include unilateral headaches, focal seizures, cranial nerve dysfunction or localized motor deficits. Signs of increased intracranial pressure (diffuse headaches, nausea and vomiting) maybe present and later meningeal signs may develop followed by meningitis and progressive loss of consciousness (from stupor to coma). This progression generally occurs over the period of two to twelve weeks [16, 10].

Treatment

Most cases of Balamuthia GAE have vague clinical profile and treated with steroids, antibacterial, antifungal and antiviral agents without any effect on course of infection. Two patients survived with combination therapy combining pentamidine isethionate, sulphadiazine, clarithromycin, fluconazole and flucytosine. In cases of Peruvian Balamuthiasis, similar types of lesions as described and few cutaneous lesions, one of the cases was recovered without treatment and other two patients improved after prolonged therapy with

Albendazole and itraconazole. *In vitro* studies have shown that pentamidine and propamidine isoethionate at a concentration of 1 µg/ ml inhibits the growth of amoeba by 82 and 80%, respectivelyboth these drugs were amoebostatic, but not amoebocidal.Among other drugs tested with little or no activity were macrolide antibiotics, gramicidin, polymyxin B, trimethoprim-sulfamethoxazole, and a combination of trimethoprim sulfamethoxazole as well as amphotericin B. Recent information based on *in vitro* data has shown that miltefosine at a concentration of more than 40 microgram lysed the amoeba. Voriconazole however, had virtually no effect on Balamuthia [10, 16]. Given the problems with the diagnosis of infection and lack of effective antimicrobial agents, the prognosis for patients with balamuthiasis is very poor.

Prevention

It usually infects immunocompromised individuals systems and presently there are no clear methods for the prevention of this infection. Acute disease symptoms require immediate attention.

Many of the recent cases that have occurred in apparently immunocompetent children suggesting subtle immunological deficiencies may also be involved in causing the disease.

SAPPINIA SPECIES

Introduction

There are two species of Sappinia known to cause CNS infections in humans, *Sappinia diplodea* and *Sappinia pedata*. Sappinia spp. is found worldwide. *S.diploidea* was first isolated from lizard faeces. Trophozoites have been found in elk and buffalo faeces and from soil contaminated with bovine faeces. Since, the lifecycle involves animal faeces, the persons handling of livestock are at risk.

Life Cycle

The Trophozoite Stage

Trophozoites measure ranging from 40-80 µm, are ovoid or oblong and occasionally surface wrinkles flatten it. There are contractile and food vacuoles in the cytoplasm.

The Mature Cyst

The mature cyst is round in shape and size ranges between 15-30 µm. It can be cultivated on non- nutrient agar plate coated with either *Escherichia coli* or *Enterobacter spp.* Sappinia has been isolated from soil, fresh water, forest litter, mammalian faeces and rectum of lizards. It has been described from Europe, North America, Egypt, the Middle east, West indies and Japan.

In 2001, it was first implicated in human disease (Gelman *et al.* 2001). Although cysts were not seen in the brain tissue, amoebic trophozoites were being isolated from the biopsy sample of human tissues. Till date, only one case of granulomatous amoebic encephalitis due to *Sappinia diploidea* has been reported. The confirmed case was a patient who was previously healthy male farmer with no evidence of immunosuppression and engaged in handling livestock. He presented with headache vomiting loss of consciousness and blurry vision following a sinus infection. The antimicrobial therapy consists of azithromycin, pentamidine, itraconazole, flucytosine, was used and the patient survived.

Laboratory Diagnosis

The diagnosis is confirmed on histopathological sample,which generally shows necrotizing haemorrhagic inflammation of infected tissue, containing trophozoites. The trophozoite of sappinia is characterized by two opposing nucleus with central flattening [21]. Diagnosis can also be done by amplifying rDNA of both Sappinia diploidea and Sappinia pedata using SSU primers. The real time PCR can also be used based on 18rRNA gene sequences.

OTHER EMERGING FREE LIVING AMOEBAE

Vahlkamphia spp.

It is emerging free living amoeba found in water and soil, belongs to same family as Naegleria spp. Till date only two reported cases of keratitis, mimicking signs and symptoms of Acanthamoeba keratitis have been reported. This amoeba was isolated from contact lens and corneal lesions of patient later identified as Vahlkamphia spp.

Paravahlkamfia Francinae

It was first isolated from CSF of young patient who presented with typical symptoms of primary amoebic meningoencephalitis, unlike Naegleria spp., Paravahlkamfia spp does not produce flagellated forms and does not grow in bacterial coated agar plates. The Trophozoites contains single vesicular nucleus and grows on cell culture, where it gets differentiated into cysts forms [3, 4].

CONFLICT OF INTEREST

The author confirms that author has no conflict of interest to declare for this publication.

ACKNOWLEDGEMENTS

Declared none.

REFERENCES

[1] FC Page A New Key to Freshwater and Soil Gymnamoebae. Cumbria, UK: Freshwater Biological Association 1988; p. 122.

[2] Hampl V, Hug L, Leigh JW, *et al.* Phylogenomic analyses support the monophyly of Excavata and resolve relationships among eukaryotic supergroups. Proc Natl Acad Sci USA 2009; 106(10): 3859-64.
[http://dx.doi.org/10.1073/pnas.0807880106] [PMID: 19237557]

[3] Rønn R, McCaig AE, Griffiths BS, Prosser JI. Impact of protozoan grazing on bacterial community structure in soil microcosms. Appl Environ Microbiol 2002; 68(12): 6094-105.
[http://dx.doi.org/10.1128/AEM.68.12.6094-6105.2002] [PMID: 12450833]

[4] Rosenberg K, Bertaux J, Krome K, Hartmann A, Scheu S, Bonkowski M. Soil amoebae rapidly change bacterial community composition in the rhizosphere of Arabidopsis thaliana. ISME J 2009; 3(6): 675-84.
[http://dx.doi.org/10.1038/ismej.2009.11] [PMID: 19242534]

[5] Stehr-Green JK, Bailey TM, Visvesvara GS. The epidemiology of Acanthamoeba keratitis in the United States. Am J Ophthalmol 1989; 107(4): 331-6.
[http://dx.doi.org/10.1016/0002-9394(89)90654-5] [PMID: 2929702]

[6] Stehr-Green JK, Bailey TM, Brandt FH, Carr JH, Bond WW, Visvesvara GS. Acanthamoeba keratitis in soft contact lens wearers. A case-control study. JAMA 1987; 258(1): 57-60.
[http://dx.doi.org/10.1001/jama.1987.03400010061028] [PMID: 3586292]

[7] Radford CF, Lehmann OJ, Dart JK. Acanthamoeba keratitis: multicentre survey in England 19926. Br J Ophthalmol 1998; 82(12): 1387-92.
[http://dx.doi.org/10.1136/bjo.82.12.1387] [PMID: 9930269]

[8] Radford CF, Minassian DC, Dart JK. Acanthamoeba keratitis in England and Wales: incidence, outcome, and risk factors. Br J Ophthalmol 2002; 86(5): 536-42.
[http://dx.doi.org/10.1136/bjo.86.5.536] [PMID: 11973250]

[9] Lalitha P, Lin CC, Srinivasan M, *et al.* Acanthamoeba keratitis in South India: a longitudinal analysis of epidemics. Ophthalmic epidemiol 2012; 19(2): 111-5.

[10] Schuster FL, Visvesvara GS. Free-living amoebae as opportunistic and non-opportunistic pathogens of humans and animals. Int J Parasitol 2004; 34(9): 1001-27.
[http://dx.doi.org/10.1016/j.ijpara.2004.06.004] [PMID: 15313128]

[11] Martinez AJ. Free-Living Amebas: Natural History, Prevention, Diagnosis, Pathology, and Treatment of Disease. Boca Raton, FL: CRC Press Inc. 1985.

[12] Pettit DA, Williamson J, Cabral GA, Marciano-Cabral F. *In vitro* destruction of nerve cell cultures by Acanthamoeba spp.: a transmission and scanning electron microscopy study. J Parasitol 1996; 82(5): 769-77.
[http://dx.doi.org/10.2307/3283890] [PMID: 8885887]

[13] Omana-Molina M, Gonzalez-Robles A, Iliana Salazar-Villatoro L, *et al.* Reevaluating the role of Acanthamoeba proteases in tissue invasion: observation of cytopathogenic mechanisms on MDCK cell monolayers and hamster corneal cells. BioMed Res Int 2013.

[14] Niederkorn JY, Alizadeh H, Leher H, McCulley JP. The pathogenesis of Acanthamoeba keratitis. Microbes Infect 1999; 1(6): 437-43.
[http://dx.doi.org/10.1016/S1286-4579(99)80047-1] [PMID: 10602676]

[15] Seal DV. Acanthamoeba keratitis update-incidence, molecular epidemiology and new drugs for treatment. Eye (Lond) 2003; 17(8): 893-905.
[http://dx.doi.org/10.1038/sj.eye.6700563] [PMID: 14631394]

[16] Jones DB, Visvesvara GS, Robinson NM. Acanthamoeba polyphaga keratitis and Acenthamoeba uveitis associated with fatal meningoencephalitis. Trans Ophthalmol Soc U K 1975; 95(2): 221-32.
[PMID: 1064220]

[17] Sharma S, Garg P, Rao GN. Patient characteristics, diagnosis, and treatment of non-contact lens related Acanthamoeba keratitis. Br J Ophthalmol 2000; 84(10): 1103-8.
[http://dx.doi.org/10.1136/bjo.84.10.1103] [PMID: 11004092]

[18] Clarke DW, Niederkorn JY. The pathophysiology of Acanthamoeba keratitis. Trends Parasitol 2006; 22(4): 175-80.
[http://dx.doi.org/10.1016/j.pt.2006.02.004] [PMID: 16500148]

[19] Chappell CL, Wright JA, Coletta M, Newsome AL. Standardized method of measuring acanthamoeba antibodies in sera from healthy human subjects. Clin Diagn Lab Immunol 2001; 8(4): 724-30.
[PMID: 11427418]

[20] Yera H, Zamfir O, Bourcier T, *et al.* Comparison of PCR, microscopic examination and culture for the early diagnosis and characterization of Acanthamoeba isolates from ocular infections. Eur J Clin Microbiol Infect Dis 2007; 26(3): 221-4.
[http://dx.doi.org/10.1007/s10096-007-0268-6] [PMID: 17393203]

[21] Marciano-Cabral FM, Patterson M, John DT, Bradley SG. Cytopathogenicity of Naegleria fowleri and Naegleria gruberi for established mammalian cell cultures. J Parasitol 1982; 68(6): 1110-6.
[http://dx.doi.org/10.2307/3281100] [PMID: 6816913]

[22] Butt CG. Primary amebic meningoencephalitis. N Engl J Med 1966; 274(26): 1473-6.
[http://dx.doi.org/10.1056/NEJM196606302742605] [PMID: 5939846]

[23] De Jonckheere JF. Origin and evolution of the worldwide distributed pathogenic amoeboflagellate Naegleria fowleri. Infect Genet Evol 2011; 11(7): 1520-8.
[http://dx.doi.org/10.1016/j.meegid.2011.07.023] [PMID: 21843657]

[24] Stevens AR, De Jonckheere J, Willaert E. Naegleria lovaniensis new species: isolation and identification of six thermophilic strains of a new species found in association with Naegleria fowleri. Int J Parasitol 1980; 10(1): 51-64.
[http://dx.doi.org/10.1016/0020-7519(80)90064-8] [PMID: 6989774]

[25] Khanna V, Khanna R, Hebbar S. Primary Amoebic Meningoencephalitis in an Infant due to Naegleria fowleri. Case Reports in Neurological Medicine 2011.

[26] De Jonckheere JF, Brown S. Defining new Naegleria spp. using ribosomal DNA sequences. Acta Protozool 1997; 36: 273-8.

[27] Mandell G, Bennett J, Dolin R, Eds. Principles and Practice of Infectious Disease Philadelphia. PA: Churchill Livingstone 2000.

[28] Visvesvara GS, Maguire JH. Pathogenic and opportunistic free-living amebas. Tropical Infectious Diseases. PA: Churchill Livingstone 2006; pp. 114-25.

[29] Seidel JS, Harmatz P, Visvesvara GS, Cohen A, Edwards J, Turner J. Successful treatment of primary amebic meningoencephalitis. N Engl J Med 1982; 306(6): 346-8.
[http://dx.doi.org/10.1056/NEJM198202113060607] [PMID: 7054710]

[30] Diffley P, Skeels MR, Sogandares-Bernal F. Delayed type hypersensitivity in guinea pig infected subcutaneously with Naegleria fowleri Carter. Z Parasitenkd 1976; 49(2): 133-7.
[http://dx.doi.org/10.1007/BF00382420] [PMID: 1274383]

[31] Newsome AL, Arnold RR. Equivalent mortality in normal and athymic mice infected with Naegleria fowleri. J Parasitol 1985; 71(5): 678-9.
[http://dx.doi.org/10.2307/3281443] [PMID: 3877155]

[32] Fischer-Stenger K, Marciano-Cabral F. The arginine-dependent cytolytic mechanism plays a role in destruction of Naegleria fowleri amoebae by activated macrophages. Infect Immun 1992; 60(12): 5126-31.
[PMID: 1452346]

[33] Visvesvara GS, Moura H, Schuster FL. Pathogenic and opportunistic free-living amoebae: Acanthamoeba spp. Balamuthia mandrillaris, Naegleria fowleri, and Sappinia diploidea FEMS Immunology & Medical Microbiology 2007; 50(1): 1-26.

[34] Sparagano O, Drouet E, Brebant R, Manet E, Denoyel GA, Pernin P. Use of monoclonal antibodies to distinguish pathogenic Naegleria fowleri (cysts, trophozoites, or flagellate forms) from other Naegleria species. J Clin Microbiol 1933; 31(10): 2758-63.

[35] Robinson BS, Monis PT, Dobson PJ. Rapid, sensitive, and discriminating identification of Naegleria spp. by real-time PCR and melting-curve analysis. Appl Environ Microbiol 2006; 72(9): 5857-63.
[http://dx.doi.org/10.1128/AEM.00113-06] [PMID: 16957204]

[36] Moura H, Ospina M, Woolfitt AR, Barr JR, Visvesvara GS. Analysis of four human microsporidian isolates by MALDI-TOF mass spectrometry. J Eukaryot Microbiol 2003; 50(3): 156-63. a
[http://dx.doi.org/10.1111/j.1550-7408.2003.tb00110.x] [PMID: 12836871]

[37] Seidel JS, Harmatz P, Visvesvara GS, Cohen A, Edwards J, Turner J. Successful treatment of primary amebic meningoencephalitis. N Engl J Med 1982; 306(6): 346-8.
[http://dx.doi.org/10.1056/NEJM198202113060607] [PMID: 7054710]

[38] Schuster FL, Mandel N. Phenothiazine compounds inhibit *in vitro* growth of pathogenic free-living amoebae. Antimicrob Agents Chemother 1984; 25(1): 109-12.
[http://dx.doi.org/10.1128/AAC.25.1.109] [PMID: 6703673]

[39] Schuster FL, Honarmand S, Visvesvara GS, Glaser CA. Detection of antibodies against free-living amoebae Balamuthia mandrillaris and Acanthamoeba species in a population of patients with encephalitis. Clin Infect Dis 2006; 42(9): 1260-5. b
[http://dx.doi.org/10.1086/503037] [PMID: 16586385]

[40] Cogo PE, Scaglia M, Gatti S. Fatal Naegleria fowleri meningoencephalitis, Italy. Emerg Infect Dis 2004; 10(10): 1835-7.

[41] Intalapaporn P, Suankratay C, Shuangshoti S, Phantumchinda K, Keelawat S, Wilde H. Balamuthia mandrillaris meningoencephalitis: the first case in southeast Asia. Am J Trop Med Hyg 2004; 70(6): 666-9.
[PMID: 15211011]

CHAPTER 7

Giardia and Giardiasis

Showgy Ma'ayeh and **Staffan Svärd***

Department of Cell and Molecular Biology, Uppsala University, Uppsala, Sweden

Abstract: *Giardia intestinalis* (syn. *G. lamblia* or *G. duodenalis*) is a unicellular protozoan parasite that infects the small intestines of humans and animals. The species *G. intestinalis* is composed of eight genotypes (called assemblages) designated from A to H. Only assemblages A and B infect humans, causing diarrhea and other associated symptoms. Giardiasis, the disease caused by the parasite, has a global distribution but is mainly endemic in developing countries where pronounced effects on children manifest in a failure to thrive condition. In adults, giardiasis might predispose for other gastrointestinal disorders such as Irritable Bowel Syndrome (IBS) after parasite clearance. The parasite exists in two forms: the cyst and the trophozoite. The cyst is the infectious form that is transmitted *via* the fecal-oral route. Infections can be asymptomatic, acute or chronic, which might be the result of interplay between parasite and host factors. The parasite is known to induce pathophysiological changes in the small intestines and trigger an immune response that result in parasite clearance. Host immune responses towards *G. intestinalis,* however, are not completely understood. Microscopy is the gold standard in clinical settings to diagnose giardiasis and treatment is usually with metronidazole or tinidazole but other drugs are also available. No human vaccines are available to date but some vaccine trials have shown promising results in animals. The spread of giardiasis is most common *via* water and is usually controlled through monitoring water treatment processes, by implementing a multi-barrier approach, and by monitoring the quality of water in recreational venues. Giardiasis adds to the global health burden and increases the costs of health care in many countries and therefore, a better knowledge of disease transmission and control is required to mitigate the risks of infections.

Keywords: Assemblages, Diarrhea, Diplomonad, Host-parasite interaction, Metronidazole, Protozoa, Van Leeuwenhoek, Variable surface proteins, Ventral disc, Virulence, Water-borne, Zoonosis.

INTRODUCTION

Giardia intestinalis, also known as *Giardia duodenalis* or *Giardia lamblia*, is a

* **Corresponding author Staffan Svärd:** Department of Cell and Molecular Biology, Uppsala University, Uppsala, Sweden; Tel/Fax: +46-18-4714558; E-mail: staffan.svard@icm.uu.se

Fabrizio Bruschi (Ed.)

unicellular protozoan parasite that infects the upper intestinal tract of humans and animals [1]. The disease, giardiasis, manifests in humans as a bout of acute diarrhea that can develop to a chronic stage but the majority of infections remain asymptomatic [2, 3]. Giardiasis has a global distribution with 280 million cases reported annually, with its impact being more pronounced in the developing world, where it is usually associated with poor socioeconomic conditions [4]. Children, elderly people and immunocompromised individuals are the most affected by the disease [5 - 7]. In children specifically, effects on growth, nutrition and cognitive function have been reported [8 - 10]. Currently, it has been suggested that giardiasis could predispose for chronic gastrointestinal disorders such as irritable bowel syndrome (IBS) [11, 12]. In 2004, giardiasis was recognized by the World Health Organization (WHO) as a neglected disease associated with poverty, impairing development and socio-cconomic improvements [13]. Because giardiasis adds to the global microbial disease burden, an initiative was instigated to implement a comprehensive approach for control and prevention [13]. One area of focus is the potential spread of giardiasis *via* food and food handling [14, 15], daycare settings [16, 17], travel to endemic areas and close human contact [18 - 20]. Potential transmission of giardiasis from animals to humans (*i.e.* zoonosis) has been also the subject of extensive research over the years. Wildlife accessibility to water used for drinking and recreational purposes, as well as living in proximity with animals, have been identified as risk factors associated with zoonosis [21 - 23]. Another body of research addressed the effect of giardiasis on livestock. Not only transmission in livestock but also economic losses associated with poor growth, weight loss, reduced productivity and even death of animals [24 - 28]. Thus, giardiasis is regarded as a disease that has a significant impact on both humans and animals.

Currently, there are eight different genotypes, also called assemblages, of *G. intestinalis*. *Giardia* assemblages are designated from A to H, the majority of which are host-specific [7]. Parasites that belong to assemblages A and B infect humans but are also common in other mammals [29]. The *Giardia* life cycle alternates between an actively dividing trophozoite and an infectious cyst. Cysts are transmitted to humans *via* the fecal-oral route and upon ingestion they break open (*i.e.* excyst) in the duodenum, releasing trophozoites [29]. Trophozoites attach to the intestinal epithelial cells (IECs), to avoid elimination by peristalsis, and start multiplication. The presence of *Giardia* in the host intestines and in close contact with IECs induce functional and structural changes in the intestinal epithelia, which overall results in diarrhea [30]. Nevertheless, given the variability in giardiasis symptoms, infection *versus* symptoms or lack of symptoms is a controversial issue where the interplay between the infecting parasite of a certain assemblage, isolate virulence and host factors could determine the outcome of infection.

Giardia does not cause overt inflammation in the host small intestines during most infections [31 - 33]. However, inflammatory responses have been seen during acute experimental and human infections [34, 35]. In line with this, IECs produce proinflammatory cytokines in response to the parasite or parasite-excreted products in an isolate-specific mode of induction [36]. At the same time, some parasite factors such as cysteine proteases (CPs) have been shown to attenuate the proinflammatory response by cytokine degradation [37]. Indeed, an accumulating body of evidence shows that cytokine production is rather inhibited during giardiasis [38 - 41]. Effector mechanisms play an early part in defense against *Giardia* and these include nitric oxide (NO) production, lactoferrins, α-defensins, phagocytes, mast cells and dendritic cells [31, 42]. The parasite, however, is able to avert some of these responses, specifically, NO production [43], dendritic cells recruitment [44] and granulocyte infiltration [45]. Patients are able to clear the infection within few weeks and re-infections manifest with reduced symptoms, indicating an acquired immunity against giardiasis [46 - 48]. Antibodies (*i.e.* B cell response) have been suggested to be important players in parasite clearance while the contribution of T cell responses is still unclear [31, 42].

Identification of cysts by microscopy is the gold standard in diagnosing human giardiasis. This is because trophozoites are only seen in stool during periods of acute diarrhea (*i.e.* watery or loose stool). Bright field microscopy coupled with morphometry, using phase contrast or differential interference contrast, is performed on wet mounts or following sample concentration [49]. Usually, multiple stool samples are submitted for analysis, especially when neither trophozoites nor cysts could be detected microscopically at initial examination. Fixation and staining can maintain trophozoite morphology and enhance visualizing trophozoite structure [49]. Antigen detection in stool (coproantigen) can be used for large-scale analyses and polymerase chain reaction (PCR) for increased sensitivity and specificity as well as genotyping [49]. For environmental samples, an initial filtration step is included to concentrate cysts in water samples. Immunomagnetic separation (IMS) followed by fluorescent microscopy is the gold standard for detecting cysts and is used for dual detection of *Giardia* cysts and *Cryptosporidium* oocysts in different water matrices [50]. The technique uses antibodies immobilized to magnetic beads to capture cysts in concentrated samples and fluorescence isothiocyanate (FITC)-linked antibodies to view cysts by fluorescent microscope [50]. The same method is adopted for food, specifically fruit and vegetables, using different wash buffers and sometimes stomaching to enhance cysts recovery [51 - 53].

Human giardiasis is usually treated with nitroheterocyclic compounds [54]. These include the 5-nitroimidazoles (*e.g.* metronidazole and tinidazole), and other non-imidazole compounds such as furazolidine and nitazoxanide. Treatment with these compounds has proven effective but side effects have been reported. Treatment failures especially with metronidazole, the most prescribed drug to treat giardiasis, have been reported in 20% of the cases and this drug is also associated with side effects and carcinogenicity [54 - 56]. Therefore, ongoing efforts are dedicated to discover new drugs to treat giardiasis. Currently, there is no vaccine against human giardiasis, but antibodies against some *Giardia* antigens have been identified in patients' sera [57], which could be used in future trials. Some antigens have been shown to provide protection against second *Giardia* infections in mice (*e.g.* α-1 giardin and cyst wall protein 2) [58, 59] and in gerbils (*e.g.* variable surface proteins, VSPs) [60] and these can be good targets for future vaccine development.

HISTORY

Discovery and Nomenclature

Giardia was first described by the Dutch microscopist van Leeuwenhoek in 1681 [61]. He was examining his own diarrheal stool under a one lens light microscope when he observed the presence of prettily moving tiny creatures; 'longer than wider with flattish belly and sundry little paws' [61, 62]. The French zoologist Alfred Mathieu Giard described the parasite following Leeuwenhoek's discovery; however, his work remained underreported in the scientific literature. Lambl (1859) was the first to describe the *Giardia* parasite in greater detail, which he named *Cercomonas intestinalis*, as many flagellates in diarrheal stool back then were assigned to this genus [62]. The parasite was then reported in rodents (*i.e. G. muris*; Grassi, 1879) [63], amphibians (*i.e. G. agilis*; Kunstler, 1882) [64] and in different hosts (*e.g.* rabbit, cat, dog, bird *etc.*...); however, the nomenclature in different hosts was controversial (*e.g. Dimorphus muris*, *Hexamita duodenalis*, *Megastoma enterica*, *Giardia intestinalis* and *Lamblia intestinalis*). Kunstler (1882) was the first to use the genus name *Giardia* (*i.e. G. intestinalis*) [64], whereas Blanchard (1888) used *Lamblia* (*i.e. Lamblia intestinalis*) to honor Lambl [65]. The species name, *intestinalis*, was dropped afterwards due to homonymy with *C. intestinalis*, the organism that Diesing renamed from *Bodo intestinalis*. The followers, thereafter, used the names *G. duodenalis* (Stiles, 1902)

[66], *G. lamblia* (1915) [63] and *G. enterica* (1920) [67] by Kofoid and Christiansen. The species name, *enterica*, was used to refer to the forms in both rodents and humans (by Grassi), and thus it was suppressed as a synonym for the rodents form as it was deemed different morphologically [68]. Dobell (1940) revised *Giardia*'s nomenclature and suggested calling the organism *Giardia* not *Lamblia* [69]. Nowadays, *G. duodenalis*, *G. intestinalis* and *G. Lamblia* are all accepted and used interchangeably, although *G. lamblia* is frequently used in medicine to refer to isolates of human origin [70, 71].

Giardia spp.

In the past, many *Giardia* species have been named based on the host of origin. This resulted in an overestimation of the number of *Giardia* species (*e.g.* 51 species have been named since 1859) [72]. Relying on morphological differences between *Giardia* isolates was a better option for speciation, which Simon was the first to use to differentiate between *G. lamblia*, the most used name at the time, and *G. muris* [73]. In 1952, Filice discriminated between *G. intestinalis*, *G. muris* and *G. agilis* based on morphological features such as the overall shape of trophozoite and median body [74]. *G. intestinalis* is pear-shaped with one or two transverse claw-shaped median bodies. *G. muris* is more rounded with a small rounded median body while *G. agilis* is longer with narrower body and a teardrop-shaped median body [74]. Filice indicated that within the three groups, there could exist some morphologically similar forms with different physiological characters. This was confirmed later with the advent of more refined and discriminatory methods such as electron microscopy and PCR. Examples are *G. pisttaci* in budgerigars (*i.e.* parakeets) [75] and *G. ardeae* in herons and ibis [76]; which were both identified by electron microscopy. *G. pisttaci* is morphologically similar to *G. intestinalis* but lacks a complete ventrolateral flange and marginal groove bordering the anterior and lateral border of adhesive disc [75]. *G. ardeae* share morphological features with both *G. intestinalis* and *G. muris*. Trophozoite shape is similar to that of *G. intestinalis* with a teardrop-shaped nuclei and a deep notch on the posterior border of adhesive disc [76]. The median bodies are pleomorphic ranging in structure from round oval (as in *G. muris*) to claw-shaped (as in *G. duodenalis*). A single caudal flagellum exists on the right side (viewed dorsally) and a rudimentary left flagellum. An exclusive feature of *G. ardeae,* in comparison to other species, is the nuclei, which is long, cylinder and has a teardrop shape [76]. Another *Giardia* species was isolated from *Microtus ochrogaster* (*i.e.* Prairie vole) and identified based on cyst morphology [77, 78]. This species is *G. microti* whose cysts contain two differentiated trophozoites with mature adhesive discs. This is unlike the *G. intestinalis* cyst that contains only one immature trophozoite (exyzoite) with four nuclei and no ventral disc [77, 78], indicating an interesting aspect of evolution. This brings the total number of *Giardia* species identified so far to six. *Giardia* spp. and host range are summarized in Table **1**.

Table 1. ***Giardia*** **species,** ***G. intestinalis*** **assemblages, proposed new species names and respective host range.**

Species	Genotype	Proposed New Species Names	Host
G. intestinalis	Assemblage A	*G. duodenalis*	Many mammals incl. humans
	Assemblage B	*G. enterica*	Many mammals incl. humans
	Assemblage C,D	*G. canis*	Dogs
	Assemblage E	*G. bovis*	Hoofed animals
	Assemblage F	*G. cati*	Cats
	Assemblage G	*G. simondi*	Rats
G. agilis			Amphibians
G. ardeae			Herons
G. microti			Rodents
G. muris			Rodents
G. psittaci			Birds

Giardia Duodenalis Assemblages

Morphological variation, specifically shape and size, within *G. intestinalis* isolates was a criterion used early in an attempt to identify subspecies with the same group. This, however, had produced limited results due to the morphological similarity of different *G. intestinalis* isolates. Analyzing the proteins (*i.e.* enzyme electrophoresis) [79 - 82] and DNA (*i.e.* DNA fingerprinting) [83 - 86] of different isolates was a more powerful approach to confirm genetic polymorphism among different *G. intestinalis* isolates, and as a result the differences among morphologically similar isolates. Thereafter, extensive studies and analyses of PCR products targeting the small-subunit ribosomal RNA genes (*ssrRNA or 18SrRNA*) as well as several housekeeping genes (*e.g.* β-giardin (*bg*), glutamate dehydrogenase (*gdh*), triosephosphate isomerase (*tpi*) and elongation factor 1 alpha (*ef-1α*)) have presented a conclusive evidence that *G. intestinalis* is a species complex, which was categorized into eight distinct subgroups or assemblages designated from A to H [87 - 96]. Among the different assemblages, only A and B have been found to infect humans. Human isolates within these two assemblages were further placed under four sub-assemblages (*e.g.* AI, AII, BIII and BIV) based on multilocus genotyping of *18SrRNA*, *bg*, *gdh*, *tpi* and analyses of single nucleotide polymorphisms [7, 71, 72]. The rest of the assemblages are host specific; C and D in canines, E in hoofed animals (*i.e.* artiodactyls), F in cats, G in rodents and H in marine mammals [96]. This has shown that host specificity might be a reliable but underestimated phenotypic trait for identifying *G. intestinalis* subspecies. The importance of this notion is represented by whether *G. intestinalis* isolates can be zoonotic but the difficulty in interpreting cross-transmission experiments results, using uncharacterized isolates in experimental infections and possibly host factors affecting the outcome of infection made host specificity unreliable factor in identifying subspecies [89, 97]. Recently, it has been proposed to rename *G. intestinalis* assemblages based on host specificity and genetic characterization. The names *G. intestinalis* has been suggested for assemblage A, *G. enterica* for B, *G. canis* for C and D, *G. bovis* for E, *G. catti* for F and *G. simondi* for G (Table **1**) [7, 72, 97].

Although many *Giardia* isolates could not be cultured *in vitro* (*i.e.* axenized), those of human origin that were axenized successfully provided valuable information on the genetic and phenotypic variation among *G. intestinalis* isolates. The human isolates, WB and GS, which belong to assemblages A and B, respectively, have been axenized for decades and their genomes have been sequenced [98, 99]. Genomic comparisons have shown that they share 77% and 78% similarity of nucleotide and amino-acid identity in the protein coding regions and this was suggested to account for the observed biological differences between the two isolates/assemblages [99] such as growth rate and infectivity [100, 101]. GS is known to be a slow grower in comparison to WB but it can readily infect mice whereas WB is often cleared before establishing infection [101]. Different *G. intestinalis* assemblages were also successfully passaged in suckling mice where a variation in growth patterns was observed. Assemblages A, B and F can adapt to growth in suckling mice easily while an erratic pattern of growth is observed for assemblages C and D [102]. Furthermore, genetic differences were found to underlie a variation in metabolism, biochemistry, duration of infection, pH susceptibility and drug sensitivity [100 - 107]. This shows that genetic makeup accounts for biological differences among assemblages and it could potentially be significant for the clinical outcomes of infection and epidemiology.

Giardia's Evolutionary Position

Excavates is a group of eukaryotes which includes several taxa (diplomonads, parabasalids, retortamonads, oxymonads, heterolobosea and euglenozoa) and *Giardia* is part of the diplomonads. Many excavate taxa, including the diplomonads, were previously believed to be "deep-branching" basal eukaryotes, which diverged prior to the appearance of the "crown-group" eukaryotes [108]. Recently, this view has been largely challenged with the appearance of more advanced and sophisticated phylogenetic tools and greater taxon sampling. Diplomonads have unusual cell organization, which includes two karyomastigonts (*i.e.* cellular cytoskeleton including flagella) and nuclei, and a recurrent flagellum that passes through a cytopharyngeal tube; however, this has been lost in the lineage leading to *Giardia*. Two subdivisions exist within the diplomonads; the Giardiinae and Hexamitinae. Giardiinae (*Giardia* and *Octomitus*) have lost the cytopharynx and recurrent flagella pass through the cellular cytoplasm whereas Hexamitinae possess a functional feeding apparatus and uses the UGA genetic code as the sole stop codon due to the utilization of UAA and UAG genetic codes to encode glutamine [109]. Members of Hexamitinae subdivision include both parasitic and free-living forms such the parasites *Hexamita* as well as *Spironucleus,* and *Trepomonas,* respectively. Curiously, there exists other diplomonads with one karyomastigont, and they resemble one half of a typical diplomonad cell. These are collectively termed 'enteromonads' emerging within the Hexamitinae subdivision but not as one group, suggesting several separate events of cellular reduction [110].

MORPHOLOGY AND LIFE CYCLE

Giardia Morphology

Giardia is a eukaryotic unicellular organism with two transcriptionally active diploid nuclei, no peroxisomes or true mitochondria [70]. The trophozoite is pyriform with four pairs of flagella and a claw-shaped median body. Flagella are named according to their location respective to trophozoite body: anterior, ventral, posterior/lateral and caudal. Trophozoite length ranges between 12 and 15 µm and width between 6 and 8 µm [70]. The adhesive disc (*i.e.* ventral disc), located on the ventral side of the parasite, is a characteristic feature of *Giardia*, distinguishing it from the rest of diplomonads [111]. It is large, rigid and concave, making it well suited for its function; that is attaching to IECs during infection. The cyst is oval and is surrounded by a thick cyst wall (CW) (0.3–0.5 µm thick). They are 8 to 12 µm long and 7 to 10 µm wide. Under the CW, there exists a double inner membrane protecting the excyzoite, which has four nuclei. CW contains glycan (~ 60%) and protein (~ 40%) in insoluble fibrils [112, 113]. The filamentous outer portion of CW is made up from the glycan β(1-3)-N-acet-l-d-galactosamine (GalNAc) that is converted into an insoluble polysaccharide [114] and three proteins constitute the main compnenets of CW, including cyst wall protein 1, 2 and 3 (CWP1, CWP2 and CWP3). The CW structure gives cysts rigidity and resistance to environmental damage.

Life Cycle

The life cycle of *Giardia* alternates between an infectious cyst and the disease-causing trophozoite (Fig. **1**). Upon ingestion and exposure to stomach acids, cysts become metabolically active and start excystation. The rapid influx of hydrogen ions from stomach acids causes a sudden drop in pH, providing the trigger to begin excystation [115]. A cysteine protease (cysteine protease 2, CP2) acts on the CW to help break open the cyst [116]. Treating cysts with N-[*N*-(L-3-*trans*-carboxyirane-2-carbonyl)-L-leucyl]-agmatine (E-64), a CP-specific irreversible inhibitor, results in excystation failure both *in vitro* and *in vivo*, showing the importance of CPs in excystation [117]. An acid phosphatase is also important for the dephosphorylation of CWPs [118]. Excystation occurs in the duodenum and jejunum, where excysted trophozoites (*i.e.* excyzoites) are away and protected from the damaging effects of gastric acids [115]. When cysts break open, flagella emerge first through an opening in the cyst pole followed by the exit of the excyzoite. Excyzoites are intermediate forms of trophozoites, which undergo cytokinesis to produce two daughter cells [119]. They are disc-less upon emergence in the small intestines but disc proteins (*i.e.* fragments) are stored inside the excyzoite and assembled directly following excystation, concurrent with the reorganization of flagella for locomotion [120]. Alpha-1 giardin, a membrane-associated protein, is believed to mediate attachment at that stage until assembly of adhesive disc is complete; otherwise the parasite will be swept away by intestinal peristaltic movement [121]. Infection establishes thereafter by parasite attachment to enterocytes, replication and colonization of the small intestines. For life cycle continuation, a fraction of the trophozoite population undergo encystation in response to changes of specific factors in the intestinal environment such as dietary lipids, bile salts, intestinal pH (pH 7.8) and lactic acid [122]. A simulation of such conditions (*i.e.* high bile and pH of 7.8) *in vitro* induces encystation in axenic cultures. When trophozoites begin encystation (Fig. **1**), fragments of the adhesive disc and flagella are internalized, causing trophozoites to round up and lose their attachment ability [120]. Internalized structures will appear in the centre of cysts as a ribbon-like microtubule structure in addition to an excyzoite with four nuclei [115, 123, 124]. The key feature of encystation is the synthesis of CW (Fig. **1**), which is composed of 50% GalNAc and 50% protein (mainly the thee CWPs). Endogenous glucose is converted into GalNAc *via* a pathway composed of five transcriptionally and allosterically regulated enzymes [125 - 127]. CWPs are transcriptionally up-regulated early in encystation, post-translationally modified and shuttled *via* the encystation specific secretory vesicles (ESV, Fig. **1**). Upon CW synthesis, transcription and metabolism decrease, preparing cells to enter dormancy [120, 128]. Infected individuals excrete mature cysts in their stool, and when cysts contaminate food and water, life cycle continues in the environment. The life cycle of *Giardia* is summarized in Fig. (**1**). It is clearly advantageous for *Giardia* to undergo encystation inside the host as this helps avoid the host immune response and once released they are protected form hostile environmental conditions [128, 129]. Therefore, encystation has been regarded as a survival mechanism aiding the transmission of giardiasis.

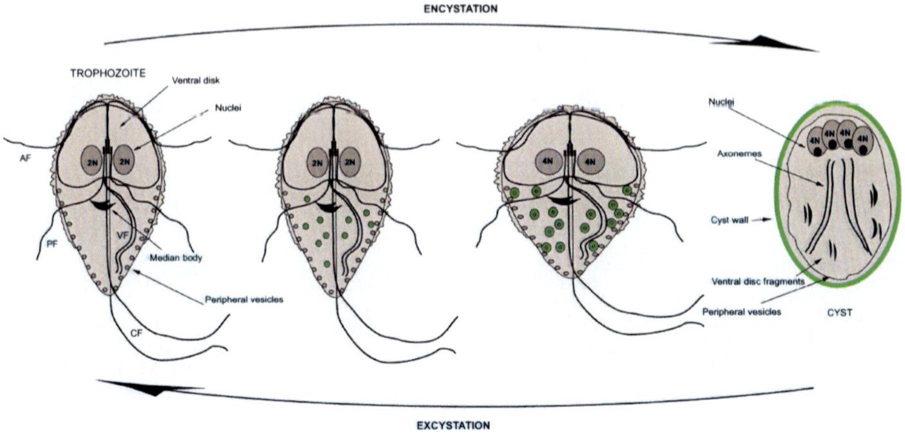

Fig. (1). *Giardia* has two major life-stages; the dividing trophozoite and the infectious cyst. Differentiation into the cyst stage is vital for the parasite to survive and transmit to new hosts. During encystation, motile trophozoites undergo gradual transformation during which they lose the ability to attach (the disc is fragmented), flagella are internalized and metabolism decreases. As a result, cells round up and enter dormancy. CWPs are transported in ESVs, here shown in green in this cartoon, and these proteins, together with the GalNAc sugar, build up the cyst wall. Mature cyst contains 4 nuclei, each harboring 4 genome copies, resulting from two rounds of replication; the last replication occurs without nuclear division. The excystation process produces 4 trophozoites from one cyst. Figure production by Elin Einarsson and Jon Jerlström-Hultqvist.

GLOBAL EPIDEMIOLOGY AND RISK FACTORS

Global Epidemiology

Human giardiasis has a global distribution with 200 million symptomatic cases reported annually in Asia, Africa and Latin America [130]. In the United States, giardiasis is the most common intestinal parasitic disease with 1.2 million cases reported annually and hospitalization costs reaching up to 34 million American dollars [131, 132]. In the latest survey of giardiasis in the US (published in 2015) [133], a total of 16,868 giardiasis cases in 2011 and 15,223 cases in 2012 were reported, 98.8% of those were confirmed. Small proportions of the cases were associated with detected outbreaks (1.5% and 1.3% in 2011 and 2012, respectively). Northwestern states had the highest incidence rates with an annual peak onset of illness during early summer through early fall. Incidence rates were reported to be 6.4 and 5.8 per a 100,000 population in 2011 and 2012, respectively, representing a slight decline from the rates reported during 2005–2010 (7.1–7.9 cases per 100,000 population). Cases exhibited an age-correlated pattern, peaking in children (1-4 years), and to a less extent for those between 5-9 years and adults aged between 45–49 years. The decline in the incidence rate of giardiasis was ascribed to changes in transmission pattern, surveillance case definition, and implementing strategies to reduce waterborne transmission. The geographical differences in incidence rates, on the other hand, were ascribed to differences in surveillance capacity across states or transmission patterns [133].

A recent survey from 31 European Union (EU) countries including the European Economic Area (EEA) reported a total of 16368 cases of confirmed giardiasis for 23 countries in 2012 [134]. Bulgaria was on top of the list with an incidence rate of 21 confirmed cases per a 100000 population, followed by Estonia (19 per 100000), Sweden (11 per 100000) and Finland (7 per 100000). The lowest rates were reported for Malta (0.24 per 100000), Luxemborg (0.38 per 100000), Lithuania (0.43 per 100000), Cyprus (0.46 per 100000) and Czech Republic (0.47 per 100000). The incidence rate of giardiasis has been relatively constant over a period of five years (2008-2012) with an average of 5.43 cases per a 100000 population. A male to female ratio of 1.19 to 1 was reported from a total of 14373 confirmed cases in EU/EEA. The incidence rate (from 12779 confirmed cases) was the highest in the age group 0-4 years for both genders, being higher for males within the same age group (0-4 years, 11.6 per 100000). No strong correlation could be established with seasonality from 14434 reported cases from 22 countries, although a slight increase in reported cases was observed in autumn compared to summer months (June, July and August).

In New Zealand, giardiasis became a notifiable disease in 1996 and since then New Zeeland has had a relatively high incidence rate (49.4 per 100000 population) [135]. Over a 10-years period (1997-2006) the average annual rate of notified giardiasis was 44.1 cases per a 100000 population with 1.7% of those being hospitalized and two fatalities reported; notification rates were slightly higher in rural areas [136]. The highest reported rates were among children (0-9 years old), adults (30-39 years old), Europeans, and those living in deprived areas. Giardiasis showed little seasonality and no correlation with farm animals density.

The distribution of cases exhibited an anthroponotic mode of transmission (*i.e.* human), with a small zoonotic contribution in rural environments and a modest contribution from overseas travel [136]. In 2014, giardiasis was the second notifiable disease in New Zealand after campylobacteriosis (6776 cases) with 1709 reported cases [137]. The highest number of cases was reported for the age group 30-39 (368 cases), followed by 1-4 years (352) and 40-49 years (241 cases). The infection rate was similar in both genders and the majority of cases were reported for the European including New Zealander ethnic group (1301 cases) [137]. A recent review has demonstrated a significant association between climate change and agricultural land use with enteric disease occurrence in humans represented by the most common pathogens in New Zealand (campylobacteriosis, salmonellosis, cryptosporidiosis and giardiasis) [138]. The study suggested that disease risk from environmental reservoirs is pathogen-specific with the presence of considerable pathogen load in rural regions, increasing the chances for zoonotic transmission. Nevertheless, demographic factors have been suggested to influence disease patterns [138]. This was apparent when disease notification data were compared across three time periods, 1997-2000, 2001-2004, and 2005-2008, where a spatio-temporal cluster for giardiasis occurred in urban areas with little seasonal influence. This was mainly related to host behavior and urban exposures [139].

Giardia is an endemic parasite in Asia [5]. A recent review looked at the current situation of giardiasis in South Asia, South East Asia and Far East from 33 published papers from 2002-2007, upon which the authors considered giardiasis the most common water-borne diarrhea-causing disease in Asia [5]. *Giardia* prevalence varied in different Asian countries (1.5% to 73.4%) being more prevalent in urban than rural areas, among poor communities and university students, in males aged between 2 and 5 years, elderly people, HIV-positive patients, and gastric carcinoma patients. The epidemiological situation of giardiasis in Asia was reflective of poor economic level and social status, represented by limited access to clean drinking water, lack of proper hygiene, sanitation and health education, and absence of appropriate surveillance [5]. Nevertheless, a recent report has indicated that the prevalence of giardiasis is declining in some countries in Asia like China [140]. While an overall across country prevalence has been estimated to be 2.52% (28.5 million infections) according to the first national survey in China [141], in the second national survey a significant decrease in giardiasis prevalence was observed, such as that in Henan Province (from 7.18% to 3.85%) and in Zhejiang Province (from 2.55% to 1.00%) [142, 143]. The previous studies acknowledged water as a vehicle for giardiasis transmission. In Malaysia, an overall steady decline in intestinal parasitic infections from 1970 to 2013 was observed in plantation communities, with prevalence rates dropping from 95.0% in the seventies to 37.0% in 2012. Giardiasis prevalence in children has been estimated to be 1.8% and a lower overall rate for intestinal parasitic infections was observed in urban societies of China [144].

Factors Affecting the Transmission of Giardiasis

Several factors in the biology of *Giardia* contribute to its successful water- and food-borne transmission and thus dissemination in the environment. *Giardia* cysts are already mature and infectious once they are excreted in the stool of infected hosts [145]. Owing to the hard CW, excreted cysts stay viable for months in cool and damp areas in the environment (*i.e.* water and soil) [146]. Cysts are also resistant to standard levels of chlorination and their small size may enable them to penetrate sand filters and therefore evade removal in a fully functional water treatment plant [147, 148]. Infected individuals release around 10^8 to 10^9 cysts daily [149, 150] and cysts shedding may last for three months or more [150]. Infected cattle excrete around 150 000 cysts per gram of feces, although an intermittent cyst-shedding pattern and variation in the numbers released were reported [151]. Therefore, the excretion of cysts in high numbers coupled with the prolonged duration of cyst shedding are factors of high significance for spreading disease in the environment. Given the low infective dose required to cause infection (10-100 cysts) [152], the presence of low amounts of cysts in water and food matrices could be sufficient to cause infection. In this scenario, it is worth noting that previous experimental infections with human volunteers presented different outcomes of the disease, depending on the isolates used (*e.g.* GS and Isr) [100], implying that the infecting isolate should be virulent enough to induce disease. The fact that human infections are basically caused by assemblage-restricted (*i.e.* either A or B) isolates, and given the broad host range of these two assemblages, the zoonotic nature of this parasite is another factor contributing to *Giardia* dissemination. Animals, birds and fish infected with the zoonotic assemblages A and B, can serve as a transmission vehicle for giardiasis in human either by direct contact or contamination of water resources. This subject was extensively reviewed over the past decades [7, 71, 89, 92, 96, 153, 154]. Therefore, the factors discussed above work collectively to aid the survival and spread of giardiasis in the environment.

Risk Factors

Transmission *via* contaminated water is currently considered the most important route for spreading giardiasis. Between 2004 and 2010, *Giardia* was responsible for 70 out of 199 reported protozoal outbreaks caused by the consumption of contaminated water [155]. The contamination is introduced to water resources during rain induced runoff events and in catchment areas accessible by wild life. Inadequate water treatment, treatment failure or distribution deficiencies may result in the spread of giardiasis to a large number of individuals, especially when a water body serves as a main water supply for a large community [7, 148, 155 - 157]. This places the consumption of inadequately treated water or contaminated water as a top risk factor for acquiring giardiasis. Foodborne transmission, on the other hand, is rather linked to either contaminated water used in food preparation and manufacturing or food handlers [15, 158 - 160]. The risk factors for giardiasis have been identified previously; however, these might vary between developed and developing countries depending on exposure routes. In developed countries, increased risks of acquiring giardiasis are usually associated with exposure to recreational waters [148, 161, 162]. The giardiasis seasonal peak together with the age-specific pattern (1-9 years) of disease might be related to increased outdoor activities (summer to early autumn) and accidental swallowing of contaminated water by kids [148, 161]. In fact, *Giardia* has been responsible for at least 18 outbreaks associated with recreational water [148]. Foreign travel to areas where *Giardia* is endemic is another risk factor for acquiring giardiasis [162 - 164]. Interestingly, a recent study looked at the prevalence of giardiasis in Swedish sentinels returning from European countries during the period 1997-2003 and identified 1112 positive cases for *Giardia*. The risk of acquiring giardiasis varied a 100-fold across Europe but this indicates that even travelling to a developed country with high incidence rate of giardiasis puts individuals at risk of acquiring the disease [165]. Children at day care centres and individuals working there are at increased risk of acquiring giardiasis, depending on the time spent and duration of attendance [166]. A recent study in day care centres investigated risk factors associated with the acquiring of five enteropathogens, including rotavirus, norovirus, astrovirus, *G. intestinalis* and *Cryptosporidium hominis/parvum* [167]. Identified risk factors were: a large day care centre capacity, crowding, having animals, nappy changing areas, sandpits, paddling pools, cleaning potties in normal sinks, cleaning vomit with paper towels and mixing of staff between children [167]. Mothers and caregivers of children infected with *Giardia* are also at increased risk for acquiring giardiasis due to close contact with sick children and the requirement of handling their waste (*e.g.* nappy changing) [16, 19]. Contact with farm animals or pets increase individual's risk of acquiring giardiasis through direct or indirect exposure to their fecal waste [7, 21 - 23]. Some sexual activities and HIV infection increase the risk for acquiring giardiasis through the oral-anal contact [168]. Immune suppression (in case of low $CD4^{+}$

count in HIV infection, chemotherapy, hypogammaglobulinemia, malnutrition and others) might increase the risk of developing chronic giardiasis [169]. Finally, some studies have identified the consumption of pre-cut lettuce or green leafs as risk factors for acquiring giardiasis [161, 170, 171]. The increased risk is likely to be associated with using reclaimed wastewater or contaminated water in irrigation [171].

In developing countries where giardiasis is endemic, a higher risk of acquiring giardiasis is largely due to the consumption of inadequately treated surface water rather than recreational activities [5, 148]. While this could be reflective of poor socioeconomic conditions, resources or infrastructure in developing countries, human-introduced faecal contamination into water and the surrounding environment is not uncommon and can occur either due to inappropriate sewage handling or some low hygienic practices. A cross-sectional study performed in South Eastern Minas Gerais State in Brazil found that risk factors associated with acquiring giardiasis and soil-transmitted helminthiasis are drinking unsafe water, inadequate sewage discharge and lack of sanitary infrastructure [172]. The aforementioned factors were pronounced in another study (*e.g.* defecation in open areas and low income families), which also touched on and highlighted the educational level of parents, especially the mother, as a risk factor for intestinal parasitism [173]. A study conducted in a rural part of Egypt showed that children living in a household without a latrine, a mud floor in the sleeping rooms, and household exposure to more than 10 chickens increase the risk for acquiring giardiasis [174]. In Sudan, a study showed that crowded and large sized families, illiterate individuals, and children especially those with malnutrition are at increased risk of acquiring giardiasis [175]. In another study targeting three different tribes in Orang Asli in Malaysia, children under 15 years of age, being male, consumption of raw vegetables and the presence of other family members infected with *G. intestinalis* were identified as the main risk factors for giardiasis [176]. In summary, risk factors for acquiring giardiasis in developing countries are mainly associated with the exposure and consumption of contaminated water and eating raw vegetables irrigated with poor quality water, socioeconomic status, malnutrition, practicing low hygiene, crowding in the same household, contact with infected family members and the educational level of parents.

PATHOPHYSIOLOGY OF HUMAN GIARDIASIS

Overview

To date, most of our knowledge about the pathophysiology of giardiasis comes from *in vitro* experiments using intestinal epithelial cell lines and, to a lesser extent, from animal infection models or human biopsies. From *in vitro* experiments, it has been shown that the attachment of trophozoites to cells induces changes in cell shape and size, compromises the integrity of IEC monolayers [177 - 180], causes digestive enzymes deficiencies [179, 180], inhibits cell cycle and proliferation [181] and induces cell apoptosis [182, 183]. Disease machnisms of giardiasis are depicted in Fig. (**2**).

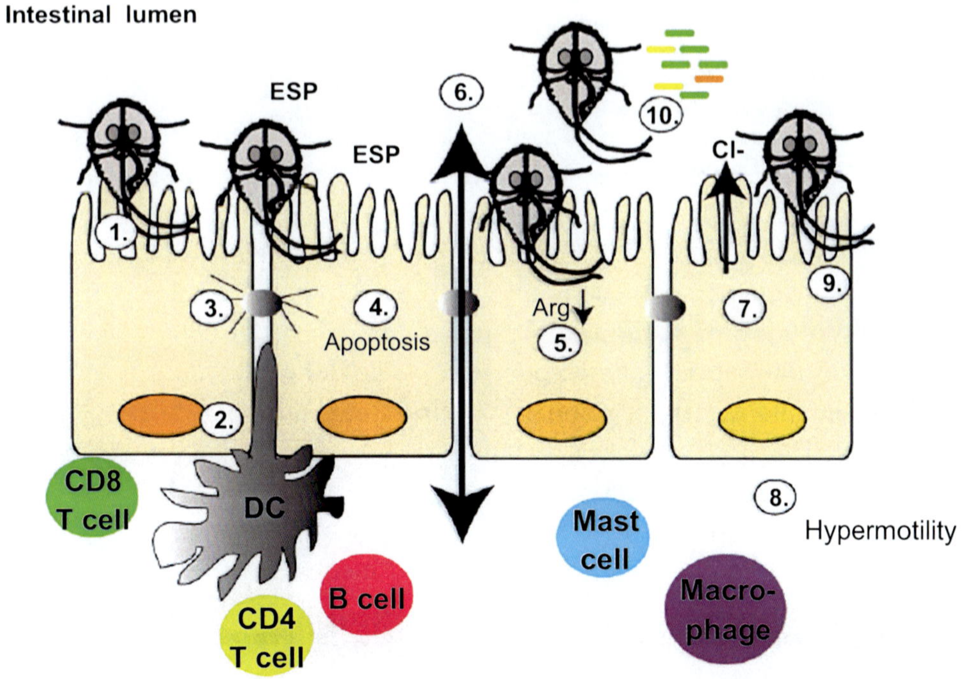

Fig. (2). Main pathophysiological effects during giardiasis. Giardiasis is a multi-factorial diarrheal disease and several different parasite-induced mechanisms are involved in disease induction. 1. Diffuse shortening of microvilli and inhibition of brush border enzymes. This involves attachment of the parasites and $CD8^+$ T cells. 2. Induction of chemokines in IECs, resulting in attraction of immune cells. 3. Disruption of tight junctions. 4. Induction of apoptosis. 5. Starvation for arginine in IECs, resulting in cell cycle arrest and apoptosis. 6. Increased intestinal permeability induced by mechanism three to five. 7. Increased chloride ion secretion. 8. Intestinal hypermotility. 9. Crypt hyperlasia and increased mucus secretion. 10. Changed composition of bacterial normal flora.

Giardia also releases some excretory/secretory products (ESPs) (*e.g.* cysteine proteases, metabolic enzymes, elongation factor-1 alpha) upon contact with IECs [184 - 187] and these ESPs might play roles in the degradation of extracellular matrix (ECM) [187] and modulation of host immune responses [37] (Fig. **2**). The results obtained from animal models or human biopsies were in agreement with those from *in vitro* experiments, and the presence of histopathological and physiological alterations inside the host intestines have been verified in addition to the induction of apoptosis [32, 188 - 192]. Some experiments, on the other hand, have provided insights into the virulence of different *Giardia* isolates, whereby isolate-specific induced pathologies have been observed [191, 193]. Nevertheless, the exact pathogenic mechanisms and the factors associated with different isolates virulence have not been completely identified but it is clear that the cause of disease is multi-factorial (Fig. **2**).

Insights from *in vitro* and *in vivo* Studies

Trophozoite Attachment

Adhesion of trophozoites to IECs is the most important step in establishing the infection. This is primarily achieved by the ventral disc and it is mediated by other cell surface molecules (*e.g.* VSPs, lipid rafts, 200 kDa protein, lectin-like proteins, glycoproteins) and proteolytic enzymes (*e.g.* cysteine proteases) [179, 184, 194 - 196]. The whole attachment machinery is believed to have a high significance in the pathophysiology of giardiasis. Trophozoites adhere to cells apically or to microvilli and basolateral membranes [197]. The adhesion force is strong and upon detachment some marks could be seen on cell surface, indicating cellular damage [194, 198]. Attachment is aided by CPs released by trophozoites during interaction with host cells [184]. Treating trophozoites with E-64, a selective CP inhibitor, before incubation with rat IECs dramatically decreased attachment *in vitro* [184]; however, how CPs promote attachment remains to be elucidated. Furthermore, the release of CPs, might be moonlighting in disease pathology. That is *Giardia* CPs have been shown to degrade a wide range of substrates, especially collagen type 1, which is an important component of ECM [199 - 201]. The ECM provides structural and biochemical support to the surrounding cells and its degradation during infection compromises the integrity of intestinal epithelial tissue [187]. *In vivo*, the oral administration of *Giardia* ESPs, including CPs, to BALB/c mice resulted in mucosal injury as well as systemic and local antibody response [202]. The histopathological changes in mouse intestines were ascribed to CPs, since the sera produced following ESP administration recognized three *Giardia* CPs. It will be interesting to see whether these effects in mice can be abolished upon ESP treatment with E-64 before oral administration.

A body of evidence has shown that *Giardia* attachment *in vitro* is partly mediated by its surface lectins that bind D-glucosyl and mannosyl residues on host cell surface. Taglin (trypsin activated *Giardia* lectin) is a *Giardia* lectin with specificity to mannose-6-phosphate residues that are present on enterocyte surface

[203, 204]. Taglin could be activated *in vivo* by host trypsin and has the ability to bind specifically to isolated brush border membranes of cells. Since taglin agglutinates the cells to which the parasite adheres, it has been suggested to contribute to microvillus shortening (Fig. **2**) during giardiasis [205]. Interestingly, monoclonal antibodies against taglin (α-L2) also recognise α1- and α2-giardins, indicating that they have similar lectin moieties [121]. Alpha-giardins are membrane-associated proteins that have the ability to bind the proteoglycans (*i.e.* glycosaminoglycans, GAGs) on IECs surface, in a Ca^{2+} dependent fashion [121, 206]. Alpha-1 giardin particularly is an immunodominant antigen, recognized by the host immune response during acute giardiasis [58]. Other than these functions, its similarity to taglin definitely warrants further investigation to identify a role in disease pathology similar to that of taglin, if any.

Dysfunctional Intestinal Epithelial Barrier and Disaccharidases Activity

Attachment of *Giardia* and secretion of ESPs affects the cellular junctions, including tight and adherens junctions (Fig. **2**), collectively called the apical junctional complexes (AJCs), which hold the cells together in order to maintain the integrity of epithelial tissue. AJCs are associated with the actin cytoskeleton and they are composed of transmembrane proteins (claudin, occludin, junctional adhesion molecules (JAMs) and tricellulin), cytosolic plaque proteins (zonula occludens (ZO) family) and cytosolic regulatory proteins (F-actin, α-actinin) [207]. Intact AJCs form a seal to prevent the paracellular diffusion of microorganisms and antigens across the epithelium [207] and the disruption or reorganization of AJC proteins affects the integrity of intestinal epithelia and lowers the transepithelial electrical resistance (TER). In patients with chronic giardiasis, claudin-1 expression is decreased and its loss is believed to be associated with epithelial barrier dysfunction in those patients [189]. *In vitro*, using different cell lines (polarized or not), *Giardia* infection was shown to result in the re-arrangement of ZO-1, α-actinin, occludin, claudin-1 and F-actin, subsequently decreasing TER and increasing the permeability of epithelial cell monolayer [178 - 180, 182, 208]. In one study, the reorganization of F-actin has been shown to induce changes in the distribution and activity of brush border-associated proteins (Fig. **2**), including sucrase-isomaltase, dipeptidylpeptidase IV and fructose transporter, GLUT5, and a decrease in sucrase activity [179]. In murine models of giardiasis and Mongolian gerbils, a significant disaccharidase enzymes deficiency (*e.g.* lactase, maltase, sucrase and trehalase) has been reported [190, 193, 209 - 213]. This has been found to be isolate-dependent, as evidenced by the ability of GS isolate but not WB to cause such effects in mice models of giardiasis [213]. In children, some reports have shown an association between giardiasis and hypolactasia as well as vitamin B-12, iron, hemoglobin and transferrin deficiencies [214, 215]. Sodium-dependent glucose uptake is also affected by giardiasis and the expression of SGLT-1, a sodium-glucose co-transporter, is decreased [212]. However, Na^+-independent glucose uptake and potassium uptake are not affected but rather increased as a compensatory mechanism for electrolyte loss during diarrhea [216]. Therefore, due to the large structural and physiological changes *Giardia* induces in the host intestines, it is not surprising to see a wide range of effects on host's nutrition and wellbeing. Indeed, this is line with the reports from developing countries where children fail to thrive due to giardiasis [5].

Diarrhea due to Structural and Functional Changes in the Small Intestines

Giardia infection reduces the total absorptive area in the intestines and this outcome can be clearly seen in chronic giardiasis patients [189]. This pathological event, however, does not occur in all *Giardia* infections and is isolate-dependent. It is also independent to the appearance of diffuse shortening (Fig. **2**) or total depletion of epithelial border microvilli [30]. Since nutrient-coupled absorption of electrolytes (*i.e.* Na/D-glucose) occurs at epithelial microvilli, their shortening impairs the absorption of water, glucose and electrolytes. This in turn causes electrolyte imbalance, including chloride hypersecretion, and results in the build-up of an osmotic gradient that draws water into the small intestinal lumen [30]. This leads to distension, rapid peristalsis (Fig. **2**) and diarrhea [192, 217, 218]. Increased levels of cholecystokinin (CKK), mast cell degranulation and the production of NO mediated by neuronal NO synthase (nNOS) [219, 220] have been also observed during giardiasis. These exert their effect by increasing smooth muscles contractility [221], leading to an increase in the gastrointestinal transit rate. In addition to malabsorption, maldigestion has been suggested as another mechanism for the induction of diarrhea. Maldigestion occurs as a result of brush border enzyme deficiencies (Fig. **2**), generating high levels of undigested carbohydrates and fatty acids by the intestinal microbiota [222 - 224]. The presence of high level of fatty acids in the intestinal lumen has a direct stimulating effect, leading to diarrhea [225]. Collectively, it can be concluded that the induction of diarrhea during giardiasis is multi-factorial, which includes brush border microvilli shortening, malabsorption, maldigestion, electrolyte imbalance and an increase in the gastrointestinal transit rate (Fig. **2**).

A study has shown that ezrin and villin, key components of actin cytoskeleton that comprises the brush border of IECs, underwent significant post-translational changes (*e.g.* phosphorylation, proteolytic cleavage) following *Giardia* infection and during the recovery phase [226]. A drop in ezrin phosphorylation but an increase in villin phosphorylation was observed at the peak of infection, correlating with the reduced brush border enzyme activity [226]. The proteolytic cleavage of ezrin and villin, specifically, has been found to be mediated by $CD4^+$ T cells (ezrin) and both $CD4^+$ and $CD8^+$ T cells in case of villin [226]. In another report, the adaptive transfer of CD8+ T lymphocytes from an infected mice resulted in diffuse shortening of microvilli, loss of brush border surface area, impaired sucrase activity, and increased crypt/villus ratios compared to respective controls [227]. Based on these findings, it appears that cellular immune responses (involving both $CD4^+$ and $CD8^+$ T cells), on top of *Giardia*-induced pathology, can be significant drivers of cytoskeletal remodeling in IECs and immunopathology (Fig. **2**).

Induction of Apoptosis

In the gastrointestinal tract, apoptotic turnover of enterocyte occurs at a high rate but the process is tightly regulated so that it does not affect the integrity of intestinal epithelia [228]. Nevertheless, various forms of gastrointestinal diseases and microbes can disrupt this homeostatic-programmed cell death, leading to a higher rate of enterocyte apoptosis and an increase in intestinal epithelial permeability [229]. Caspases, a family of cysteinyl-aspartate proteases, are mediators of apoptosis. They can be activated either intrinsically (*e.g.* caspase 9) by DNA damage and stress or extrinsically (*e.g.* caspase 8), by the binding of several factors (*e.g.* microbes) to specific receptors on cell surface (*e.g.* death receptor Fas or tumour necrosis factor (TNF)-receptor superfamily) [230]. Both pathways activate a downstream effector, caspase 3, as well as other caspases, eventually causing cells to engage in an apoptotic process characterized by membrane blebbing, cell shrinkage, chromosome condensation, DNA fragmentation and eventually cell death [183].

Apoptosis is a very complex programmed cellular event in which the balance between the expression of anti-apoptotic and pro-apoptotic regulators determines the cellular decision to presage in an apoptotic process. In *in vitro* experiments, this effect could be clearly seen in IECs incubated with *Giardia* (Fig. **2**) and is represented by the activation of caspases 3 and 9, decreased expression of the anti-apoptotic protein Bcl-2, increased expression of the pro-apoptotic protein Bax, and proteolytic cleavage of poly ADP-ribose polymerase (PARP) [183]. The proteolytic cleavage of PARP, specifically, stimulates mitochondria to release apoptosis inducing factor (AIF), leading to the induction of intrinsic apoptotic pathway. The induction of extrinsic apoptotic pathway has been also reported in cells parasitized with *Giardia*. It is unknown, however, which *Giardia* molecules activate this pathway and whether caspase 8 activation occurs as a result of caspase 6 activation, a downstream target of caspase 3 [30]. *Giardia* induces apoptosis also *in vivo* where examinations of biopsies from chronic giardiasis patients showed a slightly higher rate of enterocyte apoptosis [189]. Although the increase in apoptotic rate was found to be relatively small (0.5%), this might represent a significant contribution to intestinal pathology due to the overall effects on transepithelial conductance and permeability [231, 232] (Fig. **2**). This situation is similar to that reported in other intestinal disorders such as Crohn's disease where even low apoptotic rates contribute significantly to disease pathology [232].

Based on the accumulating information over the past decades, it is possible to point out the different factors involved in the induction of apoptosis. An example is that the induction of apoptosis in IECs coincides with the loss of epithelial barrier function induced by the disruption of ZO-1 in tight junctions, an effect that was found to be caspase-3-dependent [182]. In this scenario, caspase 3 was found to be involved in the activation of myosin light chain kinase (MLCK), a known regulator of AJCs and mediator of cytoskeleton reorganization [233], indicating an interrelationship between a dysfunctional intestinal epithelial barrier and apoptosis. Oxidative stress, hypoxia, glucose starvation and arginine depletion (Fig. **2**) are all factors known to induce apoptosis and the transcriptional data from parasitized cells are indicative of such effects [35]. Oxidative stress and hypoxia can trigger both apoptotic cascades, either through ROS-induced mitochondrial DNA damage or by up-regulating Fas expression (also called CD95 or APO-1) [234 - 238]. Glucose uptake and metabolism are associated with the regulation of the intrinsic apoptotic pathway [239], and as discussed before *Giardia* is known to impair glucose uptake through the inhibition of SGLT1 [212]. Transfection of IECs (*i.e.* Caco-2 cell line) with *sglt-1* transgene rendered the cell resistant to apoptosis when incubated with *Giardia t*rophozoites in high glucose medium [240].

While the factors above relate to host cells responses to parasitism, there are also parasite-derived factors proposed to be associated with the induction of apoptosis. Lectin-mediated binding of trophozoites to IECs has been suggested to provide the stimulus required for caspases activation (*e.g.* caspase 8, 9, and 3) [183] similar to the contact-dependent mode of apoptosis induction in cells infected with *Entamoeba histolytica* [241]. Nevertheless, since the induction of apoptosis seems to be isolate-dependent, parasite attachment itself might not provide a sufficient stimulus to induce apoptosis but it can rather work in conjunction with other parasite factors to promote this effect. The other suggested pro-apoptotic factor is *Giardia* CPs and this assumption has been based on the comparative ability of CPs in *Trichomonas vaginalis* and *E. histolytica*, *Giardia*-counterparts, to induce apoptosis in host cells [242, 243]. It has been suggested that *Giardia* CPs might activate proteinase-activated receptor 1 (PAR-1) on cell surface and once activated, PAR-1 is capable of inducing pro-apoptotic signalling [244]. Incubating WB isolate or its ESPs, containing CPs, with IECs have produced conflicting results regarding the induction of apoptosis [182, 183] and thus the available data are inconclusive. Nevertheless, since *Giardia* CPs induces tissue damage in the small intestines, this might indirectly contribute to the apoptotic process [202]. Based on the findings above, the induction of apoptosis in parasitized cells is rather multi-factorial.

Arginine Depletion and its Role in Disease Pathology

Arginine is considered to be an important energy source in *Giardia* and it is metabolized through the arginine dihydrolase pathway to produce energy. Three enzymes function consecutively to convert arginine to citrulline then to ornithine and include in order arginine deiminase (ADI), ornithine carbamoyl transferase (OCT) and carbamate kinase (CK) [245]. *Giardia* takes up external arginine rapidly [246] and has a 10-20 fold more efficient arginine transporter system in comparison to host IECs [247]. *Giardia* also releases ADI and OCT within the first 30 minutes during interaction with IECs *in vitro*, resulting in total arginine depletion in four hours [43, 185]. Arginine depletion in turn has important consequences on host cells. Arginine is the substrate required by enterocytes to produce nitric oxide (NO), a cytostatic compound to *Giardia,* and thus its depletion inhibits NO production [43, 185]. Arginine depletion also arrests the cell cycle *in vitro* through reduced polyamine levels and induces the up-regulation of cell cycle inhibitory genes [181]. It can also induce apoptosis [35, 248, 249]. *In vivo*, this overall effect might disrupt intestinal tissue homeostasis due to reduced cell proliferation and increased rate of enterocyte apoptosis (Fig. **2**). Therefore, arginine-depletion has been suggested as a disease mechanism in giardiasis and arginine metabolizing enzymes as pathogenic factors [181, 185, 250].

To conclude, *Giardia* infection induces pathophysiological effects in the host intestines through cellular attachment and the excretion of certain parasite products (Fig. **2**). This overall, causes structural damages that compromise the integrity of intestinal epithelia as well as the function of brush border enzymes and induce apoptosis in cells (Fig. **2**). These changes affect electrolytes balance and causes maldigestion, leading to the production of diarrhea. The pathological outcomes, however, are dependent on isolates' virulence and thus, they might not be seen in all *Giardia* infections.

IMMUNOLOGY AND IMMUNOPATHOLOGY

Giardia infections elicit both innate and adaptive immune responses in humans and animals [251] but the exact mechanisms controlling infection are not fully understood. Earlier in infection, operating defence mechanisms are B cell-independent and include NO, reactive oxygen species (ROS), defensins, lactoferrins, immune cells such as phagocytes, mast cells and dendritic cells, and cytokines [31, 42]. Later, an antibody-dependent phase comes into play and is responsible for parasite clearance. Hosts are able to develop memory responses to *Giardia* infection, as second infections tend to be asymptomatic [46 - 48].

Innate Immune Responses

Innate immune responses are the first lines of defence against *Giardia*, which naturally work to prevent or eliminate infection before humoral and cellular immune responses start operating. During infection, trophozoites in the intestinal lumen are exposed to duodenal fluids, whose components (*e.g.* biliary lactoferrin and α-defensins) have been found to have giardiacidal properties [252 - 254]. Biliary lactoferrin is a protein produced by the gallbladder and is released into the intestines. Lactoferrin and its derived N-terminal peptide bind to trophozoite surface *in vitro,* and shortly after binding, it induces striking and complex morphologic changes in trophozoite plasma membrane, endomembranes and cytoskeleton, as assessed by immunofluorescence. Lactoferrin binding also increases the electron density of lysosome-like peripheral vacuoles (assessed by electron microscopy) and these overall changes eventuate in cell lysis [252, 253]. *In vivo*, supplementing children diet with lactoferrin was found to lower the prevalence of colonization with *Giardia* spp. and improve their growth [255]. Therefore, lactoferrin or its N-peptide derivative might be an important innate component of host mucosal defences against *Giardia.*

Alpha-defensins are proteins secreted by Paneth cells (specialised epithelial cells) into the intestinal lumen [42]. Humans express two types of defensins, HD5 and HD6, in the intestinal tract and another four in neutrophil granules (HPN1-4) [256]. Cryptdins are mouse α-defensins that are produced in the small intestines and they are activated upon cleavage by matrix metalloproteinase-7 (matrilysin; MMP-7) [257]. Defensins of humans (HPN-2 and HPN-1 at higher dose), mice (cryptdins 2 and 3) and rabbit (NP-2) reduced trophozoites viability *in vitro* by more than 3 log units in 2 h, and their lethality was dose- and time- dependent.

These defensins were also effective against cysts [254]. *In vivo*, infections with *G. muris* in MMP-7 deficient mice, which cannot produce bioactive α-defensins, showed that they were able to control and eradicate *G. muris* similar to the wild type, indicating that defensins do not play an important role in defence against *G. muris* infection [42]. Surprisingly, parasite load was lower in MMP-7 deficient mice compared to wild type but this was suggested to be linked to alterations in microbial flora as a result of inactive defensins, affecting the ability of *G. muris* to colonize mice [42]. Another study has shown that defensins efficiency appears to be associated with the presence of a bacterial endosymbiont engulfed by *Giardia.* It has been proposed that engulfed endosymbionts alter trophozoite antigenicity and enhance trophozoite destruction and lysis [258]. However, this can be challenged by the notion that most *Giardia* isolates do not have any endosymbionts. A recent study analyzing mice transcriptome upon *G. intestinalis* infection has shown that the most abundant transcripts were antibody genes and mast cell proteases with *mmp7* gene being the highest induced [41]. In this study, mmp7-deficient mice showed only a small increase in intestinal parasite numbers and mice deficient in both MMP7 and inducible nitric oxide synthase (iNOS, Nos2) exhibited a significant increase in intestinal parasite load. Although this study has indicated a role for defensins in immune responses against *Giardia*, as inferred by *mmp7* up-regulation, α-defensins and NO have been identified as redundant mechanisms in infection clearance [41].

ROS and NO are other components of the innate immune system, which have cytotoxic effects on *Giardia* trophozoites [31]. ROS (*e.g.* hydrogen peroxide, superoxide and hydroxyl radicals) are naturally produced at low amounts in the intestinal lumen to promote signaling and tissue homeostasis, but at higher levels to fight off invading pathogens [259 - 261]. IECs and immune cells (*e.g.* macrophages and neutrophils) are the source of ROS during giardiasis. Several factors might trigger ROS production by IECs including the change in cells metabolic activity, hypoxia induced from adhering trophozoites covering cell surface and/or sensing the close presence of parasites. Different subsets of leukocytes also migrate to the site of infection, amongst which are the macrophages (Fig. **2**) as evidenced both *in vitro* and *in vivo*, which can ingest and kill the parasite by an oxidative burst [262 - 266].

ROS produced by IECs might be less effective in eliminating the parasite since both, *in vivo* and *in vitro*, trophozoites could persist on cell surface although they have been directly exposed to ROS. Indeed, this is a perplexing part of the parasite's biology because *Giardia* lacks the conventional eukaryotic oxygen/ROS detoxification system (*i.e.* catalase, superoxide dismutase and glutathione peroxidase) [245]. However, part of the explanation to *Giardia's* ability to cope with ROS could be accredited to a body of research that has identified some players in the oxidative defense machinery in *Giardia*. These include an NADH oxidase, NADH peroxidase, flavodiiron protein (FDP), a thioredoxin system (thioredoxins, peroxiredoxins, thioredoxin reductase, and NADPH), and reduced thiols (*e.g.* cysteine, thioglycolate, sulphite and CoA) [245, 267, 268].

NO is an important signaling molecule in the small intestines and is produced by nitric oxide synthases (neuronal, nNOS; inducible, iNOS; and endothelial, eNOS) using L- arginine as a substrate. Inducible NOS is usually the enzyme responsible for NO production in response to pathogens [269 - 272]. However, in murine models of giardiasis, nNOS but not iNOS has been found to be responsible for parasite clearance due to its effect on increasing intestinal motility [219]. Indeed, a body of evidence has shown that NO production, mediated by iNOS, is rather inhibited in IECs due to arginine depletion by the parasite (as discussed previously) [43, 181, 185, 250]. This, however, does not rule out the role of NO in parasite clearance but suggests that NO produced from other sources (*e.g.* neuronal and immune cells) could still contribute in the control of infection.

Mast Cell Responses

Mast cells exert their effects against *Giardia* colonization through degranulation and subsequent release of histamine. The hormone cholecystokinin (CCK) is induced during infection and it triggers mast cells to release their histamine content (*i.e.* degranulation). Released histamine acts on increasing the intestinal motility (Fig. **2**) and muscular contractions, leading to an increase in transit rate [220, 221]. These effects are believed to help *Giardia* clearance by increased peristalsis. Interestingly, due to the effect of *Giardia* infection on mast cells, represented by histamine release, an association between *Giardia* and hypersensitivity or allergy have been proposed. In gerbils, the accumulation of mast cells in small intestines following infection increased the incidence of hypersensitivity reactions [273 - 275] and in humans, infected children developed more allergic responses (*e.g.* skin test with different antigens) and higher levels of serum IgE [273, 274]. Thus, although mast cells help control *Giardia* infection, they might also induce extra-intestinal symptoms such as allergy or hypersensitivity.

Mast cells also play a role in the production of parasite-specific IgA and the proinflammatory cytokine interleukin 6 (IL-6). In mast cells, the *c-kit* gene encodes a protein responsible for the development and activation of mast cells

[276]. Mice deficient in this gene (c-kit$^{w/wv}$) fail to make parasite-specific IgA, mount a mast cell response, or eliminate the infection. Furthermore, treatment of C57BL/6 mice with antibodies against c-kit inhibited mast cell responses and reduced interleukin-6 (IL-6) mRNA levels in the small intestine. It did not, however, affect IgA production but might result in a reduction of the quantity of IgA secreted into the intestines. Anti c-kit-treated mice failed to control the infection within 10 days in comparison to wild type, suggesting that mast cells are required during the early phase of control of *Giardia* infections in mice [276]. Mast cells are an important source of IL-6 [276, 277] and this cytokine has been found to help drive B cell switching to IgA production [278]. Mast cells, however, are not the only source of this cytokine in the intestinal tract as multiple cell types contribute to the IL-6 response *in vivo* [276]. IL-6 itself, although secreted by mast cells, is not required for mast cell responses, since upon *Giardia* infection IL-6-deficient mice showed enhanced mast cell responses and a high serum level of mast cell mediators, histamine and mast cell protease 1, as well as an IgA response. The accumulation of mast cells in IL-6-deficient mice, though, is insufficient to effectively control the infection. Overall, the findings above indicate that mast cells deficiency affects IL-6 production and thus, early and rapid control of infection.

Dendritic Cells

Dendritic cells (DCs) are able to recognize microbes through a family of plasma membrane-spanning receptors called the toll-like receptors (TLRs). TLRs of DCs recognize structurally conserved molecules derived from microbes and activate TLR signaling, leading to DCs maturation, secretion of proinflammatory cytokines and activation of naïve T cells [279]. Microbes/antigens that prime DCs to secrete IL-12 induce a helper T cells response (Th1) and those that inhibit IL-12 production will induce a Th2 response, each of which leads to the production of different cytokines [280, 281]. Although *Giardia* is capable of activating DCs and inducing their maturation, inhibition of cytokines production in both murine and human derived monocyte-derived DCs (moDCs) has been observed [38, 282] but the results from bovine moDC are somehow inconclusive [283]. In one study, murine monocyte-derived DCs incubated with trophozoites or trophozoites extract expressed CD40, CD80 and CD86, and produced small amounts of TNF-α and IL-6. IL-12, but not IL-10, production was inhibited [38]. Co-incubation of DCs with lipopolysaccharides (LPS), an inducer of cytokine production, and *Giardia* resulted in enhanced IL-10 production but reduced levels of IL-12, TNF-α and IL-6. IL-12 production was significantly restored upon DCs treatment with a selective phosphoinositide 3-kinase (PI3K) inhibitor, wortmannin, and thus the inhibition of IL-12 production by *Giardia* appears to be PI3K-dependent [38]. How *Giardia* activates the PI3K pathway, leading to the inhibition of cytokine production, it is unknown. In a recent report, ADI has been identified as one parasite factor responsible for the inhibition of cytokine production (*e.g.* IL-10 and IL-12p40) by human moDCs [44]. Arginine depletion and the accumulation of ammonia have been suggested as the mechanisms underlying this inhibition [44]. The inhibition of cytokine production might also be due to *Giardia* interference with TLR signaling, and this inhibition persists even after DCs activation with different TLR agonists [38], but how the parasite interacts with different TLRs is not understood. Although *Giardia* infection has inhibitory effects on IL-12 and IL-10 production by DCs, two previous reports have shown that mice immunized with *Giardia* ESPs produced significant amounts of IL-4, IL-5 and IL-10 (Th2 response) as well as interferon gamma (Th1 response) [284, 285]. Therefore, *Giardia* ESPs stimulate a mixed Th1 and Th2 response but whether a similar response can be obtained with trophozoites (*i.e.* during infection) or whether other immune cells are involved in inducing helper T cell responses needs to be further investigated.

Although trophozoites or trophozoites extracts have inhibitory effects on cytokine production, purified antigens have the opposite effects on DCs. Bip (binding immunoglobulin protein) or its truncated form that has a heat shock protein 70 (HSP70) domain is an example of purified protein from *Giardia* lysates, which is able to activate DCs leading to the production of TNF-α, IL-6 and IL-12 [286]. The activation is induced by TLR4 signaling, involving the downstream effector MyD88, and recruiting the transcription factors NFκB (nuclear factor kappa B) and AP-1 (activator protein-1) [286]. Overall, these results show that *Giardia* or *Giardia* antigens modulate the host immune response where DCs activation (although weak) leads to IL-6 and TNF-α production, which might help control the infection while the inhibition of cytokine production prevents strong inflammatory signals.

Cytokines and Chemokines

Giardia produces little or no inflammation [32, 37] and the role of cytokines in immune response to human giardiasis is unclear. In experiments with mice, *Giardia* infection was able to attenuate acute intestinal inflammatory responses represented by granulocytes (*i.e.* polymorphonuclear leukocytes or neutrophils) recruitment into colonic tissue, even after inducing granulocyte tissue infiltration by intra-rectal instillation of *Clostridium difficile* toxins A and B [45]. This outcome was associated with a significant decrease in the amount of colonic tissue cytokines associated with PMNs recruitment (*e.g.* CXCL1, CXCL2, and IL-17) and other cytokines (CCL2, IL-6, leukocytes inhibitory factor and IL-12p70). *Giardia* trophozoites also decreased the level of cytokines in culture supernatants, including those associated with PMN recruitment such as IL-8 and CXCL1-3, when incubated with inflamed human tissues obtained from active Crohn's disease patients. IL-8, specifically, was reduced significantly in both culture supernatants and within biopsy tissue homogenates [45]. In fact, this has been seen previously in experiment with similar human tissue biopsies demonstrating the post-transcriptional degradation of CXCL-8 by *Giardia* cathepsin B proteases (*i.e.* CPs) even following cells stimulation with IL-1-β or *Salmonella typhimurium* to produce IL-8 [37]. Interestingly, the effects above were seen with the NF isolate (assemblage A) but not with GS isolate (assemblage B) [45], indicating that the attenuation of granulocytes infiltration and cytokine production is isolate-dependent. In a study that analyzed global gene transcription in the small intestines of cattle upon infection with assemblage E isolates, an inhibition of the induction of cytokine genes was observed [40]. The transcription levels of IL-6, IL-8, IL-13, IL-17, and IFN-γ was down-regulated in the jejunum of infected animals compared to controls and no immune cell recruitment, intestinal pathologies or increased levels of apoptosis could be seen [40]. In a tissue culture model incorporating IECs and murine macrophages, the production of IL-8 and GRO was abolished following incubation with trophozoites that belong to the WB/clone 6 isolate [287]. Also, in IECs (T84, HT29, Caco-2) *Giardia* infection did not induce the transcription of IL-2, IL-4, IL-5, IL-6, IL-12p40, IFN-gamma, IL-1 or IL-10 [33]. Therefore, in many studies *Giardia* infection appears to have inhibitory effects on cytokine production.

In mice infected with the GS isolate, elevated intestinal IFN-γ, TNF-α, and IL-1-β levels were detected on day 35 post-infection concurrent with the observed damage of tight junctions in intestinal mucosa, mucosal influx of bacteria, neutrophil infiltration and myeloperoxidase activity (*i.e.* marker of granulocytes infiltration) [288]. Furthermore, the GS isolate or its ESPs has been found to enhance cytokine production (*e.g.* IL-8, IL-1-β, TNF-α and CXCL-10) in *Clostridium*-toxins induced colitis mice model [45] or even without prior immune stimulation with microbial agents or proinflammatory cytokines [36]. Weaned mice with persistent giardiasis resulting from infection with H3 isolate (assemblage B) also produced elevated levels of IL-4 and IL-5 [289]. In C57BL/6J mice infected with either WB or GS isolate, analysis of cytokine production by spleen and mesenteric lymph node cells showed production of IL-4, IL-10, IL-13, IL-17, IL-22, TNF-α, and IFN-γ post-infection, with IFN-γ being the dominant cytokine for both isolates [213]. In humans, one study showed that patients infected with *G. intestinalis* had significant increase of IFN-γ and IL-13 in their sera [39] and in another study, IL-5 was twice as high compared to the control group, IL-6 was two and a half times higher and IFN-γ was almost four times higher compared to healthy controls [290]. In a study in Turkey, children infected with *Giardia* had elevated levels of TNF-α and soluble IL-2 receptor (sIL-2R) in their sera [291]. Another study found a correlation between fecal concentrations of cytokines and duration of infection [292]. Particularily, the increased levels of monocyte chemoattractant protein-1 (MCP-1), IFN-γ, IL-4, and IL-5 were associated with an increased duration of infection whereas higher amounts of IL-8 demonstrated an opposite effect [292]. Based on the results from all of these studies, it is inconclusive how cytokines are involved in immune response and whether *Giardia* infections tend to develop a specific polarized immune response (*i.e.* either Th-1 or Th-2). Nevertheless, studies in murine models produced clearer data on the involvement of specific cytokines in the immune response during giardiasis. For example, mice lacking tumour necrosis factor-α (TNF-α) or IL-6 showed significant increase in trophozoite numbers in the intestines and delayed parasite clearance and thus both cytokines were shown to be required for early parasite clearance [293, 294].

Even in *in vitro* experiments, the incubation of differentiated Caco-2 cells (*i.e.* polarized IECs) with the WB isolate induced the transcription of genes encoding chemokine ligand 20 (CCL-20), chemokine ligand 2 (CCL-2) and the chemokines CXCL1-3 [35]. CCL20 was secreted by parasitized IECs. The role of these chemokines in giardiasis has not been examined directly in this study, but their transcription and production at protein level suggests the recruitment of immune cells such as T and B cells, DCs, macrophages and neutrophils to the site of infection [35]. Nevertheless, more research is required to identify the induction of chemokines production in human giardiasis and their function in immune response.

B Cell Responses and Antibody Production

Giardia infection results in the development of a humoral immune response in humans [42, 290, 295]. Among the different immunoglobulin populations, IgA is of paramount importance in *Giardia* elimination due to its secretion into the intestinal lumen and its role in mucosal immunity [296]. Nevertheless, IgA is more efficient in eliminating *Giardia* at later stages of infection because of antigenic variation (*i.e.* VSP switching) which occurs every 6-13 generations during parasite proliferation [297]. In fact, IgAs produced in the first 10 days of infection only recognize a small subset of trophozoites *ex vivo* but those produced at later stages (*i.e.* 60 days post-infection) recognize all trophozoites. This indicates that VSP switching during early stage of infection renders the IgAs produced at that time ineffective [297] and those produced after 60 days of infection can either recognize invariant epitopes or all the potential trophozoites variant antigens [297]. Therefore, *Giardia*'s ability to undergo antigenic variation might delay *Giardia* elimination until the majority of antigens including VSPs are recognized by host antibodies.

The development of chronic giardiasis in patients with hypogammaglobulinemia is an evidence to the involvement of humoral immune response in parasite clearance at a later stage of infection [298, 299]. The total absence of antibodies due to B cell deficiency, however, does not compromise mice ability to control infection. In murine giardiasis, μMT mice that lack B-cells due to the deletion of μ exon of Ig heavy chain and those lacking the poly-Ig-receptor required to transport IgM and IgA across the epithelial barrier into the lumen were both able to eliminate infection with kinetics similar to the wild type [300, 301]. This does not imply that antibodies cannot control giardiasis but rather indicate that B cell/antibody-independent mechanisms are responsible for eliminating infection [297].

Cellular Immune Response to *Giardia* Infection

The role of T cells in controlling giardiasis in humans is unclear. In murine models of giardiasis, it has been shown that $CD4^+$ T cell responses, but not $CD8^+$, are required to control infection [251]. However, the total absence, depletion and defects in T cells (*e.g.* athymic nude mice, mice treated with anti $CD4^+$, and mice lacking the β T cell receptor (TCR-β)) drive the disease to the chronic stage. In this scenario, an association between $CD4^+$ T cells role in supporting B cells function (*e.g.* IgA production), and chronicity of infection has been suggested [251, 297, 300, 302]. $CD8^+$ T cells, on the other hand, were found to be involved in disease pathology and this was discussed in a previous section [227]. In humans, T cell deficiencies do not increase individual's susceptibility to *Giardia* infection since infections are able to resolve normally [297]. This does not either increase the susceptibility of HIV positive individuals to *Giardia* infections, and infection rate is as low as 6.2% [303 - 305]. In other immunodeficiencies such as thymic dysplasia (DiGeorge syndrome), patients do not develop severe diarrhea from giardiasis [298]. In both diseases, $CD4^+$ T cell count is extremely low but these findings do not rule out T cell function in clearing *Giardia* infection but indicate that even low $CD4^+$ T cell count could still be sufficient to clear the infection normally [297].

Giardia Antigens: Possible Targets for the Development of Vaccines and Diagnostics

VSPs can be problematic in the development of vaccines and diagnostics due to antigenic variation. As a result, non-variant *Giardia* antigens that are capable of inducing an immune response can be a more suitable option for this purpose. Some *Giardia* antigens have been found to be present in different isolates and could be recognized by the host immune responses. These include surface proteins (82 kDa and 65 kDa polypeptides) [306, 307], giardins [308], tubulins [309], heat shock proteins [310], and lectins [205] but their suitability for vaccine development or diagnostics has not been assessed yet. Sera collected from patients after an outbreak of giardiasis in Sweden were able to recognize some immunodominant antigens, many of which have been found to be immunodominant in other studies and they include OCT, ADI, fructose bisphosphate aldolase, α-enolase, giardins (β, α 1, α -2, α -7.1 and α -11), tubulins (β and α-2), *Giardia* trophozoite antigen-2 (GTA-2) and uridine phosphorylase-like protein 1 (UPL-1) [57, 301, 311].

So far the efficacy of some vaccines have been tested in mice and Mongolian gerbils. An α-1 giardin-based heterologous vaccine was developed and showed protection against *G. duodenalis* in a murine model [58]. Also, a vaccine against CWP2 was tested with results showing fewer cysts shedding by BALB/c mice. This vaccine has been suggested for use in the development of a transmission-blocking vaccine [59]. An attempt also aimed at developing a vaccine against the VSPs covering parasite surface. In Mongolian gerbils, infection with transgenic trophozoite lines simultaneously expressing many VSPs (as a result of dicer knockout) or immunization with purified VSPs from transgenic cells protected gerbils from subsequent *Giardia* infections [60]. Nevertheless, the efficiency of these vaccines requires evaluation in humans and could help control infection in endemic areas.

SYMPTOMS

Clinical manifestations of giardiasis are variable ranging from asymptomatic carriage to acute or chronic disease. The incubation period is around 1-2 weeks during which symptoms arise 7-10 days post infection (on average, range 3-25 days). Prepatent period, which is the time between infection and detection of cysts in stool, is around 9.1 days [152] and symptoms can start without cysts being detectable in stool. In asymptomatic infections, hosts still pass cysts with feces and excreted cysts are infectious to susceptible hosts [312]. When giardiasis manifests in an acute phase, the most prominent feature is diarrhea and this is usually associated with distention, flatulence, cramps, nausea and vomiting [2]. Diarrhea might be frequent and watery with the passage of bulky, greasy and offensive stool or protracted and mild with semi solid stool [313]. Usually, infected individuals clear the infection within 3-6 weeks but when it persists longer, the disease might become chronic. It is not clear, though, what timeframe defines the onset of chronic phase in giardiasis. In one study, the authors used a six months period as a criterion for the disease to presage to a chronic phase [314] but in another study, the authors suggested that the continuation of symptoms or parasite presence in host stool beyond two months could define a chronic stage during which the parasite have had established a permanent colonization in host gut [169]. At this stage, infected hosts continue to have diarrhea, sometimes interrupted by episodes of constipation, and stools are pale or yellow, frequent and have small volume [313]. Individuals with chronic giardiasis suffer from malaise, weight loss and malabsorption [34, 223]. Malabsorption in particular arises from disaccharidase enzymes deficiency, mainly lactase, with effects extending to involve the absorption of micronutrients such as vitamins A and B12, folate, zinc and D-xylose [3, 313, 315]. Steatorrhea can also occur as a result of malabsorption and maldigestion [30]. Some reports have shown that inflammation might or might not be present in the host intestines in chronic giardiasis and sometimes the severity of diarrhea correlated with mucosal changes [30, 32, 192, 315, 316]. Recent research has reported the emergence of some disorders after *Giardia* infection, which includes IBS, functional dysplasia, chronic fatigue, arthritis, ocular pathologies, allergies and muscular complications [12, 315, 317]. The etiology of these disorders remains obscure. Nevertheless, in children with giardiasis it can be well reasoned that malabsorption and malnutrition could be the leading cause for stunting, failure to thrive and poor cognitive functions [317].

The factors underlying the variability in clinical manifestations of giardiasis are poorly understood. In many studies both assemblages A and B were capable of producing symptomatic disease and differing lengths of disease have been reported in infections with the same assemblage [34]. In an early study in the Netherlands, Homan and Mank [318] found a strong correlation between a mild intermittent diarrhea with assemblage A and severe persistent diarrhea accompanied with duodenal inflammation with assemblage B. In a study conducted in Southwest London, assemblages A and B were found to cause similar illness but assemblage A was significantly more associated with fever [319]. In other studies conducted in Bangladesh, Spain, Australia, Peru and Turkey, symptomatic giardiasis was significantly associated with assemblage A [320 - 323]. On the other hand, in Cuba, Ethiopia, Malaysia, and India assemblage B was significantly associated with clinical symptoms [324 - 327]. In Argentina, assemblage B was found to be strongly associated with abdominal pain but neither assemblage was significantly associated with diarrhea [328]. In Saudi Arabia assemblage B was found to produce symptomatic infection whereas assemblage A infections were asymptomatic [329]. Therefore, studies of the possible association of *G. intestinalis* assemblages with diarrhea and other clinical symptoms have produced inconsistent results and overall are inconclusive. The infective dose and dual infection with different genotypes/assemblages have been suggested as other factors involved in the variability of symptoms during giardiasis or causing a synergistic increase in the disease pathology (as in the case of co-infections) [7].

Host factors might also contribute to the variability in symptoms during giardiasis. Individuals previously infected with *Giardia* might not develop a symptomatic disease unless infected with a new genotype [46]. Age, nutrition and immune status of the host might also play a role in the outcome of disease as evidenced in some studies where symptomatic disease is confined mainly to young children or elderly people [322, 330]. The increased susceptibility of these two groups is believed to be due to the immaturity of immune system or malnutrition in case of children and immune incompetence in case of the elderly [331]. Recent studies have also suggested that host microbiota might have stimulatory effects on the pathogenicity of *G. intestinalis* [332, 333], hence the outcome of disease. Based on the discussion above, it can be concluded that both parasite and host factors are involved in the outcome of disease upon *Giardia* infection.

DIAGNOSIS AND DETECTION METHODS

Microscopy is the gold standard practiced in diagnostic parasitology laboratories incorporating both morphology and morphometry as definitive criteria for the diagnosis of human giardiasis. In acute phase of giardiasis, stool is usually loose and watery and may contain large amounts of trophozoites. A wet mount performed on a hot stool sample can detect motile trophozoites, as they might be present in large numbers, also cysts [49]. Wet mounts are prepared as fecal suspensions (in warm 150 mM NaCl or saline) that are smeared on a microscope slide for examination. In case of formed stool, cysts are more abundant than trophozoites, but due to their excretion in low numbers a concentration step generally using formalin-ether or formalin ethyl-acetate is performed in clinical diagnostic settings to enhance cysts recovery [49]. Trophozoites, however, cannot withstand concentration solutions due to their fragile nature. Delays in sample processing, due to large number of samples submitted for analysis or transport, necessitate the use of fixative to preserve trophozoite and cyst morphology. Some of the preservatives used include sodium acetate-acetic acid-formalin (SAF), formalin and buffered formalin but formalin and buffered formalin might reduce PCR sensitivity if required for further analysis [49]. At low magnification (10x) internal details of trophozoites and cysts are not readily visible by bright field microscopy. Phase contrast and Nomarski differential interference contract at higher magnification (20x and 40x) might provide a sharper image of trophozoite and cyst perimeter. Staining, whether temporary or permanent, can enhance viewing internal structures and help keep a permanent record of positive samples [49]. Some of the stains used include Lugol's iodine, iron-haematoxylin, Giemsa, trichrome and Chlorazol black dye [334]. The absence of cysts in stool samples is not definitive to exclude infection simply because of the intermittent nature of cysts shedding, their excretion in low numbers and the possible appearance of symptoms before cysts or trophozoites are present in stool [49]. This in turn warrants taking multiple samples (at least three) on a weekly basis (1-2 weeks) to avoid reporting false negative results [335]. Sometimes when symptoms persist in the absence of laboratory-confirmed giardiasis, more invasive methods such as duodenal or jejunal fluids aspirate, Enterotest (gelatine capsule with a string swallowed but remains attached to patients' mouth with a tape) or tissue biopsy are required [49].

Some immunological methods have been implemented in the diagnosis of human giardiasis as well as in environmental samples. For example, fluorescent microscopy and FITC-conjugated monoclonal anti-*Giardia* antibody can be used directly or indirectly to detect surface epitopes on cysts in fecal or environmental samples with increased sensitivity (93%- 100%) and specificity (99.8%-100%) [334]. A prior concentration step for cysts, however, is recommended, making immunofluorescence more laborious as a diagnostic method for human infections [49]. The same holds true for environmental and food samples, where immunofluorescence microscopy is preceded with filtering large volumes of water or excessive washing of food samples with buffers, followed by IMS, antibody binding (direct or indirect) and immunofluorescence microscopy [50 - 53]. Nevertheless, this method compared to bright field microscopy can enhance the detection of distorted cysts or those with poor morphology that might be missed during microscopic examinations [49]. The detection of *Giardia* antigens in stool is another option for diagnosis of human giardiasis, where screening large number of samples in a short time is possible. This is very practical in case of an outbreak, where immediate actions for treatment and disease control are required. The detection of *Giardia* antigens implies an enzyme linked immunosorbent assay (ELISA) where antibodies against *Giardia*-specific antigen, for example GSA (a

65 kD protein), are immobilized on a solid phase (wells in a plate) to which a stool sample is added followed by the development of coloured reaction product [49, 334]. The use of ELISA is very helpful as it can detect the presence of the parasite when microscopy has failed to do so. Indeed, it has been reported that ELISA based on GSA has greater sensitivity than microscopy (per single specimen), with a sensitivity range between 95%-100% and 100% specificity. Some laboratories use lateral flow immunochromatographic assays (*i.e.* Dipstick) to detect *Giardia* antigens in stool [49]. This is a rapid and practical method through which results are made available within 15 minutes [49]. Despite the ease of use and the high sensitivity of antigen detection methods mentioned above, the general opinion is that they cannot replace or substitute routine microscopic examination.

PCR-based methods are more sensitive than microscopic and immunological tests to detect *G. intestinalis* in human stool. However, because of the refined and discriminatory ability of PCR to identify *G. intestinalis* at assemblage and sub-assemblage levels, they are restricted to specialist laboratories where epidemiological and transmission studies are performed. The technique relies on the specific amplification of one or many genes loci from small amounts of DNA extracted from cysts [49, 334]. Following PCR, restriction fragment length polymorphism and/or sequencing are used to identify *G. intestinalis* at sub-assemblage level. Some of the most common genes used in genotyping include *ssrRNA*, *bg*, *gdh*, *tpi*, and *ef-1α* [49, 71, 85 - 91, 334]. Central to this process is the techniques used for DNA extraction from biological and environmental samples, some of which selectively remove constituents that are inhibitory to PCR amplification (*e.g.* humic acid). Some of these DNA extraction methods adopted by research laboratories are sonication, freeze/thaw cycles, glass beads, phenol-chloroform extraction and commercially available kits [49, 334]. The use of real time PCR and multiplex PCR is gaining a wide acceptance by many research laboratories, since these techniques can reduce labour, cross contamination, and can amplify multiple targets in the same reaction [49]. Indeed, a recent study has shown that multiplex real time PCR can be also used as a diagnostic test in clinical laboratories to simultaneously detect *G. intestinalis*, *E. histolytica* and *Cryptosporidium hominis/parvum* in patients suffering from gastrointestinal disorders [336 - 338].

TREATMENT

Nitroheterocyclic drugs are commonly used to treat human giardiasis. These include the 5-nitroimidazoles (5-NIs) family, metronidazole (MTZ), ornidazole, tinidazole (TNZ) and secnidazole and the non-imidazole compounds, nitazoxanide and furazolidine [245, 339]. Nitroheterocyclic compounds have a nitro group extending from a heterocyclic backbone and utilize the antioxidant machinery of anaerobic/microaerophilic parasites, including oxireductases (pyruvate:ferredoxin oxireductase, PFOR), nitroreductases, thioredoxin reductases and NADH oxidases, for drug activation [340]. Nitroheterocyclic compounds are 'redox-active' with low redox potential, and thus can be reduced within the parasitic cell into toxic intermediates. The mode of action of many nitroheterocyclic compounds is not completely elucidated but for reduced MTZ, it involves the induction of DNA damage upon DNA binding, inhibition of respiration and oxidation of biomolecules [245, 341].

Among 5-NIs, MTZ and TNZ are the most commonly used drugs to treat human giardiasis worldwide. MTZ, known as Flagyl, is also used to treated *E. histolytica* and *T. vaginalis* infections [341]. MTZ is given in three oral doses (250 mg) for 5 to 10 days with treatment efficiency ranging between 60-95% [339]. This treatment schedule could usually be tolerated well by patients although they might experience some side effects such as headache, vertigo, nausea, and metallic taste in the mouth. MTZ can be also administered in high doses (2-2.4 g/dose) for shorter periods (3 days) to improve compliance without sacrificing efficacy (93%-100%) [341]. High doses of MTZ, however, might be accompanied with bad side effects and thus are less desirable in treatment regimens. TNZ (Fasigyn) is becoming the drug of choice in the US to treat human giardiasis due to its longer half-life and high efficacy when a single dose is compared to MTZ [339, 341]. Furthermore, side effects reported for MTZ are less common with TNZ. A single dose of 2 g of TNZ has proven a clinical efficacy of 80 to 100% and as a result compliance is improved with this treatment schedule [341]. The same has been reported for ornidazole (efficacy 92-100%) and secnidazole (> 85% efficacy) as single-dose therapies [342]. Side effects have been reported for secnidazole but this does not usually require discontinuation of drug treatment [341].

Furazolidone (Furoxone) is a nitrofuran drug used in the treatment of many parasitic and bacterial infections. Its mode of action is not completely elucidated but might involve inducing oxidative damage of many cellular components including biomolecules in *Giardia* [341]. Furazolidone is used worldwide to treat giardiasis and because of its availability in liquid form it is widely used in the treatment of pediatric giardiasis [341]. Four doses (100 mg) of furazolidone are given to adults on daily bases for 10 days but achieving high curative efficacy (80-96%) requires compliance with the long treatment schedule (*i.e.* 10 days) [341]. Side effects also arise with furazolidine treatment and include gastrointestinal upsets, brown discoloration of urine and hemolysis in case of patients with glucose-6-phosphate dehydrogenase (G6PD) deficiency. Nitazoxanide is another broad-spectrum anti-parasitic/anti-bacterial drug and its mode of action involves the inactivation of giardial PFOR similar to that of MTZ [339, 341]. Nitazoxanide has been shown to reduce the duration of symptoms in children giardiasis and in symptomatic individuals with unconfirmed giardiasis or gastroenteritis [339]. This drug is given for 3 days and side effects are similar to MTZ [341].

Quinacrine, the drug originally used to treat malaria infection, has been successfully used to treat human giardiasis with an efficacy exceeding 90% [341]. The mode of action of is not fully elucidated but it is believed to inhibit nucleic acid synthesis. A selective advantage of quinacrine is that it can reduce cysts viability and excystation rate whereas this cannot be achieved with MTZ and TNZ due to poor diffusion rates in cysts [341]. In one report, quinacrine was successfully used to treat a family that failed to clear infection with TNZ [343]. Although using quinacrine have achieved efficacies exceeding 95% upon 5-10 days treatment of 100 mg dose (three times daily), the side effects especially in children makes this drug less favourable in giardiasis treatment [341]. Some of the side effects include gastrointestinal upsets, discoloration of skin and hemolysis in individuals with glucose-6-phosphate dehydrogenase deficiency.

Benzimidazoles such as albendazole (Albenza) and mebendazole (Vermox) are anti-helminthic drugs that can be also used to treat giardiasis. The drugs can bind *Giardia* β-tubulin in cytoskeleton, inhibiting polymerization and glucose uptake [341]. Albendazole efficacy in treating giardiasis is similar to that of MTZ when given for 5-7 days (400mg daily dose) and it can be used in combination with MTZ with 100% efficacy in the treatment of drug resistant strains [339, 341]. Mebendazole efficacy, on the other hand, is variable (60-90%) and decreases against *Giardia* during mass treatment for helminths [341]. Therefore, albendazole is more favourable in treating giardiasis and co-infecting helminths with an added advantage of being less prone to giving side effects. Long-term treatment as in the case of cestode larval infections, however, might cause reversible neutropenia and elevated hepatic enzymes level [341].

Paromomycin (Humatin) is an aminoglycoside used to treat infections with *E. histolytica* and *Trichomonas*. In *Giardia*, the drug is capable of inhibiting protein synthesis by interfering with 50S and 30S ribosomal RNA subunits, causing misreading of mRNA codons [341]. The drug is poorly absorbed in the intestines, and thus despite its low anti-protozoal activity compared to MTZ, quinacrine, or furazolidine, the presence of high concentrations therein compensates for its lower anti-protozoal activity [339, 341]. As a result, paromomycin is the drug prescribed to treat giardiasis during pregnancy. The drug has a 55-90% efficacy and is administered in 500 mg concentration three times daily for 10 days. Caution should be practiced when administering the drug to patients with kidney problems as it might be absorbed systemically and cause ototoxicity and nephrotoxicity [341]. The drug is less effective in adults with MTZ-refractory giardiasis and is uncommonly used in clinical practice [339].

Finally, treatment regimens for refractory giardiasis might involve administering drugs for longer periods, using higher doses, switching to a different class of drug or combination therapy. Usually, using different families of drugs (*e.g.* 5-NIs with paromomycin and/or benzimidazoles) for 7-10 days showed high efficacy in treating refractory giardiasis with no serious side effects [339, 341, 343]. Ongoing efforts are also directed towards developing new drugs to treat giardiasis, mainly to overcome problems associated with drug resistance and ensure better compliance with treatment regimens.

PERSPECTIVES OF CONTROL

Water serves as the primary vehicle for the spread of human giardiasis. In the 1970s and 1980s, a peak of giardiasis-water related outbreaks (more than 30) has been reported, most of which took place in the US [148]. Outbreaks, during these periods and in the last decade, were also common in other developed countries such as Canada, UK, Finland, Sweden, Germany, New Zealand and Norway [148]. Although, one expects that in these countries detection and monitoring systems of water are in place, inadequate and poorly operated treatment and disinfection systems were responsible for more than 75% of these outbreaks. Around 12% of the total outbreaks were also related to deficiencies in distribution systems [148]. Nevertheless, a marked decrease in the number of outbreaks (11 in total) has been reported since 2000. This was ascribed, particularly in the US and UK, to the implementation of regulations that involve monitoring for protozoan parasites such as *Cryptosporidium* in the UK and *Giardia* (both in US and UK) in raw water resources and as a result the appropriate treatments to the water source were applied [148, 344].

The establishment of surveillance systems in the US over the past decade has provided useful information for the implementation of effective control measures and identifying treatment deficiencies. Monitoring for *Giardia* and *Cryptosporidium* was first performed in the US between 1997-1998 (18 months period) under the Information Collection Rule (ICR) and was followed by supplemental surveys (ICRSS, US 1623) [50, 148]. According to ICR, if raw water (surface and ground water) supplying 10 000 people contained more than 10 cysts/oocysts per liter during any month in a 12 months period, then treated water should be monitored too. The ICR rule was followed by US EPA method 1623 to provide supplementary data for large and medium-sized water systems. The US EPA 1623 method, involved analyzing two water samples per month at a participating site in a total of 12 months period (1999-2000) [50]. The results from assessment and monitoring studies formed the basis for later regulations such as the Final Long term 2 Enhanced Surface Water Treatment Rule (LT2) [345]. In LT2, testing of the source water (24 months) is required to identify treatment requirement to achieve at least a 4-log removal for viruses, 3-log removal for *Giardia* and 2-log removal of *Cryptosporidium*. The rule was further expanded to stipulate that systems that store their treated water in open reservoirs must either cover the reservoir or treat the reservoir discharge to achieve the previous log reduction criteria. Systems must also review their current level of microbial treatment before they incorporate any significant change in their treatment practice [345]. Owing to better monitoring and enhanced regulations, only two water outbreaks (one *Cryptosporidium*, one *Giardia*) were recorded by the Center of Disease Control (CDC) in 2005 and these were due to a point of use contamination at a gymnasium [148].

A preventive risk-based approach has been developed to revise the WHO Guidelines of Drinking Water Quality [148]. This aimed to provide an expansion of the current approach to help protect consumers against the adverse health effects resulting from consuming contaminated water [346]. Although this assessment attempted at identifying risk factors associated with the presence of *Cryptosporidium* in surface water, the necessity for monitoring catchment and surface waters, especially during peak events, was re-emphasized therein despite methodological difficulties [346]. This is because relying on regular sampling schemes implemented by monitoring programs might miss the presence of important pathogens in water (surface and ground water) especially during peak events represented by heavy rainfall, melting of snow and increased farming activities. In this case, a site-specific pre-sampling survey incorporating relatively simple indicators such as turbidity or fecal indicator bacteria could underlie the basis for more comprehensive and complex monitoring schemes [346]. As a result, an optimum sampling scheme can be adopted to better assess and mitigate the risks of contamination in a specific site. Furthermore, knowledge of the pathogens in a water catchment can help identify the contamination source and thus implementing certain measures to prevent contamination by human or animal waste [346]. Although the study above was mainly focusing on *Cryptosporidium*, the fact that water contamination of both *Giardia* and *Cryptosporidium* is common and both cysts/oocysts (*i.e.* (oo)cysts) are robust and resistant to disinfection imply that these recommendations are largely relevant to *Giardia*.

Since 1999, the UK set a standard of regulatory monitoring of treated drinking water for *Cryptosporidium* under the Water Quality Regulation SI No.1524 [347]. The standard implied continuous monitoring for oocysts at certain water treatment works, so the number of oocysts does not exceed 1-oocyst/10 liters. In 2007, water undertakers were obliged to carry out risk assessments on all their water supply sites to avoid any risks associated with *Cryptosporidium* presence in final treated water [347]. When certain sites were identified as high risk, additional treatments in the form of properly controlled coagulation/flocculation filtration systems, membrane or UV treatment systems were installed. The UK regulations also request water companies to design and operate adequate treatment and disinfection continuously. A proven failure to comply with these requirements is now an offence [347] and the same standard is also practiced in Northern Ireland. Although this legislation does not control for the presence of *Giardia* in treated water, it can be well justified that meeting the requirements for *Cryptosporidium* elimination from drinking water should also provide effective risk control of giardiasis transmission.

In countries like Canada, Australia and New Zealand a routine monitoring of treated water for enteric protozoa is not practiced [348 - 350]. Usually, the costs associated with such monitoring is high and in Canada it is reasoned that the current methods for the detection of (oo)cysts have low recovery rates and they do not provide information on (oo)cysts viability. Also, because of the lag time between sample collection and test result, individuals' health is already at risk by the time a warning signal is received by water authorities as in the case of an outbreak [348]. The current Canadian Drinking Water Quality Guidelines, on the other hand, encourage the adoption of a multi-barrier approach to produce clean, safe and reliable drinking water [348]. In this approach, a minimum 3-log reduction and/or inactivation of *Cryptosporidium* and *Giardia* have been established as a health-based treatment goal with the implementation of indicator microorganisms, turbidity and disinfectant residuals as parameters for the multi-barrier approach to determine the quality of the treated drinking water [348]. Therefore, the application of multi-barrier approach aims at assessing the entire drinking water system, from water source, treatment, distribution and at point of use to identify risks and apply appropriate measures to prevent illness. It also uses site-specific information of water quality to add further treatment or optimize existing barriers to enhance the microbiological drinking water quality. The multi-barrier approach is also adopted in countries like Australia and New Zealand [349, 350].

Protozoan waterborne outbreaks could still occur even when detection and monitoring systems are in place as witnessed in many developed countries. This might be due to an incidental contamination of treatment plant process as a result of filter breakthrough, improper coagulation, use of recycled, concentrated backwash water, process overload, or improper maintenance [351]. Distribution system contamination can also take place as a result of cross-connection, broken or leaking waterlines, or back-siphonage [351]. If a high number of cases of giardiasis during a certain time in a certain area are reported, this defines an outbreak situation. In this case, health care professionals should notify water authorities upon which an investigation is initiated to identify the source of the outbreak and prevent further exposure to the source. In many countries, in case of a water outbreak and until contamination source is determined, public health authorities will issue a 'boil water advisory' to help control the spread of illness or advise using household bleach under certain instructions to disinfect water (see Table **2**) [348 - 352]. Once the contamination source is determined, it is either that another water source is sought, or water treatment processes are fixed or modified to include another treatment barrier and disinfection steps. *Giardia* cysts can be readily removed by conventional water treatment processes such as coagulation, sedimentation, rapid filtration and slow sand filtration [353]. Following filtration, the incorporation of some disinfection methods such as chlorination, ozonation and UV light irradiation can inactivate/destruct both *Giardia* and *Cryptosporidium* (oo)cysts [353]. Chlorine is a cheap disinfection method that is usually effective against *Giardia* but at high doses (5-15 mg/L) and longer contact time (30 min-2h) than what is usually used in drinking water treatment [354]. Chlorine dioxide has been demonstrated to be reasonably effective in achieving 90% deactivation with practicable dosing regimes [355]. Ozonation has been shown to destruct *Giardia* cysts [356] while irradiation with ultraviolet (UV) light is capable of inactivating cysts [354, 357, 358]. Sometimes the incorporation of two disinfection methods (chlorination and UV irradiation or UV irradiation and ozonation) may potentially be credited towards meeting the overall disinfection efficiency and inactivation of both *Giardia* and *Cryptosporidium* (oo)cysts [353].

Recreational waters (swimming pools, hot tubs, water parks, water play areas, interactive fountains, lakes, rivers, or oceans) have been recognized as an important source for the spread of human diseases including giardiasis and cryptosporidiosis [148]. *Giardia* has been identified as an etiologic agent for at least 18 outbreaks associated with recreational waters in countries like the US,

Table 2. Guidelines for the control and prevention of human giardiasis infections [348, 352].

Measures for Control and Prevention (CDC Guidelines)

Contaminated drinking and recreational water

- Public water supplies should undergo flocculation, sedimentation and filtration. Filtration is really the only method that guarantees cyst removal.
- Watershed management to control both sewage inputs and the populations of aquatic mammals in the vicinity of water intakes and adherence to drinking water regulations and well placement.
- Drinking water

  If the safety of drinking water is in doubt (for example, during or after an outbreak, in a place with poor sanitation or lack of water treatment systems), do one of the following:

  - Drink bottled water.
  - Disinfect tap water by heating it to a rolling boil for 1 minute.
  - If boiling is not possible, 2 to 4 drops of household bleach or 0.5 ml of 2% tincture of iodine can be added to each liter of water and hold water for 20 minutes before drinking (longer if water is cold or turbid)
  - Use a filter that has been tested and rated by National Safety Foundation (NSF) Standard 53 or NSF Standard 58 for *Giardia* cyst and *Cryptosporidium* oocyst reduction (< 1.0 μM pore size).
  - Do not drink poorly treated water or ice made from water during community outbreaks caused by contaminated drinking water.
  - Do not use or drink poorly treated water or use ice when traveling in countries where the water supply might be unsafe.

- Recreational water
  - Do not swallow water while swimming in pools, hot tubs, interactive fountains, lakes, rivers, springs, ponds, streams or the ocean.
  - Do not drink untreated water from lakes, rivers, springs, ponds, streams, or shallow wells.
  - Proper management of fecal accidents in recreational water facilities.

Practicing good hygiene

- Wash hands with soap and clean, running water for at least 20 seconds; rub your hands together to make a lather and be sure to scrub the backs of your hands, between your fingers, and under your nails.

  - Before, during, and after preparing food
  - Before eating food
  - Before and after caring for someone who is sick
  - After using the toilet
  - After changing diapers or cleaning up a child who has used the toilet
  - After touching an animal or animal waste
  - After handling pet food or pet treats

- Help young children and other people you are caring for with hand washing as needed.
- Exclusion of infected individuals handling food until disease is treated and symptoms disappeared.
- Infected persons should avoid using recreational water venues (*e.g.*, swimming pools, wading pools, whirlpools) until two weeks after resolution of symptoms.
- Education on preventive measures should be provided to cases and their contacts.

(Table 2) contd.....

Measures for Control and Prevention (CDC Guidelines)
At childcare facilities • To reduce the risk of spreading the disease, children with diarrhea should be removed from childcare settings until the diarrhea has stopped. • Childcare facility workers who work directly with children should be excluded from child care settings until diarrhea has resolved. • Early recognition and treatment of cases in childcare facilities to prevent spread to the community. • At recreational water venues (*e.g.* pools, beaches, fountains) ▪ Protect others by not swimming if you have diarrhea (this is most important for children in diapers). ▪ Shower before entering the water. ▪ Wash children thoroughly (especially their bottoms) with soap and water after they use the bathroom or after their diapers are changed and before they enter the water. ▪ Take children on frequent bathroom breaks and check their diapers often. ▪ Change diapers in the bathroom, not by the water.
Around animals • Minimize contact with the feces (poop) of all animals, especially young animals. ▪ When cleaning up animal feces (poop), wear disposable gloves and always wash hands when finished. ▪ Wash hands after any contact with animals or their living areas.
Eating food that may be contaminated ▪ Use safe, uncontaminated water to wash all food that is to be eaten raw. ▪ After washing vegetables and fruit in safe, uncontaminated water, peel them if you plan to eat them raw. ▪ Avoid eating raw or uncooked foods when traveling in countries with poor food and water treatment.
Contact and contamination with feces (poop) during sex. ▪ Use a barrier during oral-anal sex. ▪ Wash hands right after handling a condom used during anal sex and after touching the anus or rectal area.

UK and Canada [148]. The largest recreational water outbreak happened in a water park in Georgia, US involving an estimated 5449 cases after a fecal accident in a children's pool [359]. While this can be the primary reason for the spread of giardiasis through swimming pools, poor construction or maintenance of plumbing, poor filtration systems and insufficient use of disinfectants can be added factors contributing to the spread of disease [360]. In 2008, the CDC analyzed data from 121,020 routine pool inspections conducted by a convenience sample of 15 state and local agencies. This had resulted in an immediate closure of 12.1% of the treated recreational water venues due to violations of pool codes (*e.g.* disinfectant level, pH values, water chemistry, filtration and circulation, improperly maintained pool log, unapproved water test kit used and operator training documentation) [361]. Thereafter, a Model Aquatic Health Code (MAHC) (released online in 2014) has been developed and includes national standards for pool design, construction, operation, and maintenance, and guidance for inspections, which are based on scientific evidence or best practices to reduce the risk for illness from recreational water and injury at public venues [362]. Since the health of many individuals practicing water activities is already at risk by the time an illness is reported and an outbreak situation is identified, the CDC has also published some facts and guidelines for swimmers so they can keep themselves and their family healthy every time they swim (Table **2**) [352]. Contamination of recreational natural bodies of water could occur as result of urban and non-urban runoff, industrial pollution, storm waters, and human or animal wastes. Holidaymakers, bathers and campers are usually advised to minimize swallowing water while practicing water activities (See Table **2**) [352].

Implementing good prevention and control measures for giardiasis requires a full knowledge of the factors that aid in parasite transmission. As a result, and based on a large body of research, many health authorities worldwide have complied a list of practices to mitigate risks associated with acquiring infection through water or food as well as through contact with infected humans and animals. These practices are summarized in Table **2** and are derived from CDC guidelines, Health Canada and published research studies [348, 352]. Overall, the summarized practices touch upon and emphasize the need to practice a good personal hygiene and require individual's awareness of disease routes of transmission in order to protect themselves and their families.

CONCLUSIONS

Gastrointestinal parasites currently infect over 1 billion people and most of our farm animals. *Giardia intestinalis* is an intestinal protozoan parasite that causes the diarrheal disease giardiasis. Giardiasis has a global distribution with 280 million cases reported annually, with its impact being more pronounced in the developing world. Water serves as the primary vehicle for the spread of human giardiasis. Giardiasis is characterized by watery diarrhea, epigastric pain, nausea, vomiting and weight loss. The cause of disease is most likely multi-factorial. Chronic infections are common but about half of the infections during epidemics are asymptomatic and infection frequently resolve spontaneously. Minimizing fecal contamination and transmission in water and food can control infections. More research is needed in order to understand how *Giardia* affects other intestinal infections and which parts of the immune response are important for parasite clearance. Several drugs are available for the treatment of giardiasis but resistance is emerging, making it important to identify new drugs.

CONFLICT OF INTEREST

The authors confirm that they have no conflict of interest to declare for this publication.

ACKNOWLEDGEMENTS

Declared none.

REFERENCES

[1] Thompson RC. Giardiasis as a re-emerging infectious disease and its zoonotic potential. Int J Parasitol 2000; 30(12-13): 1259-67.
[http://dx.doi.org/10.1016/S0020-7519(00)00127-2] [PMID: 11113253]

[2] Wolfe MS. Giardiasis. N Engl J Med 1978; 298(6): 319-21.
[http://dx.doi.org/10.1056/NEJM197802092980606] [PMID: 622089]

[3] Farthing MJ. Giardiasis. Gastroenterol Clin North Am 1996; 25(3): 493-515.
[http://dx.doi.org/10.1016/S0889-8553(05)70260-0] [PMID: 8863037]

[4] Lane S, Lloyd D. Current trends in research into the waterborne parasite *Giardia*. Crit Rev Microbiol 2002; 28(2): 123-47.
[http://dx.doi.org/10.1080/1040-840291046713] [PMID: 12109771]

[5] Dib HH, Lu SQ, Wen SF. Prevalence of *Giardia lamblia* with or without diarrhea in South East, South East Asia and the Far East. Parasitol Res 2008; 103(2): 239-51.
[http://dx.doi.org/10.1007/s00436-008-0968-6] [PMID: 18425689]

[6] Stark D, Barratt JL, van Hal S, Marriott D, Harkness J, Ellis JT. Clinical significance of enteric protozoa in the immunosuppressed human population. Clin Microbiol Rev 2009; 22(4): 634-50.
[http://dx.doi.org/10.1128/CMR.00017-09] [PMID: 19822892]

[7] Feng Y, Xiao L. Zoonotic potential and molecular epidemiology of *Giardia* species and giardiasis. Clin Microbiol Rev 2011; 24(1): 110-40.
[http://dx.doi.org/10.1128/CMR.00033-10] [PMID: 21233509]

[8] Nematian J, Gholamrezanezhad A, Nematian E. Giardiasis and other intestinal parasitic infections in relation to anthropometric indicators of malnutrition: a large, population-based survey of schoolchildren in Tehran. Ann Trop Med Parasitol 2008; 102(3): 209-14.
[http://dx.doi.org/10.1179/136485908X267876] [PMID: 18348775]

[9] Simsek Z, Zeyrek FY, Kurcer MA. Effect of *Giardia* infection on growth and psychomotor development of children aged 05 years. J Trop Pediatr 2004; 50(2): 90-3.
[http://dx.doi.org/10.1093/tropej/50.2.90] [PMID: 15088797]

[10] Prado MS, Cairncross S, Strina A, Barreto ML, Oliveira-Assis AM, Rego S. Asymptomatic giardiasis and growth in young children; a longitudinal study in Salvador, Brazil. Parasitology 2005; 131(Pt 1): 51-6.
[http://dx.doi.org/10.1017/S0031182005007353] [PMID: 16038396]

[11] Hanevik K, Dizdar V, Langeland N, Hausken T. Development of functional gastrointestinal disorders

after *Giardia lamblia* infection. BMC Gastroenterol 2009; 9: 27.
[http://dx.doi.org/10.1186/1471-230X-9-27] [PMID: 19383162]

[12] Morken MH, Lind RA, Valeur J, Wilhelmsen I, Berstad A. Subjective health complaints and quality of life in patients with irritable bowel syndrome following *Giardia lamblia* infection: a case control study. Scand J Gastroenterol 2009; 44(3): 308-13.
[http://dx.doi.org/10.1080/00365520802588091] [PMID: 19031266]

[13] Savioli L, Smith H, Thompson A. *Giardia* and *Cryptosporidium* join the Neglected Diseases Initiative. Trends Parasitol 2006; 22(5): 203-8.
[http://dx.doi.org/10.1016/j.pt.2006.02.015] [PMID: 16545611]

[14] Sanchez JL, Rios C, Hernandez-Fragoso I, Ho CK. Parasitological evaluation of a foodhandler population cohort in Panama: risk factors for intestinal parasitism. Mil Med 1990; 155(6): 250-5.
[PMID: 2122299]

[15] Pozio E. Epidemiology and control prospects of foodborne parasitic zoonoses in the European Union. Parassitologia 2008; 50(1-2): 17-24.
[PMID: 18693552]

[16] Hoque ME, Hope VT, Scragg R, Kjellström T, Lay-Yee R. Nappy handling and risk of giardiasis. Lancet 2001; 357(9261): 1017-8.
[http://dx.doi.org/10.1016/S0140-6736(00)04251-3] [PMID: 11293600]

[17] Ang LH. Outbreak of giardiasis in a daycare nursery. Commun Dis Public Health 2000; 3(3): 212-3.
[PMID: 11014038]

[18] Chute CG, Smith RP, Baron JA. Risk factors for endemic giardiasis. Am J Public Health 1987; 77(5): 585-7.
[http://dx.doi.org/10.2105/AJPH.77.5.585] [PMID: 3565652]

[19] Hoque ME, Hope VT, Kjellström T, Scragg R, Lay-Yee R. Risk of giardiasis in Aucklanders: a case-control study. Int J Infect Dis 2002; 6(3): 191-7.
[http://dx.doi.org/10.1016/S1201-9712(02)90110-4] [PMID: 12718834]

[20] Faustini A, Marinacci C, Fabrizi E, *et al.* The impact of the Catholic Jubilee in 2000 on infectious diseases. A case-control study of giardiasis, Rome, Italy 20002001. Epidemiol Infect 2006; 134(3): 649-58.
[http://dx.doi.org/10.1017/S0950268805005327] [PMID: 16255833]

[21] Appelbee AJ, Thompson RC, Olson ME. *Giardia* and *Cryptosporidium* in mammalian wildlifecurrent status and future needs. Trends Parasitol 2005; 21(8): 370-6.
[http://dx.doi.org/10.1016/j.pt.2005.06.004] [PMID: 15982929]

[22] Thompson RC, Kutz SJ, Smith A. Parasite zoonoses and wildlife: emerging issues. Int J Environ Res Public Health 2009; 6(2): 678-93.
[http://dx.doi.org/10.3390/ijerph6020678] [PMID: 19440409]

[23] Plutzer J, Ongerth J, Karanis P. *Giardia* taxonomy, phylogeny and epidemiology: Facts and open questions. Int J Hyg Environ Health 2010; 213(5): 321-33.
[http://dx.doi.org/10.1016/j.ijheh.2010.06.005] [PMID: 20619729]

[24] O'Handley RM, Buret AG, McAllister TA, Jelinski M, Olson ME. Giardiasis in dairy calves: effects of fenbendazole treatment on intestinal structure and function. Int J Parasitol 2001; 31(1): 73-9.
[http://dx.doi.org/10.1016/S0020-7519(00)00148-X] [PMID: 11165274]

[25] Geurden T, Claerebout E, Vercruysse J. [Protozoan infection causes diarrhea in calves]. Tijdschr Diergeneeskd 2005; 130(23): 734-7.
[PMID: 16363207]

[26] Aloisio F, Filippini G, Antenucci P, *et al.* Severe weight loss in lambs infected with *Giardia duodenalis* assemblage B. Vet Parasitol 2006; 142(1-2): 154-8. [http://dx.doi.org/10.1016/j.vetpar.2006.06.023] [PMID: 16891057]

[27] O'Handley RM, Olson ME. Giardiasis and cryptosporidiosis in ruminants. Vet Clin North Am Food Anim Pract 2006; 22(3): 623-43. [http://dx.doi.org/10.1016/j.cvfa.2006.07.002] [PMID: 17071357]

[28] Sweeny JP, Ryan UM, Robertson ID, Jacobson C. *Cryptosporidium* and *Giardia* associated with reduced lamb carcase productivity. Vet Parasitol 2011; 182(2-4): 127-39. [http://dx.doi.org/10.1016/j.vetpar.2011.05.050] [PMID: 21719199]

[29] Thompson RC. The zoonotic significance and molecular epidemiology of *Giardia* and giardiasis. Vet Parasitol 2004; 126(1-2): 15-35. [http://dx.doi.org/10.1016/j.vetpar.2004.09.008] [PMID: 15567577]

[30] Cotton JA, Beatty JK, Buret AG. Host parasite interactions and pathophysiology in *Giardia* infections. Int J Parasitol 2011; 41(9): 925-33. [http://dx.doi.org/10.1016/j.ijpara.2011.05.002] [PMID: 21683702]

[31] Roxström-Lindquist K, Palm D, Reiner D, Ringqvist E, Svärd SG. *Giardia* immunityan update. Trends Parasitol 2006; 22(1): 26-31. [http://dx.doi.org/10.1016/j.pt.2005.11.005] [PMID: 16303332]

[32] Oberhuber G, Kastner N, Stolte M. Giardiasis: a histologic analysis of 567 cases. Scand J Gastroenterol 1997; 32(1): 48-51. [http://dx.doi.org/10.3109/00365529709025062] [PMID: 9018766]

[33] Jung HC, Eckmann L, Yang SK, *et al.* A distinct array of proinflammatory cytokines is expressed in human colon epithelial cells in response to bacterial invasion. J Clin Invest 1995; 95(1): 55-65. [http://dx.doi.org/10.1172/JCI117676] [PMID: 7814646]

[34] Hanevik K, Hausken T, Morken MH, *et al.* Persisting symptoms and duodenal inflammation related to *Giardia duodenalis* infection. J Infect 2007; 55(6): 524-30. [http://dx.doi.org/10.1016/j.jinf.2007.09.004] [PMID: 17964658]

[35] Roxström-Lindquist K, Ringqvist E, Palm D, Svärd S. *Giardia lamblia*-induced changes in gene expression in differentiated Caco-2 human intestinal epithelial cells. Infect Immun 2005; 73(12): 8204-8. [http://dx.doi.org/10.1128/IAI.73.12.8204-8208.2005] [PMID: 16299316]

[36] Lee HY, Hyung S, Lee NY, Yong TS, Han SH, Park SJ. Excretory-secretory products of *Giardia lamblia* induce interleukin-8 production in human colonic cells *via* activation of p38, ERK1/2, NF-κB and AP-1. Parasite Immunol 2012; 34(4): 183-98. [http://dx.doi.org/10.1111/j.1365-3024.2012.01354.x] [PMID: 22224945]

[37] Cotton JA, Bhargava A, Ferraz JG, Yates RM, Beck PL, Buret AG. *Giardia duodenalis* cathepsin B proteases degrade intestinal epithelial interleukin-8 and attenuate interleukin-8-induced neutrophil chemotaxis. Infect Immun 2014; 82(7): 2772-87. [http://dx.doi.org/10.1128/IAI.01771-14] [PMID: 24733096]

[38] Kamda JD, Singer SM. Phosphoinositide 3-kinase-dependent inhibition of dendritic cell interleukin-12 production by *Giardia lamblia.* Infect Immun 2009; 77(2): 685-93. [http://dx.doi.org/10.1128/IAI.00718-08] [PMID: 19047410]

[39] Matowicka-Karna J, Kralisz M, Kemona H. Assessment of the levels of nitric oxide (NO) and

cytokines (IL-5, IL-6, IL-13, TNF, IFN-gamma) in giardiosis. Folia Histochem Cytobiol 2011; 49(2): 280-4.
[http://dx.doi.org/10.5603/FHC.2011.0039] [PMID: 21744329]

[40] Dreesen L, Rinaldi M, Chiers K, *et al.* Microarray analysis of the intestinal host response in *Giardia duodenalis* assemblage E infected calves. PLoS One 2012; 7(7): e40985.
[http://dx.doi.org/10.1371/journal.pone.0040985] [PMID: 22848418]

[41] Tako EA, Hassimi MF, Li E, Singer SM. Transcriptomic analysis of the host response to *Giardia duodenalis* infection reveals redundant mechanisms for parasite control. MBio 2013; 4(6): e00660-13.
[http://dx.doi.org/10.1128/mBio.00660-13] [PMID: 24194537]

[42] Eckmann L. Mucosal defences against *Giardia.* Parasite Immunol 2003; 25(5): 259-70.
[http://dx.doi.org/10.1046/j.1365-3024.2003.00634.x] [PMID: 12969444]

[43] Eckmann L, Laurent F, Langford TD, *et al.* Nitric oxide production by human intestinal epithelial cells and competition for arginine as potential determinants of host defense against the lumen-dwelling pathogen *Giardia lamblia.* J Immunol 2000; 164(3): 1478-87.
[http://dx.doi.org/10.4049/jimmunol.164.3.1478] [PMID: 10640765]

[44] Banik S, Renner Viveros P, Seeber F, Klotz C, Ignatius R, Aebischer T. *Giardia duodenalis* arginine deiminase modulates the phenotype and cytokine secretion of human dendritic cells by depletion of arginine and formation of ammonia. Infect Immun 2013; 81(7): 2309-17.
[http://dx.doi.org/10.1128/IAI.00004-13] [PMID: 23589577]

[45] Cotton JA, Motta JP, Schenck LP, Hirota SA, Beck PL, Buret AG. *Giardia duodenalis* infection reduces granulocyte infiltration in an *in vivo* model of bacterial toxin-induced colitis and attenuates inflammation in human intestinal tissue. PLoS One 2014; 9(10): e109087.
[http://dx.doi.org/10.1371/journal.pone.0109087] [PMID: 25289678]

[46] Istre GR, Dunlop TS, Gaspard GB, Hopkins RS. Waterborne giardiasis at a mountain resort: evidence for acquired immunity. Am J Public Health 1984; 74(6): 602-4.
[http://dx.doi.org/10.2105/AJPH.74.6.602] [PMID: 6721017]

[47] Isaac-Renton JL, Lewis LF, Ong CS, Nulsen MF. A second community outbreak of waterborne giardiasis in Canada and serological investigation of patients. Trans R Soc Trop Med Hyg 1994; 88(4): 395-9.
[http://dx.doi.org/10.1016/0035-9203(94)90397-2] [PMID: 7570815]

[48] Kohli A, Bushen OY, Pinkerton RC, *et al. Giardia duodenalis* assemblage, clinical presentation and markers of intestinal inflammation in Brazilian children. Trans R Soc Trop Med Hyg 2008; 102(7): 718-25.
[http://dx.doi.org/10.1016/j.trstmh.2008.03.002] [PMID: 18485429]

[49] Smith HW, Mank TG. Diagnosis of human giardiasis. In: Lujan HD, Svärd S, Eds. *Giardia*: a Model Organism. Austria: SpringerWienNew York 2011; pp. 353-77.
[http://dx.doi.org/10.1007/978-3-7091-0198-8_22]

[50] United States Environmental Protection Agency (USEPA). Method 1623.1: *Cryptosporidium* and *Giardia* in Water by Filtration/IMS/FA, EPA 815-R-05-002. Ohio: Office of Water (4607) 2012.

[51] Robertson LJ, Gjerde B. Isolation and enumeration of *Giardia* cysts, *cryptosporidium* oocysts, and *Ascaris* eggs from fruits and vegetables. J Food Prot 2000; 63(6): 775-8.
[http://dx.doi.org/10.4315/0362-028X-63.6.775] [PMID: 10852573]

[52] Robertson LJ, Gjerde B. Factors affecting recovery efficiency in isolation of *Cryptosporidium* oocysts and *Giardia* cysts from vegetables for standard method development. J Food Prot 2001; 64(11): 1799-805.
[http://dx.doi.org/10.4315/0362-028X-64.11.1799] [PMID: 11726162]

[53] Cook N, Nichols RA, Wilkinson N, Paton CA, Barker K, Smith HV. Development of a method for detection of *Giardia duodenalis* cysts on lettuce and for simultaneous analysis of salad products for the presence of *Giardia* cysts and *Cryptosporidium* oocysts. Appl Environ Microbiol 2007; 73(22): 7388-91.
[http://dx.doi.org/10.1128/AEM.00552-07] [PMID: 17890337]

[54] Abboud P, Lemée V, Gargala G, *et al.* Successful treatment of metronidazole- and albendazole-resistant giardiasis with nitazoxanide in a patient with acquired immunodeficiency syndrome. Clin Infect Dis 2001; 32(12): 1792-4.
[http://dx.doi.org/10.1086/320751] [PMID: 11360222]

[55] Upcroft P, Upcroft JA. Drug targets and mechanisms of resistance in the anaerobic protozoa. Clin Microbiol Rev 2001; 14(1): 150-64.
[http://dx.doi.org/10.1128/CMR.14.1.150-164.2001] [PMID: 11148007]

[56] Tejman-Yarden N, Millman M, Lauwaet T, *et al.* Impaired parasite attachment as fitness cost of metronidazole resistance in *Giardia lamblia.* Antimicrob Agents Chemother 2011; 55(10): 4643-51.
[http://dx.doi.org/10.1128/AAC.00384-11] [PMID: 21825286]

[57] Palm JE, Weiland ME, Griffiths WJ, Ljungström I, Svärd SG. Identification of immunoreactive proteins during acute human giardiasis. J Infect Dis 2003; 187(12): 1849-59.
[http://dx.doi.org/10.1086/375356] [PMID: 12792861]

[58] Jenikova G, Hruz P, Andersson MK, *et al.* A1-giardin based live heterologous vaccine protects against *Giardia lamblia* infection in a murine model. Vaccine 2011; 29(51): 9529-37.
[http://dx.doi.org/10.1016/j.vaccine.2011.09.126] [PMID: 22001876]

[59] Larocque R, Nakagaki K, Lee P, Abdul-Wahid A, Faubert GM. Oral immunization of BALB/c mice with *Giardia duodenalis* recombinant cyst wall protein inhibits shedding of cysts. Infect Immun 2003; 71(10): 5662-9.
[http://dx.doi.org/10.1128/IAI.71.10.5662-5669.2003] [PMID: 14500486]

[60] Rivero FD, Saura A, Prucca CG, Carranza PG, Torri A, Lujan HD. Disruption of antigenic variation is crucial for effective parasite vaccine. Nat Med 2010; 16(5): 551-557, 1p, 557.
[http://dx.doi.org/10.1038/nm.2141] [PMID: 20418884]

[61] Lambl W. Mikroskopische untersuchungen der Darmexcrete. Vierteljahrsschr. Prakst Heikunde 1859; 61: 1-58.

[62] Dobell C. The Discovery of the Intestinal Protozoa of Man. Proc R Soc Med 1920; 13(Sect Hist Med): 1-15.

[63] Kofoid CA, Christensen EB. On binary and multiple fission in *Giardia muris* (Grassi). Univ Calif Publ Zool 1915; 16: 30-54.

[64] Kunstler J. Sur cinq protozaires parasites nouveaus. C R Seances Soc Biol Filiales 1882; 95: 347-9.

[65] Blanchard R. Remarques sur le megastome intestinal. Bull Soc Zool Fr 1888; 13: 18.

[66] Stiles CW. The type species of certain genera of parasitic flagellates, particularly Grassi's genera of 1879 and 1881. Zool Anz 1902; 25: 689.

[67] Kofoid CA, Christensen ED. A critical review of the nomenclature of human intestinal flagellates. Univ Calif Publ Zool 1920; 20: 160.

[68] Corradetti A. *Giardia enterica* (Grassi), 1881, the correct name for the *Giardia* of man. J Parasitol 1935; 21: 310-1.
[http://dx.doi.org/10.2307/3271363]

[69] Dobell C. Vilem Lambl (1824–1895): A portrait and a biographical note. Parasitol 1940; 32: 122-5.
[http://dx.doi.org/10.1017/S003118200001564X]

[70] Adam RD. Biology of *Giardia lamblia.* Clin Microbiol Rev 2001; 14(3): 447-75. [http://dx.doi.org/10.1128/CMR.14.3.447-475.2001] [PMID: 11432808]

[71] Xiao L, Fayer R. Molecular characterisation of species and genotypes of *Cryptosporidium* and *Giardia* and assessment of zoonotic transmission. Int J Parasitol 2008; 38(11): 1239-55. [http://dx.doi.org/10.1016/j.ijpara.2008.03.006] [PMID: 18479685]

[72] Thompson RC, Monis PT. Variation in *Giardia*: implications for taxonomy and epidemiology. Adv Parasitol 2004; 58: 69-137. [http://dx.doi.org/10.1016/S0065-308X(04)58002-8] [PMID: 15603762]

[73] Simon CE. A critique of the supposed rodent origin of human giardiasis. Am J Hyg 1922; 2: 406-34.

[74] Filice FP. Studies on the cytology and life history of a *Giardia* from the laboratory rat. Univ Calif Publ Zool 1952; 57: 53-146.

[75] Erlandsen SL, Bemrick WJ. SEM evidence for a new species, *Giardia psittaci.* J Parasitol 1987; 73(3): 623-9. [http://dx.doi.org/10.2307/3282146] [PMID: 3598809]

[76] Erlandsen SL, Bemrick WJ, Wells CL, *et al.* Axenic culture and characterization of *Giardia ardeae* from the great blue heron (*Ardea herodias*). J Parasitol 1990; 76(5): 717-24. [http://dx.doi.org/10.2307/3282988] [PMID: 2213415]

[77] Feely DE. Morphology of the cyst of *Giardia microti* by light and electron microscopy. J Protozool 1988; 35(1): 52-4. [http://dx.doi.org/10.1111/j.1550-7408.1988.tb04075.x] [PMID: 3285001]

[78] Erlandsen SL, Sherlock LA, Januschka M, *et al.* Cross-species transmission of *Giardia* spp.: inoculation of beavers and muskrats with cysts of human, beaver, mouse, and muskrat origin. Appl Environ Microbiol 1988; 54(11): 2777-85. [PMID: 3063208]

[79] Nash TE, Keister DB. Differences in excretory-secretory products and surface antigens among 19 isolates of *Giardia.* J Infect Dis 1985; 152(6): 1166-71. [http://dx.doi.org/10.1093/infdis/152.6.1166] [PMID: 4067331]

[80] Andrews RH, Adams M, Boreham PF, Mayrhofer G, Meloni BP. *Giardia intestinalis*: electrophoretic evidence for a species complex. Int J Parasitol 1989; 19(2): 183-90. [http://dx.doi.org/10.1016/0020-7519(89)90006-4] [PMID: 2722391]

[81] Andrews RH, Mayrhofer G, Chilton NB, Boreham PF, Grimmond TR. Changes in allozyme pattern of the protozoan parasite *Giardia intestinalis.* Int J Parasitol 1992; 22(3): 403-6. [http://dx.doi.org/10.1016/S0020-7519(05)80022-0] [PMID: 1639578]

[82] Mayrhofer G, Andrews RH, Ey PL, Chilton NB. Division of *Giardia isolates* from humans into two genetically distinct assemblages by electrophoretic analysis of enzymes encoded at 27 loci and comparison with *Giardia muris.* Parasitology 1995; 111(Pt 1): 11-7. [http://dx.doi.org/10.1017/S0031182000064556] [PMID: 7609985]

[83] Upcroft P, Mitchell R, Boreham PF. DNA fingerprinting of the intestinal parasite *Giardia duodenalis* with the M13 phage genome. Int J Parasitol 1990; 20(3): 319-23. [http://dx.doi.org/10.1016/0020-7519(90)90146-E] [PMID: 2358315]

[84] Archibald SC, Mitchell RW, Upcroft JA, Boreham PF, Upcroft P. Variation between human and animal isolates of *Giardia* as demonstrated by DNA fingerprinting. Int J Parasitol 1991; 21(1): 123-4. [http://dx.doi.org/10.1016/0020-7519(91)90131-P] [PMID: 2040561]

[85] Homan WL, Gilsing M, Bentala H, Limper L, van Knapen F. Characterization of *Giardia duodenalis* by polymerase-chain-reaction fingerprinting. Parasitol Res 1998; 84(9): 707-14.

[http://dx.doi.org/10.1007/s004360050474] [PMID: 9766898]

[86] Hopkins RM, Constantine CC, Groth DA, Wetherall JD, Reynoldson JA, Thompson RC. PCR-based DNA fingerprinting of *Giardia duodenalis* isolates using the intergenic rDNA spacer. Parasitology 1999; 118(Pt 6): 531-9.
[http://dx.doi.org/10.1017/S003118209900428X] [PMID: 10406031]

[87] Monis PT, Andrews RH, Mayrhofer G, Ey PL. Molecular systematics of the parasitic protozoan *Giardia intestinalis*. Mol Biol Evol 1999; 16(9): 1135-44.
[http://dx.doi.org/10.1093/oxfordjournals.molbev.a026204] [PMID: 10486969]

[88] Monis PT, Andrews RH, Mayrhofer G, Ey PL. Genetic diversity within the morphological species *Giardia intestinalis* and its relationship to host origin. Infect Genet Evol 2003; 3(1): 29-38.
[http://dx.doi.org/10.1016/S1567-1348(02)00149-1] [PMID: 12797970]

[89] Monis PT, Thompson RC. *Cryptosporidium* and *Giardia*-zoonoses: fact or fiction? Infect Genet Evol 2003; 3(4): 233-44.
[http://dx.doi.org/10.1016/j.meegid.2003.08.003] [PMID: 14636685]

[90] Cacciò SM, Thompson RC, McLauchlin J, Smith HV. Unravelling *Cryptosporidium* and *Giardia* epidemiology. Trends Parasitol 2005; 21(9): 430-7.
[http://dx.doi.org/10.1016/j.pt.2005.06.013] [PMID: 16046184]

[91] Lalle M, Pozio E, Capelli G, Bruschi F, Crotti D, Cacciò SM. Genetic heterogeneity at the beta-giardin locus among human and animal isolates of *Giardia duodenalis* and identification of potentially zoonotic subgenotypes. Int J Parasitol 2005; 35(2): 207-13.
[http://dx.doi.org/10.1016/j.ijpara.2004.10.022] [PMID: 15710441]

[92] Cacciò SM, Ryan U. Molecular epidemiology of giardiasis. Mol Biochem Parasitol 2008; 160(2): 75-80.
[http://dx.doi.org/10.1016/j.molbiopara.2008.04.006] [PMID: 18501440]

[93] Sprong H, Cacciò SM, van der Giessen JW. Identification of zoonotic genotypes of *Giardia duodenalis*. PLoS Negl Trop Dis 2009; 3(12): e558.
[http://dx.doi.org/10.1371/journal.pntd.0000558] [PMID: 19956662]

[94] Lasek-Nesselquist E, Bogomolni AL, Gast RJ, *et al.* Molecular characterization of *Giardia intestinalis* haplotypes in marine animals: variation and zoonotic potential. Dis Aquat Organ 2008; 81(1): 39-51.
[http://dx.doi.org/10.3354/dao01931] [PMID: 18828561]

[95] Wielinga C, Ryan U, Andrew Thompson RC, Monis P. Multi-locus analysis of *Giardia duodenalis* intra-Assemblage B substitution patterns in cloned culture isolates suggests sub-Assemblage B analyses will require multi-locus genotyping with conserved and variable genes. Int J Parasitol 2011; 41(5): 495-503.
[http://dx.doi.org/10.1016/j.ijpara.2010.11.007] [PMID: 21176781]

[96] Ryan U, Cacciò SM. Zoonotic potential of *Giardia.* Int J Parasitol 2013; 43(12-13): 943-56.
[http://dx.doi.org/10.1016/j.ijpara.2013.06.001] [PMID: 23856595]

[97] Monis PT, Caccio SM, Thompson RC. Variation in *Giardia*: towards a taxonomic revision of the genus. Trends Parasitol 2009; 25(2): 93-100.
[http://dx.doi.org/10.1016/j.pt.2008.11.006] [PMID: 19135417]

[98] Morrison HG, McArthur AG, Gillin FD, *et al.* Genomic minimalism in the early diverging intestinal parasite *Giardia lamblia.* Science 2007; 317(5846): 1921-6.

[http://dx.doi.org/10.1126/science.1143837] [PMID: 17901334]

[99] Franzén O, Jerlström-Hultqvist J, Castro E, *et al.* Draft genome sequencing of *giardia intestinalis* assemblage B isolate GS: is human giardiasis caused by two different species? PLoS Pathog 2009; 5(8): e1000560.
[http://dx.doi.org/10.1371/journal.ppat.1000560] [PMID: 19696920]

[100] Nash TE, Herrington DA, Losonsky GA, Levine MM. Experimental human infections with *Giardia lamblia.* J Infect Dis 1987; 156(6): 974-84.
[http://dx.doi.org/10.1093/infdis/156.6.974] [PMID: 3680997]

[101] Byrd LG, Conrad JT, Nash TE. *Giardia lamblia* infections in adult mice. Infect Immun 1994; 62(8): 3583-5.
[PMID: 8039934]

[102] Mayrhofer G, Andrews RH, Ey PL, Albert MJ, Grimmond TR, Merry DJ. The use of suckling mice to isolate and grow *Giardia* from mammalian faecal specimens for genetic analysis. Parasitology 1992; 105(Pt 2): 255-63.
[http://dx.doi.org/10.1017/S0031182000074187] [PMID: 1454424]

[103] Andrews RH, Chilton NB, Mayrhofer G. Selection of specific genotypes of *Giardia intestinalis* by growth *in vitro* and *in vivo.* Parasitology 1992; 105(Pt 3): 375-86.
[http://dx.doi.org/10.1017/S0031182000074540] [PMID: 1461678]

[104] Binz N, Thompson RC, Lymbery AJ, Hobbs RP. Comparative studies on the growth dynamics of two genetically distinct isolates of *Giardia duodenalis in vitro.* Int J Parasitol 1992; 22(2): 195-202.
[http://dx.doi.org/10.1016/0020-7519(92)90101-P] [PMID: 1587683]

[105] Hall ML, Costa ND, Thompson RC, Lymbery AJ, Meloni BP, Wales RG. Genetic variants of *Giardia duodenalis* differ in their metabolism. Parasitol Res 1992; 78(8): 712-4.
[http://dx.doi.org/10.1007/BF00931528] [PMID: 1480612]

[106] Farbey MD, Reynoldson JA, Thompson RC. *In vitro* drug susceptibility of 29 isolates of *Giardia duodenalis* from humans as assessed by an adhesion assay. Int J Parasitol 1995; 25(5): 593-9.
[http://dx.doi.org/10.1016/0020-7519(94)00174-M] [PMID: 7635637]

[107] Thompson RC, Lymbery AJ, Pearce DA, Finn KC, Reynoldson JA, Meloni BP. *Giardia duodenalis*: exposure to metronidazole inhibits competitive interactions between isolates of the parasite *in vitro.* J Parasitol 1996; 82(4): 679-83.
[http://dx.doi.org/10.2307/3283808] [PMID: 8691391]

[108] Simpson AG. Cytoskeletal organization, phylogenetic affinities and systematics in the contentious taxon Excavata (Eukaryota). Int J Syst Evol Microbiol 2003; 53(Pt 6): 1759-77.
[http://dx.doi.org/10.1099/ijs.0.02578-0] [PMID: 14657103]

[109] Keeling PJ, Doolittle WF. A non-canonical genetic code in an early diverging eukaryotic lineage. EMBO J 1996; 15(9): 2285-90.
[PMID: 8641293]

[110] Kolisko M, Cepicka I, Hampl V, *et al.* Molecular phylogeny of diplomonads and enteromonads based on SSU rRNA, alpha-tubulin and HSP90 genes: implications for the evolutionary history of the double karyomastigont of diplomonads. BMC Evol Biol 2008; 8: 205.
[http://dx.doi.org/10.1186/1471-2148-8-205] [PMID: 18627633]

[111] Thompson AR, Monis PT. Taxonomy of *Giardia* species. In: Luján HD, Svärd S, Eds. *Giardia* A Model Organism. Austria: SpringerWienNew York 2011; pp. 3-15.
[http://dx.doi.org/10.1007/978-3-7091-0198-8_1]

[112] Erlandsen SL, Macechko PT, van Keulen H, Jarroll EL. Formation of the *Giardia* cyst wall: studies on extracellular assembly using immunogold labeling and high resolution field emission SEM. J Eukaryot

Microbiol 1996; 43(5): 416-29.
[http://dx.doi.org/10.1111/j.1550-7408.1996.tb05053.x] [PMID: 8822813]

[113] Svärd SG, Hagblom P, Palm JE. *Giardia lamblia* a model organism for eukaryotic cell differentiation. FEMS Microbiol Lett 2003; 218(1): 3-7.
[PMID: 12583890]

[114] Gerwig GJ, van Kuik JA, Leeflang BR, *et al.* The *Giardia intestinalis* filamentous cyst wall contains a novel beta(13)-N-acetyl-D-galactosamine polymer: a structural and conformational study. Glycobiology 2002; 12(8): 499-505.
[http://dx.doi.org/10.1093/glycob/cwf059] [PMID: 12145190]

[115] Hetsko ML, McCaffery JM, Svärd SG, Meng TC, Que X, Gillin FD. Cellular and transcriptional changes during excystation of *Giardia lamblia in vitro.* Exp Parasitol 1998; 88(3): 172-83.
[http://dx.doi.org/10.1006/expr.1998.4246] [PMID: 9562420]

[116] Ward W, Alvarado L, Rawlings ND, Engel JC, Franklin C, McKerrow JH. A primitive enzyme for a primitive cell: the protease required for excystation of *Giardia.* Cell 1997; 89(3): 437-44.
[http://dx.doi.org/10.1016/S0092-8674(00)80224-X] [PMID: 9150143]

[117] Hussein EM, Dawoud HA, Salem AM, Atwa MM. Antiparasitic activity of cystine protease inhibitor E-64 against *Giardia lamblia* excystation *in vitro* and *in vivo.* J Egypt Soc Parasitol 2009; 39(1): 111-9.
[PMID: 19530614]

[118] Slavin I, Saura A, Carranza PG, Touz MC, Nores MJ, Luján HD. Dephosphorylation of cyst wall proteins by a secreted lysosomal acid phosphatase is essential for excystation of *Giardia lamblia.* Mol Biochem Parasitol 2002; 122(1): 95-8.
[http://dx.doi.org/10.1016/S0166-6851(02)00065-8] [PMID: 12076774]

[119] Bernander R, Palm JE, Svärd SG. Genome ploidy in different stages of the *Giardia lamblia* life cycle. Cell Microbiol 2001; 3(1): 55-62.
[http://dx.doi.org/10.1046/j.1462-5822.2001.00094.x] [PMID: 11207620]

[120] Palm D, Weiland M, McArthur AG, *et al.* Developmental changes in the adhesive disk during *Giardia* differentiation. Mol Biochem Parasitol 2005; 141(2): 199-207.
[http://dx.doi.org/10.1016/j.molbiopara.2005.03.005] [PMID: 15850703]

[121] Weiland ME, Palm JE, Griffiths WJ, McCaffery JM, Svärd SG. Characterisation of alpha-1 giardin: an immunodominant *Giardia lamblia* annexin with glycosaminoglycan-binding activity. Int J Parasitol 2003; 33(12): 1341-51.
[http://dx.doi.org/10.1016/S0020-7519(03)00201-7] [PMID: 14527517]

[122] Gillin FD, Reiner DS, McCaffery JM. Cell biology of the primitive eukaryote *Giardia lamblia.* Annu Rev Microbiol 1996; 50: 679-705.
[http://dx.doi.org/10.1146/annurev.micro.50.1.679] [PMID: 8905095]

[123] Sheffield HG, Bjorvat B. Ultrastructure of the cyst of *Giardia lamblia.* Am J Trop Med Hyg 1977; 26(1): 23-30.
[PMID: 842783]

[124] Chávez-Munguía B, Omaña-Molina M, González-Lázaro M, *et al.* Ultrastructure of cyst differentiation in parasitic protozoa. Parasitol Res 2007; 100(6): 1169-75.
[http://dx.doi.org/10.1007/s00436-006-0447-x] [PMID: 17252271]

[125] Knodler LA, Svärd SG, Silberman JD, Davids BJ, Gillin FD. Developmental gene regulation in *Giardia lamblia*: first evidence for an encystation-specific promoter and differential 5' mRNA processing. Mol Microbiol 1999; 34(2): 327-40.
[http://dx.doi.org/10.1046/j.1365-2958.1999.01602.x] [PMID: 10564476]

[126] Jarroll EL, Macechko PT, Steimle PA, *et al.* Regulation of carbohydrate metabolism during *Giardia* encystment. J Eukaryot Microbiol 2001; 48(1): 22-6.
[http://dx.doi.org/10.1111/j.1550-7408.2001.tb00412.x] [PMID: 11249189]

[127] Karr CD, Jarroll EL. Cyst wall synthase: N-acetylgalactosaminyltransferase activity is induced to form the novel N-acetylgalactosamine polysaccharide in the *Giardia* cyst wall. Microbiology 2004; 150(Pt 5): 1237-43.
[http://dx.doi.org/10.1099/mic.0.26922-0] [PMID: 15133086]

[128] Lauwaet T, Davids BJ, Reiner DS, Gillin FD. Encystation of *Giardia lamblia*: a model for other parasites. Curr Opin Microbiol 2007; 10(6): 554-9.
[http://dx.doi.org/10.1016/j.mib.2007.09.011] [PMID: 17981075]

[129] Carranza PG, Lujan HD. New insights regarding the biology of *Giardia lamblia.* Microbes Infect 2010; 12(1): 71-80.
[http://dx.doi.org/10.1016/j.micinf.2009.09.008] [PMID: 19772929]

[130] World Health Organization (WHO). Fighting Disease Fostering Development. Geneva: World Health Organization 1996. http://www.who.int/whr/1996/en/whr96_en.pdf

[131] Kappus KD, Lundgren RG Jr, Juranek DD, Roberts JM, Spencer HC. Intestinal parasitism in the United States: update on a continuing problem. Am J Trop Med Hyg 1994; 50(6): 705-13.
[PMID: 8024063]

[132] Collier SA, Stockman LJ, Hicks LA, Garrison LE, Zhou FJ, Beach MJ. Direct healthcare costs of selected diseases primarily or partially transmitted by water. Epidemiol Infect 2012; 140(11): 2003-13.
[http://dx.doi.org/10.1017/S0950268811002858] [PMID: 22233584]

[133] Painter JE, Gargano JW, Collier SA, Yoder JS. Giardiasis surveillance United States, 20112012. MMWR Surveill Summ 2015; 64(3): 15-25.
[PMID: 25928582]

[134] European Centre for Disease Prevention and Control (ECDC). Annual epidemiological report 2014- food- and waterborne diseases and zoonoses. Stockholm: ECDC 2014. http://ecdc.europa.eu/en/publications/Publications/food-waterborne-diseases-annual-epidemiological-report-2014.pdf

[135] Hoque E, Hope V, Scragg R, Baker M, Shrestha R. A descriptive epidemiology of giardiasis in New Zealand and gaps in surveillance data. N Z Med J 2004; 117(1205): U1149.
[PMID: 15570332]

[136] Snel SJ, Baker MG, Venugopal K. The epidemiology of giardiasis in New Zealand, 1997-2006. N Z Med J 2009; 122(1290): 62-75.
[PMID: 19319169]

[137] New Zealand Ministry of Health [Home page on the Internet]. Annual Notifiable Disease Tables: Notifiable diseases tables by age, sex, ethnic group, 2014.Wellington: NZ Public Health Surveillance 2014; Available from: https://surv.esr.cri.nz/surveillance/annual_diseasetables.php.

[138] Lal A, Lill AW, Mcintyre M, Hales S, Baker MG, French NP. Environmental change and enteric zoonoses in New Zealand: a systematic review of the evidence. Aust N Z J Public Health 2015; 39(1): 63-8.
[http://dx.doi.org/10.1111/1753-6405.12274] [PMID: 25307352]

[139] Lal A, Hales S. Heterogeneity in hotspots: spatio-temporal patterns in neglected parasitic diseases. Epidemiol Infect 2015; 143(3): 631-9.
[http://dx.doi.org/10.1017/S0950268814001101] [PMID: 24819745]

[140] Lv S, Tian LG, Liu Q, *et al.* Water-related parasitic diseases in China. Int J Environ Res Public Health 2013; 10(5): 1977-2016.
[http://dx.doi.org/10.3390/ijerph10051977] [PMID: 23685826]

[141] Xu LQ, Yu SH, Xu SM. Distribution and Pathogenic Impact of Human Parasites in China. Beijing: People's Medical Publishing House 1999.

[142] Tu XG, Yao LN, Huang XM, *et al.* [Second sampled survey on the distribution of human parasites in Zhejiang Province]. Zhongguo Ji Sheng Chong Xue Yu Ji Sheng Chong Bing Za Zhi 2001; 19(3): 145-8.
[PMID: 12571938]

[143] Xu BL, Zhao XD, Su YP, *et al.* Survey on present epidemic status of main human parasites infection in Henan province. Chin J Parasitic Dis Contr 2005; 18: 454-7.

[144] Sinniah B, Hassan A KR, Sabaridah I, Soe MM, Ibrahim Z, Ali O. Prevalence of intestinal parasitic infections among communities living in different habitats and its comparison with one hundred and one studies conducted over the past 42 years (1970 to 2013) in Malaysia. Trop Biomed 2014; 31(2): 190-206.
[PMID: 25134888]

[145] Huang DB, White AC. An updated review on *Cryptosporidium* and *Giardia.* Gastroenterol Clin North Am 2006; 35(2): 291-314, viii.
[http://dx.doi.org/10.1016/j.gtc.2006.03.006] [PMID: 16880067]

[146] deRegnier DP, Cole L, Schupp DG, Erlandsen SL. Viability of *Giardia* cysts suspended in lake, river, and tap water. Appl Environ Microbiol 1989; 55(5): 1223-9.
[PMID: 2757381]

[147] Erickson MC, Ortega YR. Inactivation of protozoan parasites in food, water, and environmental systems. J Food Prot 2006; 69(11): 2786-808.
[http://dx.doi.org/10.4315/0362-028X-69.11.2786] [PMID: 17133829]

[148] Robertson LJ, Ai Lian Lim Y. Waterborne and Environmentally-Borne Giardiasis. In: Lujan HD, Svärd S, Eds. Giardia a Model Organism. Austria: Springer Wein New York 2011; pp. 29-69.
[http://dx.doi.org/10.1007/978-3-7091-0198-8_3]

[149] Pickering LK, Woodward WE, DuPont HL, Sullivan P. Occurrence of *Giardia lamblia* in children in day care centers. J Pediatr 1984; 104(4): 522-6.
[http://dx.doi.org/10.1016/S0022-3476(84)80540-5] [PMID: 6707812]

[150] Danciger M, Lopez M. Numbers of *Giardia* in the feces of infected children. Am J Trop Med Hyg 1975; 24(2): 237-42.
[PMID: 1119665]

[151] Geurden T, Vandenhoute E, Pohle H, *et al.* The effect of a fenbendazole treatment on cyst excretion and weight gain in calves experimentally infected with *Giardia duode*nalis. Vet Parasitol 2010; 169(1-2): 18-23.
[http://dx.doi.org/10.1016/j.vetpar.2009.12.027] [PMID: 20089358]

[152] Rendtorff RC. The experimental transmission of human intestinal protozoan parasites. II. *Giardia lamblia* cysts given in capsules. Am J Hyg 1954; 59(2): 209-20.
[PMID: 13138586]

[153] Esch KJ, Petersen CA. Transmission and epidemiology of zoonotic protozoal diseases of companion animals. Clin Microbiol Rev 2013; 26(1): 58-85.
[http://dx.doi.org/10.1128/CMR.00067-12] [PMID: 23297259]

[154] Kasprzak W, Pawlowski Z. Zoonotic aspects of giardiasis: a review. Vet Parasitol 1989; 32(2-3): 101-8.
[http://dx.doi.org/10.1016/0304-4017(89)90110-6] [PMID: 2672545]

[155] Baldursson S, Karanis P. Waterborne transmission of protozoan parasites: review of worldwide outbreaks - an update 2004-2010. Water Res 2011; 45(20): 6603-14.
[http://dx.doi.org/10.1016/j.watres.2011.10.013] [PMID: 22048017]

[156] Karanis P, Kourenti C, Smith H. Waterborne transmission of protozoan parasites: a worldwide review of outbreaks and lessons learnt. J Water Health 2007; 5(1): 1-38.
[http://dx.doi.org/10.2166/wh.2006.002] [PMID: 17402277]

[157] Braeye T, DE Schrijver K, Wollants E, van Ranst M, Verhaegen J. A large community outbreak of gastroenteritis associated with consumption of drinking water contaminated by river water, Belgium, 2010. Epidemiol Infect 2015; 143(4): 711-9.
[http://dx.doi.org/10.1017/S0950268814001629] [PMID: 25062494]

[158] Porter JD, Gaffney C, Heymann D, Parkin W. Food-borne outbreak of *Giardia lamblia.* Am J Public Health 1990; 80(10): 1259-60.
[http://dx.doi.org/10.2105/AJPH.80.10.1259] [PMID: 2400040]

[159] Rose JB, Slifko TR. *Giardia, Cryptosporidium*, and *Cyclospora* and their impact on foods: a review. J Food Prot 1999; 62(9): 1059-70.
[http://dx.doi.org/10.4315/0362-028X-62.9.1059] [PMID: 10492484]

[160] Greig JD, Todd EC, Bartleson CA, Michaels BS. Outbreaks where food workers have been implicated in the spread of foodborne disease. Part 1. Description of the problem, methods, and agents involved. J Food Prot 2007; 70(7): 1752-61.
[http://dx.doi.org/10.4315/0362-028X-70.7.1752] [PMID: 17685355]

[161] Stuart JM, Orr HJ, Warburton FG, *et al.* Risk factors for sporadic giardiasis: a case-control study in southwestern England. Emerg Infect Dis 2003; 9(2): 229-33.
[http://dx.doi.org/10.3201/eid0902.010488] [PMID: 12603995]

[162] Gray SF, Gunnell DJ, Peters TJ. Risk factors for giardiasis: a case-control study in Avon and Somerset. Epidemiol Infect 1994; 113(1): 95-102.
[http://dx.doi.org/10.1017/S0950268800051505] [PMID: 8062884]

[163] Alexander C, Jones B, Inverarity D, Pollock KG. Genotyping of *Giardia* isolates in Scotland: a descriptive epidemiological study. Epidemiol Infect 2014; 142(8): 1636-9.
[http://dx.doi.org/10.1017/S0950268813002604] [PMID: 24125434]

[164] Shlim DR, Hoge CW, Rajah R, Scott RM, Pandy P, Echeverria P. Persistent high risk of diarrhea among foreigners in Nepal during the first 2 years of residence. Clin Infect Dis 1999; 29(3): 613-6.
[http://dx.doi.org/10.1086/598642] [PMID: 10530456]

[165] Ekdahl K, Giesecke J. Travellers returning to Sweden as sentinels for comparative disease incidence in other European countries, *Campylobacter* and *Giardia* infection as examples. Euro Surveill 2004; 9(9): 6-9.
[PMID: 15381837]

[166] Novotny TE, Hopkins RS, Shillam P, Janoff EN. Prevalence of *Giardia lamblia* and risk factors for infection among children attending day-care facilities in Denver. Public Health Rep 1990; 105(1): 72-5.
[PMID: 2106707]

[167] Enserink R, Mughini-Gras L, Duizer E, Kortbeek T. VAN Pelt W. Risk factors for gastroenteritis in

child day care. Epidemiol Infect 2015; 16: 1-14.

[168] Escobedo AA, Almirall P, Alfonso M, Cimerman S, Chacín-Bonilla L. Sexual transmission of giardiasis: a neglected route of spread? Acta Trop 2014; 132: 106-11. [http://dx.doi.org/10.1016/j.actatropica.2013.12.025] [PMID: 24434784]

[169] Escobedo AA, Hanevik K, Almirall P, Cimerman S, Alfonso M. Management of chronic *Giardia* infection. Expert Rev Anti Infect Ther 2014; 12(9): 1143-57. [http://dx.doi.org/10.1586/14787210.2014.942283] [PMID: 25059638]

[170] Dixon B, Parrington L, Cook A, Pollari F, Farber J. Detection of *Cyclospora, Cryptosporidium*, and *Giardia* in ready-to-eat packaged leafy greens in Ontario, Canada. J Food Prot 2013; 76(2): 307-13. [http://dx.doi.org/10.4315/0362-028X.JFP-12-282] [PMID: 23433379]

[171] Amahmid O, Asmama S, Bouhoum K. The effect of waste water reuse in irrigation on the contamination level of food crops by *Giardia* cysts and *Ascaris* eggs. Int J Food Microbiol 1999; 49(1-2): 19-26. [http://dx.doi.org/10.1016/S0168-1605(99)00058-6] [PMID: 10477066]

[172] Pinheiro Ide O, de Castro MF, Mitterofhe A, *et al.* Prevalence and risk factors for giardiasis and soil-transmitted helminthiasis in three municipalities of Southeastern Minas Gerais State, Brazil: risk factors for giardiasis and soil-transmitted helminthiasis. Parasitol Res 2011; 108(5): 1123-30. [http://dx.doi.org/10.1007/s00436-010-2154-x] [PMID: 21243507]

[173] Quihui L, Valencia ME, Crompton DW, *et al.* Role of the employment status and education of mothers in the prevalence of intestinal parasitic infections in Mexican rural schoolchildren. BMC Public Health 2006; 6: 225. [http://dx.doi.org/10.1186/1471-2458-6-225] [PMID: 16956417]

[174] Mahmud MA, Chappell C, Hossain MM, Habib M, Dupont HL. Risk factors for development of first symptomatic *Giardia* infection among infants of a birth cohort in rural Egypt. Am J Trop Med Hyg 1995; 53(1): 84-8. [PMID: 7625540]

[175] Karrar ZA, Rahim FA. Prevalence and risk factors of parasitic infections among under-five Sudanese children: a community based study. East Afr Med J 1995; 72(2): 103-9. [PMID: 7796747]

[176] Anuar TS, Al-Mekhlafi HM, Ghani MK, *et al.* Giardiasis among different tribes of Orang Asli in Malaysia: highlighting the presence of other family members infected with *Giardia intestinalis* as a main risk factor. Int J Parasitol 2012; 42(9): 871-80. [http://dx.doi.org/10.1016/j.ijpara.2012.07.003] [PMID: 22846786]

[177] Chávez B, González-Mariscal L, Cedillo-Rivera R, Martínez-Palomo A. *Giardia lamblia*: *in vitro* cytopathic effect of human isolates. Exp Parasitol 1995; 80(1): 133-8. [http://dx.doi.org/10.1006/expr.1995.1015] [PMID: 7821402]

[178] Teoh DA, Kamieniecki D, Pang G, Buret AG. *Giardia lamblia* rearranges F-actin and alpha-actinin in human colonic and duodenal monolayers and reduces transepithelial electrical resistance. J Parasitol 2000; 86(4): 800-6. [PMID: 10958459]

[179] Humen MA, Pérez PF, Liévin-Le Moal V. Lipid raft-dependent adhesion of *Giardia intestinalis* trophozoites to a cultured human enterocyte-like Caco-2/TC7 cell monolayer leads to cytoskeleton-dependent functional injuries. Cell Microbiol 2011; 13(11): 1683-702. [http://dx.doi.org/10.1111/j.1462-5822.2011.01647.x] [PMID: 21790940]

[180] Maia-Brigagão C, Morgado-Díaz JA, De Souza W. *Giardia* disrupts the arrangement of tight,

adherens and desmosomal junction proteins of intestinal cells. Parasitol Int 2012; 61(2): 280-7. [http://dx.doi.org/10.1016/j.parint.2011.11.002] [PMID: 22146155]

[181] Stadelmann B, Merino MC, Persson L, Svärd SG. Arginine consumption by the intestinal parasite *Giardia intestinalis* reduces proliferation of intestinal epithelial cells. PLoS One 2012; 7(9): e45325. [http://dx.doi.org/10.1371/journal.pone.0045325] [PMID: 23028934]

[182] Chin AC, Teoh DA, Scott KG, Meddings JB, Macnaughton WK, Buret AG. Strain-dependent induction of enterocyte apoptosis by *Giardia lamblia* disrupts epithelial barrier function in a caspase-3-dependent manner. Infect Immun 2002; 70(7): 3673-80. [http://dx.doi.org/10.1128/IAI.70.7.3673-3680.2002] [PMID: 12065509]

[183] Panaro MA, Cianciulli A, Mitolo V, *et al.* Caspase-dependent apoptosis of the HCT-8 epithelial cell line induced by the parasite *Giardia intestinalis*. FEMS Immunol Med Microbiol 2007; 51(2): 302-9. [http://dx.doi.org/10.1111/j.1574-695X.2007.00304.x] [PMID: 17714487]

[184] Rodríguez-Fuentes GB, Cedillo-Rivera R, Fonseca-Liñán R, *et al. Giardia duodenalis*: analysis of secreted proteases upon trophozoite-epithelial cell interaction *in vitro*. Mem Inst Oswaldo Cruz 2006; 101(6): 693-6. [http://dx.doi.org/10.1590/S0074-02762006000600020] [PMID: 17072486]

[185] Ringqvist E, Palm JE, Skarin H, *et al.* Release of metabolic enzymes by *Giardia* in response to interaction with intestinal epithelial cells. Mol Biochem Parasitol 2008; 159(2): 85-91. [http://dx.doi.org/10.1016/j.molbiopara.2008.02.005] [PMID: 18359106]

[186] Skarin H, Ringqvist E, Hellman U, Svärd SG. Elongation factor 1-alpha is released into the culture medium during growth of *Giardia intestinalis* trophozoites. Exp Parasitol 2011; 127(4): 804-10. [http://dx.doi.org/10.1016/j.exppara.2011.01.006] [PMID: 21276445]

[187] Piña-Vázquez C, Reyes-López M, Ortíz-Estrada G, de la Garza M, Serrano-Luna J. Host-parasite interaction: parasite-derived and -induced proteases that degrade human extracellular matrix. J Parasitol Res 2012. 2012: 748206.

[188] Kaur H, Ghosh S, Samra H, Vinayak VK, Ganguly NK. Identification and characterization of an excretory-secretory product from *Giardia lamblia*. Parasitology 2001; 123(Pt 4): 347-56. [http://dx.doi.org/10.1017/S0031182001008629] [PMID: 11676366]

[189] Troeger H, Epple HJ, Schneider T, *et al.* Effect of chronic *Giardia lamblia* infection on epithelial transport and barrier function in human duodenum. Gut 2007; 56(3): 328-35. [http://dx.doi.org/10.1136/gut.2006.100198] [PMID: 16935925]

[190] Belosevic M, Faubert GM, MacLean JD. Disaccharidase activity in the small intestine of gerbils (*Meriones unguiculatus*) during primary and challenge infections with *Giardia lamblia*. Gut 1989; 30(9): 1213-9. [http://dx.doi.org/10.1136/gut.30.9.1213] [PMID: 2806988]

[191] Bénéré E, Van Assche T, Van Ginneken C, Peulen O, Cos P, Maes L. Intestinal growth and pathology of *Giardia duodenalis* assemblage subtype A(I), A(II), B and E in the gerbil model. Parasitology 2012; 139(4): 424-33. [http://dx.doi.org/10.1017/S0031182011002137] [PMID: 22216783]

[192] Koot BG, ten Kate FJ, Juffrie M, Rosalina I, Taminiau JJ, Benninga MA. Does *Giardia lamblia* cause villous atrophy in children?: A retrospective cohort study of the histological abnormalities in giardiasis. J Pediatr Gastroenterol Nutr 2009; 49(3): 304-8. [http://dx.doi.org/10.1097/MPG.0b013e31818de3c4] [PMID: 19525873]

[193] Cevallos A, Carnaby S, James M, Farthing JG. Small intestinal injury in a neonatal rat model of giardiasis is strain dependent. Gastroenterology 1995; 109(3): 766-73.
[http://dx.doi.org/10.1016/0016-5085(95)90383-6] [PMID: 7657104]

[194] Sousa MC, Gonçalves CA, Bairos VA, Poiares-Da-Silva J. Adherence of *Giardia lamblia* trophozoites to Int-407 human intestinal cells. Clin Diagn Lab Immunol 2001; 8(2): 258-65.
[PMID: 11238205]

[195] Bermúdez-Cruz RM, Ortega-Pierres G, Ceja V, *et al.* A 63 kDa VSP9B10A-like protein expressed in a C-8 *Giardia duodenalis* Mexican clone. Arch Med Res 2004; 35(3): 199-208.
[http://dx.doi.org/10.1016/j.arcmed.2003.12.005] [PMID: 15163460]

[196] Hernández-Sánchez J, Liñan RF, Salinas-Tobón MdelR, Ortega-Pierres G. *Giardia duodenalis*: adhesion-deficient clones have reduced ability to establish infection in Mongolian gerbils. Exp Parasitol 2008; 119(3): 364-72.
[http://dx.doi.org/10.1016/j.exppara.2008.03.010] [PMID: 18456259]

[197] Inge PM, Edson CM, Farthing MJ. Attachment of *Giardia lamblia* to rat intestinal epithelial cells. Gut 1988; 29(6): 795-801.
[http://dx.doi.org/10.1136/gut.29.6.795] [PMID: 3384364]

[198] Magne D, Favennec L, Chochillon C, *et al.* Role of cytoskeleton and surface lectins in *Giardia duodenalis* attachment to Caco2 cells. Parasitol Res 1991; 77(8): 659-62.
[http://dx.doi.org/10.1007/BF00928679] [PMID: 1805208]

[199] Knaippe F. *Giardia lamblia* attachment to biological and inert substrates. Microsc Electron Biol Celular 1990; 14(1): 35-43.
[PMID: 2102989]

[200] Coradi ST, Guimarães S. *Giardia duodenalis*: protein substrates degradation by trophozoite proteases. Parasitol Res 2006; 99(2): 131-6.
[http://dx.doi.org/10.1007/s00436-005-0124-5] [PMID: 16521040]

[201] de Carvalho TB, David EB, Coradi ST, Guimarães S. Protease activity in extracellular products secreted *in vitro* by trophozoites of *Giardia duodenalis.* Parasitol Res 2008; 104(1): 185-90.
[http://dx.doi.org/10.1007/s00436-008-1185-z] [PMID: 18797927]

[202] Jiménez JC, Fontaine J, Grzych JM, Dei-Cas E, Capron M. Systemic and mucosal responses to oral administration of excretory and secretory antigens from *Giardia intestinalis.* Clin Diagn Lab Immunol 2004; 11(1): 152-60.
[PMID: 14715563]

[203] Lev B, Ward H, Keusch GT, Pereira ME. Lectin activation in *Giardia lamblia* by host protease: a novel host-parasite interaction. Science 1986; 232(4746): 71-3.
[http://dx.doi.org/10.1126/science.3513312] [PMID: 3513312]

[204] Ward HD, Lev BI, Kane AV, Keusch GT, Pereira ME. Identification and characterization of taglin, a mannose 6-phosphate binding, trypsin-activated lectin from *Giardia lamblia.* Biochemistry 1987; 26(26): 8669-75.
[http://dx.doi.org/10.1021/bi00400a027] [PMID: 3442682]

[205] Farthing MJ, Pereira ME, Keusch GT. Description and characterization of a surface lectin from *Giardia lamblia.* Infect Immun 1986; 51(2): 661-7.
[PMID: 3943906]

[206] Bauer B, Engelbrecht S, Bakker-Grunwald T, Scholze H. Functional identification of alpha 1-giardin

as an annexin of *Giardia lamblia.* FEMS Microbiol Lett 1999; 173(1): 147-53.
[PMID: 10220891]

[207] Ulluwishewa D, Anderson RC, McNabb WC, Moughan PJ, Wells JM, Roy NC. Regulation of tight junction permeability by intestinal bacteria and dietary components. J Nutr 2011; 141(5): 769-76.
[http://dx.doi.org/10.3945/jn.110.135657] [PMID: 21430248]

[208] Buret AG, Mitchell K, Muench DG, Scott KG. *Giardia lamblia* disrupts tight junctional ZO-1 and increases permeability in non-transformed human small intestinal epithelial monolayers: effects of epidermal growth factor. Parasitology 2002; 125(Pt 1): 11-9.
[http://dx.doi.org/10.1017/S0031182002001853] [PMID: 12166516]

[209] Buret A, Gall DG, Olson ME. Effects of murine giardiasis on growth, intestinal morphology, and disaccharidase activity. J Parasitol 1990; 76(3): 403-9.
[http://dx.doi.org/10.2307/3282675] [PMID: 2191103]

[210] Buret A, Gall DG, Olson ME. Growth, activities of enzymes in the small intestine, and ultrastructure of microvillous border in gerbils infected with *Giardia duodenalis.* Parasitol Res 1991; 77(2): 109-14.
[http://dx.doi.org/10.1007/BF00935423] [PMID: 2027878]

[211] Singh KD, Bhasin DK, Rana SV, *et al.* Effect of *Giardia lamblia* on duodenal disaccharidase levels in humans. Trop Gastroenterol 2000; 21(4): 174-6.
[PMID: 11194577]

[212] Mahmood S, Sodhi CP, Ganguly NK. Expression of sodium-glucose co-transporter and brush border disaccharidases in *Giardia lamblia* infected rat intestine. Indian J Biochem Biophys 2002; 39(3): 185-90.
[PMID: 22905389]

[213] Solaymani-Mohammadi S, Singer SM. Host immunity and pathogen strain contribute to intestinal disaccharidase impairment following gut infection. J Immunol 2011; 187(7): 3769-75.
[http://dx.doi.org/10.4049/jimmunol.1100606] [PMID: 21873528]

[214] Sackey ME, Weigel MM, Armijos RX. Predictors and nutritional consequences of intestinal parasitic infections in rural Ecuadorian children. J Trop Pediatr 2003; 49(1): 17-23.
[http://dx.doi.org/10.1093/tropej/49.1.17] [PMID: 12630715]

[215] Monajemzadeh SM, Monajemzadeh M. Comparison of iron and hematological indices in *Giardia lamblia* infection before and after treatment in 102 children in Ahwaz, Iran. Med Sci Monit 2008; 14(1): CR19-23.
[PMID: 18160940]

[216] Gomes MA, de Oliveira DR, de Freitas SE, de Pinho Viana M, Borges EL. Effect of giardiasis combined with low-protein diet on intestinal absorption of glucose and electrolytes in gerbils. Exp Parasitol 2012; 131(4): 448-51.
[http://dx.doi.org/10.1016/j.exppara.2012.04.016] [PMID: 22659230]

[217] Gorowara S, Ganguly NK, Mahajan RC, Walia BN. Study on the mechanism of *Giardia lamblia* induced diarrhoea in mice. Biochim Biophys Acta 1992; 1138(2): 122-6.
[http://dx.doi.org/10.1016/0925-4439(92)90051-N] [PMID: 1540658]

[218] Buret AG. Pathophysiology of enteric infections with *Giardia duodenalis.* Parasite 2008; 15(3): 261-5.
[http://dx.doi.org/10.1051/parasite/2008153261] [PMID: 18814692]

[219] Li E, Zhou P, Singer SM. Neuronal nitric oxide synthase is necessary for elimination of *Giardia lamblia* infections in mice. J Immunol 2006; 176(1): 516-21.

[http://dx.doi.org/10.4049/jimmunol.176.1.516] [PMID: 16365445]

[220] Li E, Zhao A, Shea-Donohue T, Singer SM. Mast cell-mediated changes in smooth muscle contractility during mouse giardiasis. Infect Immun 2007; 75(9): 4514-8. [http://dx.doi.org/10.1128/IAI.00596-07] [PMID: 17620354]

[221] Deselliers LP, Tan DT, Scott RB, Olson ME. Effects of *Giardia lamblia* infection on gastrointestinal transit and contractility in Mongolian gerbils. Dig Dis Sci 1997; 42(12): 2411-9. [http://dx.doi.org/10.1023/A:1018879621272] [PMID: 9440613]

[222] Robayo-Torres CC, Quezada-Calvillo R, Nichols BL. Disaccharide digestion: clinical and molecular aspects. Clin Gastroenterol Hepatol 2006; 4(3): 276-87. [http://dx.doi.org/10.1016/j.cgh.2005.12.023] [PMID: 16527688]

[223] Morken MH, Nysaeter G, Strand EA, Hausken T, Berstad A. Lactulose breath test results in patients with persistent abdominal symptoms following *Giardia lamblia* infection. Scand J Gastroenterol 2008; 43(2): 141-5. [http://dx.doi.org/10.1080/00365520701673960] [PMID: 17943632]

[224] Morken MH, Valeur J, Norin E, Midtvedt T, Nysaeter G, Berstad A. Antibiotic or bacterial therapy in post-giardiasis irritable bowel syndrome. Scand J Gastroenterol 2009; 44(11): 1296-303. [http://dx.doi.org/10.3109/00365520903274401] [PMID: 19821794]

[225] Phillips SF. Fatty Acid Diarrhoea: Relationships between Diarrhoea and Steatorrhoea. In: Rommel K, Goebell H, Böhmer R, Eds. Lipid Absorption: Biochemical and Clinical Aspects. Netherlands: Springer Netherlands 1976; pp. 241-54. [http://dx.doi.org/10.1007/978-94-011-7176-2_12]

[226] Solaymani-Mohammadi S, Singer SM. Regulation of intestinal epithelial cell cytoskeletal remodeling by cellular immunity following gut infection. Mucosal Immunol 2013; 6(2): 369-78. [http://dx.doi.org/10.1038/mi.2012.80] [PMID: 22910215]

[227] Scott KG, Yu LC, Buret AG. Role of CD8+ and CD4+ T lymphocytes in jejunal mucosal injury during murine giardiasis. Infect Immun 2004; 72(6): 3536-42. [http://dx.doi.org/10.1128/IAI.72.6.3536-3542.2004] [PMID: 15155662]

[228] Hall PA, Coates PJ, Ansari B, Hopwood D. Regulation of cell number in the mammalian gastrointestinal tract: the importance of apoptosis. J Cell Sci 1994; 107(Pt 12): 3569-77. [PMID: 7706406]

[229] Flynn AN, Wang A, McKay DM, Buret AG. Apoptosis-inducing factor contributes to epithelial cell apoptosis induced by enteropathogenic *Escherichia coli.* Can J Physiol Pharmacol 2011; 89(2): 143-8. [http://dx.doi.org/10.1139/Y11-002] [PMID: 21326346]

[230] Slee EA, Adrain C, Martin SJ. Serial killers: ordering caspase activation events in apoptosis. Cell Death Differ 1999; 6(11): 1067-74. [http://dx.doi.org/10.1038/sj.cdd.4400601] [PMID: 10578175]

[231] Bojarski C, Gitter AH, Bendfeldt K, *et al.* Permeability of human HT-29/B6 colonic epithelium as a function of apoptosis. J Physiol 2001; 535(Pt 2): 541-52. [http://dx.doi.org/10.1111/j.1469-7793.2001.00541.x] [PMID: 11533143]

[232] Zeissig S, Bojarski C, Buergel N, *et al.* Downregulation of epithelial apoptosis and barrier repair in active Crohns disease by tumour necrosis factor alpha antibody treatment. Gut 2004; 53(9): 1295-302. [http://dx.doi.org/10.1136/gut.2003.036632] [PMID: 15306588]

[233] Scott KG, Meddings JB, Kirk DR, Lees-Miller SP, Buret AG. Intestinal infection with *Giardia* spp.

reduces epithelial barrier function in a myosin light chain kinase-dependent fashion. Gastroenterology 2002; 123(4): 1179-90.
[http://dx.doi.org/10.1053/gast.2002.36002] [PMID: 12360480]

[234] Murphy BJ, Robin ED, Tapper DP, Wong RJ, Clayton DA. Hypoxic coordinate regulation of mitochondrial enzymes in mammalian cells. Science 1984; 223(4637): 707-9.
[http://dx.doi.org/10.1126/science.6320368] [PMID: 6320368]

[235] Denning TL, Takaishi H, Crowe SE, Boldogh I, Jevnikar A, Ernst PB. Oxidative stress induces the expression of Fas and Fas ligand and apoptosis in murine intestinal epithelial cells. Free Radic Biol Med 2002; 33(12): 1641-50.
[http://dx.doi.org/10.1016/S0891-5849(02)01141-3] [PMID: 12488132]

[236] Weinmann M, Jendrossek V, Handrick R, Güner D, Goecke B, Belka C. Molecular ordering of hypoxia-induced apoptosis: critical involvement of the mitochondrial death pathway in a FADD/caspase-8 independent manner. Oncogene 2004; 23(21): 3757-69.
[http://dx.doi.org/10.1038/sj.onc.1207481] [PMID: 15034549]

[237] Rollwagen FM, Madhavan S, Singh A, Li YY, Wolcott K, Maheshwari R. IL-6 protects enterocytes from hypoxia-induced apoptosis by induction of bcl-2 mRNA and reduction of fas mRNA. Biochem Biophys Res Commun 2006; 347(4): 1094-8.
[http://dx.doi.org/10.1016/j.bbrc.2006.07.016] [PMID: 16870148]

[238] Circu ML, Aw TY. Reactive oxygen species, cellular redox systems, and apoptosis. Free Radic Biol Med 2010; 48(6): 749-62.
[http://dx.doi.org/10.1016/j.freeradbiomed.2009.12.022] [PMID: 20045723]

[239] Malhotra R, Lin Z, Vincenz C, Brosius FC III. Hypoxia induces apoptosis *via* two independent pathways in Jurkat cells: differential regulation by glucose. Am J Physiol Cell Physiol 2001; 281(5): C1596-603.
[PMID: 11600423]

[240] Yu LC, Huang CY, Kuo WT, Sayer H, Turner JR, Buret AG. SGLT-1-mediated glucose uptake protects human intestinal epithelial cells against *Giardia duodenalis*-induced apoptosis. Int J Parasitol 2008; 38(8-9): 923-34.
[http://dx.doi.org/10.1016/j.ijpara.2007.12.004] [PMID: 18281046]

[241] Boettner DR, Petri WA. *Entamoeba histolytica* activates host cell caspases during contact-dependent cell killing. Curr Top Microbiol Immunol 2005; 289: 175-84.
[http://dx.doi.org/10.1007/3-540-27320-4_8] [PMID: 15791956]

[242] Singh D, Naik SR, Naik S. Role of cysteine proteinase of *Entamoeba histolytica* in target cell death. Parasitology 2004; 129(Pt 2): 127-35.
[http://dx.doi.org/10.1017/S0031182004005451] [PMID: 15376772]

[243] Sommer U, Costello CE, Hayes GR, *et al.* Identification of *Trichomonas vaginalis* cysteine proteases that induce apoptosis in human vaginal epithelial cells. J Biol Chem 2005; 280(25): 23853-60.
[http://dx.doi.org/10.1074/jbc.M501752200] [PMID: 15843376]

[244] Chin AC, Vergnolle N, MacNaughton WK, Wallace JL, Hollenberg MD, Buret AG. Proteinase-activated receptor 1 activation induces epithelial apoptosis and increases intestinal permeability. Proc Natl Acad Sci USA 2003; 100(19): 11104-9.
[http://dx.doi.org/10.1073/pnas.1831452100] [PMID: 12960392]

[245] Brown DM, Upcroft JA, Edwards MR, Upcroft P. Anaerobic bacterial metabolism in the ancient eukaryote *Giardia duodenalis*. Int J Parasitol 1998; 28(1): 149-64.
[http://dx.doi.org/10.1016/S0020-7519(97)00172-0] [PMID: 9504342]

[246] Edwards MR, Schofield PJ, OSullivan WJ, Costello M. Arginine metabolism during culture of *Giardia intestinalis*. Mol Biochem Parasitol 1992; 53(1-2): 97-103.
[http://dx.doi.org/10.1016/0166-6851(92)90011-8] [PMID: 1501649]

[247] Rópolo AS, Touz MC. A lesson in survival, by *Giardia lamblia*. ScientificWorldJournal 2010; 10: 2019-31.
[http://dx.doi.org/10.1100/tsw.2010.200] [PMID: 20953552]

[248] Philip R, Campbell E, Wheatley DN. Arginine deprivation, growth inhibition and tumour cell death: 2. Enzymatic degradation of arginine in normal and malignant cell cultures. Br J Cancer 2003; 88(4): 613-23.
[http://dx.doi.org/10.1038/sj.bjc.6600681] [PMID: 12592378]

[249] Potoka DA, Upperman JS, Zhang XR, *et al.* Peroxynitrite inhibits enterocyte proliferation and modulates Src kinase activity *in vitro.* Am J Physiol Gastrointest Liver Physiol 2003; 285(5): G861-9.
[http://dx.doi.org/10.1152/ajpgi.00412.2002] [PMID: 12842830]

[250] Stadelmann B, Hanevik K, Andersson MK, Bruserud O, Svärd SG. The role of arginine and arginine-metabolizing enzymes during *Giardia* - host cell interactions *in vitro.* BMC Microbiol 2013; 13: 256.
[http://dx.doi.org/10.1186/1471-2180-13-256] [PMID: 24228819]

[251] Solaymani-Mohammadi S, Singer SM. *Giardia duodenalis*: the double-edged sword of immune responses in giardiasis. Exp Parasitol 2010; 126(3): 292-7.
[http://dx.doi.org/10.1016/j.exppara.2010.06.014] [PMID: 20599999]

[252] Turchany JM, McCaffery JM, Aley SB, Gillin FD. Ultrastructural effects of lactoferrin binding on *Giardia lamblia* trophozoites. J Eukaryot Microbiol 1997; 44(1): 68-72.
[http://dx.doi.org/10.1111/j.1550-7408.1997.tb05694.x] [PMID: 9172835]

[253] Turchany JM, Aley SB, Gillin FD. Giardicidal activity of lactoferrin and N-terminal peptides. Infect Immun 1995; 63(11): 4550-2.
[PMID: 7591103]

[254] Aley SB, Zimmerman M, Hetsko M, Selsted ME, Gillin FD. Killing of *Giardia lamblia* by cryptdins and cationic neutrophil peptides. Infect Immun 1994; 62(12): 5397-403.
[PMID: 7960119]

[255] Ochoa TJ, Chea-Woo E, Campos M, *et al.* Impact of lactoferrin supplementation on growth and prevalence of *Giardia* colonization in children. Clin Infect Dis 2008; 46(12): 1881-3.
[http://dx.doi.org/10.1086/588476] [PMID: 18462105]

[256] Ouellette AJ. Paneth cells and innate immunity in the crypt microenvironment. Gastroenterology 1997; 113(5): 1779-84.
[http://dx.doi.org/10.1053/gast.1997.v113.pm9352884] [PMID: 9352884]

[257] Wilson CL, Ouellette AJ, Satchell DP, *et al.* Regulation of intestinal alpha-defensin activation by the metalloproteinase matrilysin in innate host defense. Science 1999; 286(5437): 113-7.
[http://dx.doi.org/10.1126/science.286.5437.113] [PMID: 10506557]

[258] El-Shewy KA, Eid RA. *In vivo* killing of *Giardia* trophozoites harbouring bacterial endosymbionts by intestinal Paneth cells: an ultrastructural study. Parasitology 2005; 130(Pt 3): 269-74.
[http://dx.doi.org/10.1017/S0031182004006547] [PMID: 15796009]

[259] Spooner R, Yilmaz O. The role of reactive-oxygen-species in microbial persistence and inflammation. Int J Mol Sci 2011; 12(1): 334-52.
[http://dx.doi.org/10.3390/ijms12010334] [PMID: 21339989]

[260] Jones RM, Mercante JW, Neish AS. Reactive oxygen production induced by the gut microbiota:

pharmacotherapeutic implications. Curr Med Chem 2012; 19(10): 1519-29. [http://dx.doi.org/10.2174/092986712799828283] [PMID: 22360484]

[261] Neish AS, Jones RM. Redox signaling mediates symbiosis between the gut microbiota and the intestine. Gut Microbes 2014; 5(2): 250-3. [http://dx.doi.org/10.4161/gmic.27917] [PMID: 24637602]

[262] Owen RL, Allen CL, Stevens DP. Phagocytosis of *Giardia muris* by macrophages in Peyers patch epithelium in mice. Infect Immun 1981; 33(2): 591-601. [PMID: 7275318]

[263] Hill DR, Pearson RD. Ingestion of *Giardia lamblia* trophozoites by human mononuclear phagocytes. Infect Immun 1987; 55(12): 3155-61. [PMID: 3679547]

[264] Hill DR, Pohl R. Ingestion of *Giardia lamblia* trophozoites by murine Peyers patch macrophages. Infect Immun 1990; 58(10): 3202-7. [PMID: 2401561]

[265] Belosevic M, Daniels CW. Phagocytosis of *Giardia lamblia* trophozoites by cytokine-activated macrophages. Clin Exp Immunol 1992; 87(2): 304-9. [http://dx.doi.org/10.1111/j.1365-2249.1992.tb02992.x] [PMID: 1735194]

[266] Goyal R, Mahajan RC, Ganguly NK, Sehgal R, Gorowara S, Singh K. Macrophage-mediated enterocyte damage in BALB/c mice infected with different strains of *Giardia lamblia.* Scand J Gastroenterol 1993; 28(9): 845-8. [http://dx.doi.org/10.3109/00365529309104020] [PMID: 8235442]

[267] Mastronicola D, Falabella M, Testa F, *et al.* Functional characterization of peroxiredoxins from the human protozoan parasite *Giardia intestinalis.* PLoS Negl Trop Dis 2014; 8(1): e2631. [http://dx.doi.org/10.1371/journal.pntd.0002631] [PMID: 24416465]

[268] Di Matteo A, Scandurra FM, Testa F, *et al.* The O2-scavenging flavodiiron protein in the human parasite *Giardia intestinalis.* J Biol Chem 2008; 283(7): 4061-8. [http://dx.doi.org/10.1074/jbc.M705605200] [PMID: 18077462]

[269] Pavanelli WR, Gutierrez FR, Silva JJ, *et al.* The effects of nitric oxide on the immune response during giardiasis. Braz J Infect Dis 2010; 14(6): 606-12. [http://dx.doi.org/10.1016/S1413-8670(10)70119-7] [PMID: 21340302]

[270] Lefer AM, Lefer DJ. Nitric oxide. II. Nitric oxide protects in intestinal inflammation. Am J Physiol 1999; 276(3 Pt 1): G572-5. [PMID: 10070031]

[271] Potoka DA, Nadler EP, Upperman JS, Ford HR. Role of nitric oxide and peroxynitrite in gut barrier failure. World J Surg 2002; 26(7): 806-11. [http://dx.doi.org/10.1007/s00268-002-4056-2] [PMID: 11948371]

[272] Marc Rhoads J, Wu G. Glutamine, arginine, and leucine signaling in the intestine. Amino Acids 2009; 37(1): 111-22. [http://dx.doi.org/10.1007/s00726-008-0225-4] [PMID: 19130170]

[273] Hardin JA, Buret AG, Olson ME, Kimm MH, Gall DG. Mast cell hyperplasia and increased macromolecular uptake in an animal model of giardiasis. J Parasitol 1997; 83(5): 908-12. [http://dx.doi.org/10.2307/3284287] [PMID: 9379297]

[274] Di Prisco MC, Hagel I, Lynch NR, Barrios RM, Alvarez N, López R. Possible relationship between allergic disease and infection by *Giardia lamblia.* Ann Allergy 1993; 70(3): 210-3. [PMID: 8452315]

[275] Di Prisco MC, Hagel I, Lynch NR, *et al.* Association between giardiasis and allergy. Ann Allergy

Asthma Immunol 1998; 81(3): 261-5.
[http://dx.doi.org/10.1016/S1081-1206(10)62823-2] [PMID: 9759805]

[276] Li E, Zhou P, Petrin Z, Singer SM. Mast cell-dependent control of *Giardia lamblia* infections in mice. Infect Immun 2004; 72(11): 6642-9.
[http://dx.doi.org/10.1128/IAI.72.11.6642-6649.2004] [PMID: 15501797]

[277] Muñoz-Cruz S, Gómez-García A, Millán-Ibarra J, Giono-Cerezo S, Yépez-Mulia L. *Giardia lamblia*: interleukin 6 and tumor necrosis factor-alpha release from mast cells induced through an Ig-independent pathway. Exp Parasitol 2010; 126(3): 298-303.
[http://dx.doi.org/10.1016/j.exppara.2010.06.013] [PMID: 20600000]

[278] Merluzzi S, Frossi B, Gri G, Parusso S, Tripodo C, Pucillo C. Mast cells enhance proliferation of B lymphocytes and drive their differentiation toward IgA-secreting plasma cells. Blood 2010; 115(14): 2810-7.
[http://dx.doi.org/10.1182/blood-2009-10-250126] [PMID: 20101023]

[279] Hemmi H, Akira S. TLR signalling and the function of dendritic cells. Chem Immunol Allergy 2005; 86: 120-35.
[http://dx.doi.org/10.1159/000086657] [PMID: 15976491]

[280] Sher A, Pearce E, Kaye P. Shaping the immune response to parasites: role of dendritic cells. Curr Opin Immunol 2003; 15(4): 421-9.
[http://dx.doi.org/10.1016/S0952-7915(03)00072-4] [PMID: 12900274]

[281] Pulendran B. Modulating TH1/TH2 responses with microbes, dendritic cells, and pathogen recognition receptors. Immunol Res 2004; 29(1-3): 187-96.
[http://dx.doi.org/10.1385/IR:29:1-3:187] [PMID: 15181281]

[282] Obendorf J, Renner Viveros P, Fehlings M, Klotz C, Aebischer T, Ignatius R. Increased expression of CD25, CD83, and CD86, and secretion of IL-12, IL-23, and IL-10 by human dendritic cells incubated in the presence of Toll-like receptor 2 ligands and *Giardia duodenalis*. Parasit Vectors 2013; 6(1): 317.
[http://dx.doi.org/10.1186/1756-3305-6-317] [PMID: 24499474]

[283] Grit GH, Devriendt B, Van Coppernolle S, *et al.* *Giardia duodenalis* stimulates partial maturation of bovine dendritic cells associated with altered cytokine secretion and induction of T-cell proliferation. Parasite Immunol 2014; 36(4): 157-69.
[http://dx.doi.org/10.1111/pim.12095] [PMID: 24372144]

[284] Jimenez JC, Fontaine J, Grzych JM, Capron M, Dei-Cas E. Antibody and cytokine responses in BALB/c mice immunized with the excreted/secreted proteins of *Giardia intestinalis*: the role of cysteine proteases. Ann Trop Med Parasitol 2009; 103(8): 693-703.
[http://dx.doi.org/10.1179/000349809X12502035776351] [PMID: 20030993]

[285] Jiménez JC, Fontaine J, Creusy C, *et al.* Antibody and cytokine responses to *Giardia* excretory/secretory proteins in *Giardia intestinalis*-infected BALB/c mice. Parasitol Res 2014; 113(7): 2709-18.
[http://dx.doi.org/10.1007/s00436-014-3927-4] [PMID: 24867815]

[286] Lee HY, Kim J, Noh HJ, Kim HP, Park SJ. *Giardia lamblia* binding immunoglobulin protein triggers maturation of dendritic cells *via* activation of TLR4-MyD88-p38 and ERK1/2 MAPKs. Parasite Immunol 2014; 36(12): 627-46.
[http://dx.doi.org/10.1111/pim.12119] [PMID: 24871487]

[287] Fisher BS, Estraño CE, Cole JA. Modeling long-term host cell-*Giardia lamblia* interactions in an *in vitro* co-culture system. PLoS One 2013; 8(12): e81104.
[http://dx.doi.org/10.1371/journal.pone.0081104] [PMID: 24312526]

[288] Chen TL, Chen S, Wu HW, *et al.* Persistent gut barrier damage and commensal bacterial influx following eradication of *Giardia* infection in mice. Gut Pathog 2013; 5(1): 26. [http://dx.doi.org/10.1186/1757-4749-5-26] [PMID: 23991642]

[289] Bartelt LA, Roche J, Kolling G, *et al.* Persistent *G. lamblia* impairs growth in a murine malnutrition model. J Clin Invest 2013; 123(6): 2672-84. [http://dx.doi.org/10.1172/JCI67294] [PMID: 23728173]

[290] Matowicka-Karna J, Dymicka-Piekarska V, Kemona H. IFN-gamma, IL-5, IL-6 and IgE in patients infected with *Giardia intestinalis*. Folia Histochem Cytobiol 2009; 47(1): 93-7. [http://dx.doi.org/10.2478/v10042-009-0013-3] [PMID: 19419945]

[291] Bayraktar MR, Mehmet N, Durmaz R. Serum cytokine changes in Turkish children infected with *Giardia lamblia* with and without allergy: Effect of metronidazole treatment. Acta Trop 2005; 95(2): 116-22. [http://dx.doi.org/10.1016/j.actatropica.2005.05.006] [PMID: 15979046]

[292] Long KZ, Rosado JL, Santos JI, *et al.* Associations between mucosal innate and adaptive immune responses and resolution of diarrheal pathogen infections. Infect Immun 2010; 78(3): 1221-8. [http://dx.doi.org/10.1128/IAI.00767-09] [PMID: 20038536]

[293] Zhou P, Li E, Zhu N, Robertson J, Nash T, Singer SM. Role of interleukin-6 in the control of acute and chronic *Giardia lamblia* infections in mice. Infect Immun 2003; 71(3): 1566-8. [http://dx.doi.org/10.1128/IAI.71.3.1566-1568.2003] [PMID: 12595478]

[294] Zhou P, Li E, Shea-Donohue T, Singer SM. Tumour necrosis factor alpha contributes to protection against *Giardia lamblia* infection in mice. Parasite Immunol 2007; 29(7): 367-74. [http://dx.doi.org/10.1111/j.1365-3024.2007.00953.x] [PMID: 17576366]

[295] Ljungström I, Castor B. Immune response to *Giardia lamblia* in a water-borne outbreak of giardiasis in Sweden. J Med Microbiol 1992; 36(5): 347-52. [http://dx.doi.org/10.1099/00222615-36-5-347] [PMID: 1588586]

[296] Singer SM, Kamda J. Immune Response to *Giardia* Infection: Lessons from Animal Models. In: Ortega Pierres P, Caccio SM, Fayer R, Mank TG, Smith HV, Thompson RCA, Eds. *Giardia* and *Cryptosporidium*: from molecules to disease. Wallingford, UK: Cabi Publishing 2009; pp. 451-62.

[297] Singer SM. Immunology of giardiasis. In: Luján HD, Svärd S, Eds. *Giardia*: A Model Organism. Austria: Springer Wien New York 2011; pp. 319-31. [http://dx.doi.org/10.1007/978-3-7091-0198-8_20]

[298] Webster AD. Giardiasis and immunodeficiency diseases. Trans R Soc Trop Med Hyg 1980; 74(4): 440-3. [http://dx.doi.org/10.1016/0035-9203(80)90047-4] [PMID: 7445039]

[299] Perlmutter DH, Leichtner AM, Goldman H, Winter HS. Chronic diarrhea associated with hypogammaglobulinemia and enteropathy in infants and children. Dig Dis Sci 1985; 30(12): 1149-55. [http://dx.doi.org/10.1007/BF01314049] [PMID: 4064865]

[300] Singer SM, Nash TE. T-cell-dependent control of acute *Giardia lamblia* infections in mice. Infect Immun 2000; 68(1): 170-5. [http://dx.doi.org/10.1128/IAI.68.1.170-175.2000] [PMID: 10603384]

[301] Davids BJ, Palm JE, Housley MP, *et al.* Polymeric immunoglobulin receptor in intestinal immune defense against the lumen-dwelling protozoan parasite *Giardia*. J Immunol 2006; 177(9): 6281-90. [http://dx.doi.org/10.4049/jimmunol.177.9.6281] [PMID: 17056558]

[302] Roberts-Thomson IC, Mitchell GF. Giardiasis in mice. I. Prolonged infections in certain mouse strains and hypothymic (nude) mice. Gastroenterology 1978; 75(1): 42-6. [PMID: 401094]

[303] Janoff EN, Smith PD, Blaser MJ. Acute antibody responses to *Giardia lamblia* are depressed in patients with AIDS. J Infect Dis 1988; 157(4): 798-804.
[http://dx.doi.org/10.1093/infdis/157.4.798] [PMID: 3346571]

[304] Barrett DM, Steel-Duncan J, Christie CD, Eldemire-Shearer D, Lindo JF. Absence of opportunistic parasitic infestations in children living with HIV/AIDS in childrens homes in Jamaica: pilot investigations. West Indian Med J 2008; 57(3): 253-6.
[PMID: 19583124]

[305] Viriyavejakul P, Nintasen R, Punsawad C, Chaisri U, Punpoowong B, Riganti M. High prevalence of *Microsporidium* infection in HIV-infected patients. Southeast Asian J Trop Med Public Health 2009; 40(2): 223-8.
[PMID: 19323005]

[306] Einfeld DA, Stibbs HH. Identification and characterization of a major surface antigen of *Giardia lamblia.* Infect Immun 1984; 46(2): 377-83.
[PMID: 6389343]

[307] Moss DM, Mathews HM, Visvesvara GS, Dickerson JW, Walker EM. Antigenic variation of *Giardia lamblia* in the feces of Mongolian gerbils. J Clin Microbiol 1990; 28(2): 254-7.
[PMID: 2312672]

[308] Crossley R, Holberton D. Assembly of 2.5 nm filaments from giardin, a protein associated with cytoskeletal microtubules in *Giardia.* J Cell Sci 1985; 78: 205-31.
[PMID: 4093472]

[309] Crossley R, Holberton DV. Characterization of proteins from the cytoskeleton of *Giardia lamblia.* J Cell Sci 1983; 59: 81-103.
[PMID: 6863412]

[310] Char S, Cevallos AM, Farthing MJ. An immunodominant antigen of *Giardia lamblia* is a heat shock protein. Biotechnol Ther 1992; 3(3-4): 151-7.
[PMID: 1303728]

[311] Téllez A, Palm D, Weiland M, *et al.* Secretory antibodies against *Giardia intestinalis* in lactating Nicaraguan women. Parasite Immunol 2005; 27(5): 163-9.
[http://dx.doi.org/10.1111/j.1365-3024.2005.00758.x] [PMID: 15987339]

[312] Ish-Horowicz M, Korman SH, Shapiro M, *et al.* Asymptomatic giardiasis in children. Pediatr Infect Dis J 1989; 8(11): 773-9.
[http://dx.doi.org/10.1097/00006454-198911000-00009] [PMID: 2512565]

[313] Faubert G. Immune response to *Giardia duodenalis.* Clin Microbiol Rev 2000; 13(1): 35-54.
[http://dx.doi.org/10.1128/CMR.13.1.35-54.2000] [PMID: 10627490]

[314] Chester AC, MacMurray FG, Restifo MD, Mann O. Giardiasis as a chronic disease. Dig Dis Sci 1985; 30(3): 215-8.
[http://dx.doi.org/10.1007/BF01347886] [PMID: 2578921]

[315] Buret A, Cotton J. Pathophysiological processes and clinical manifestations of giardiasis. In: Luján HD, Svärd S, Eds. *Giardia*: A Model Organism. Austria: SpringerWienNew York 2011; pp. 301-17.
[http://dx.doi.org/10.1007/978-3-7091-0198-8_19]

[316] Arévalo F, Aragón V, Morales L D, Morales Caramutti D, Arandia J, Alcocer G. [Duodenal villous atrophy, an unexpectedly common finding in *Giardia lamblia* infestation]. Rev Gastroenterol Peru 2010; 30(4): 272-6.
[PMID: 21263755]

[317] Halliez MC, Buret AG. Extra-intestinal and long term consequences of *Giardia duodenalis* infections.

World J Gastroenterol 2013; 19(47): 8974-85.
[http://dx.doi.org/10.3748/wjg.v19.i47.8974] [PMID: 24379622]

[318] Homan WL, Mank TG. Human giardiasis: genotype linked differences in clinical symptomatology. Int J Parasitol 2001; 31(8): 822-6.
[http://dx.doi.org/10.1016/S0020-7519(01)00183-7] [PMID: 11403774]

[319] Breathnach AS, McHugh TD, Butcher PD. Prevalence and clinical correlations of genetic subtypes of *Giardia lamblia* in an urban setting. Epidemiol Infect 2010; 138(10): 1459-67.
[http://dx.doi.org/10.1017/S0950268810000208] [PMID: 20144251]

[320] Peréz Cordón G, Cordova Paz Soldan O, Vargas Vásquez F, *et al.* Prevalence of enteroparasites and genotyping of *Giardia lamblia* in Peruvian children. Parasitol Res 2008; 103(2): 459-65.
[http://dx.doi.org/10.1007/s00436-008-1007-3] [PMID: 18470699]

[321] Read C, Walters J, Robertson ID, Thompson RC. Correlation between genotype of *Giardia duodenalis* and diarrhoea. Int J Parasitol 2002; 32(2): 229-31.
[http://dx.doi.org/10.1016/S0020-7519(01)00340-X] [PMID: 11812501]

[322] Sahagún J, Clavel A, Goñi P, *et al.* Correlation between the presence of symptoms and the *Giardia duodenalis* genotype. Eur J Clin Microbiol Infect Dis 2008; 27(1): 81-3.
[http://dx.doi.org/10.1007/s10096-007-0404-3] [PMID: 17943329]

[323] Haque R, Mondal D, Karim A, *et al.* Prospective case-control study of the association between common enteric protozoal parasites and diarrhea in Bangladesh. Clin Infect Dis 2009; 48(9): 1191-7.
[http://dx.doi.org/10.1086/597580] [PMID: 19323634]

[324] Pelayo L, Nuñez FA, Rojas L, *et al.* *Giardia* infections in Cuban children: the genotypes circulating in a rural population. Ann Trop Med Parasitol 2008; 102(7): 585-95.
[http://dx.doi.org/10.1179/136485908X355247] [PMID: 18817599]

[325] Gelanew T, Lalle M, Hailu A, Pozio E, Cacciò SM. Molecular characterization of human isolates of *Giardia duodenalis* from Ethiopia. Acta Trop 2007; 102(2): 92-9.
[http://dx.doi.org/10.1016/j.actatropica.2007.04.003] [PMID: 17498637]

[326] Mohammed Mahdy AK, Surin J, Wan KL, Mohd-Adnan A, Al-Mekhlafi MS, Lim YA. *Giardia intestinalis* genotypes: Risk factors and correlation with clinical symptoms. Acta Trop 2009; 112(1): 67-70.
[http://dx.doi.org/10.1016/j.actatropica.2009.06.012] [PMID: 19560431]

[327] Tak V, Mirdha BR, Yadav P, Vyas P, Makharia GK, Bhatnagar S. Molecular characterisation of *Giardia intestinalis* assemblages from human isolates at a tertiary care centre of India. Indian J Med Microbiol 2014; 32(1): 19-25. [B].
[http://dx.doi.org/10.4103/0255-0857.124290] [PMID: 24399382]

[328] Molina N, Minvielle M, Grenóvero S, Salomón C, Basualdo J. High prevalences of infection with *Giardia intestinalis* genotype B among children in urban and rural areas of Argentina. Ann Trop Med Parasitol 2011; 105(4): 299-309.
[http://dx.doi.org/10.1179/136485911X12987676649665] [PMID: 21871166]

[329] Al-Mohammed HI. Genotypes of *Giardia intestinalis* clinical isolates of gastrointestinal symptomatic and asymptomatic Saudi children. Parasitol Res 2011; 108(6): 1375-81.
[http://dx.doi.org/10.1007/s00436-010-2033-5] [PMID: 20838811]

[330] Tungtrongchitr A, Sookrung N, Indrawattana N, Kwangsi S, Ongrotchanakun J, Chaicumpa W. *Giardia intestinalis* in Thailand: identification of genotypes. J Health Popul Nutr 2010; 28(1): 42-52.
[http://dx.doi.org/10.3329/jhpn.v28i1.4522] [PMID: 20214085]

[331] Gorczynski RM, Terzioglu E. Aging and the immune system. Int Urol Nephrol 2008; 40(4): 1117-25. [http://dx.doi.org/10.1007/s11255-008-9412-1] [PMID: 18683074]

[332] Torres MF, Uetanabaro AP, Costa AF, *et al.* Influence of bacteria from the duodenal microbiota of patients with symptomatic giardiasis on the pathogenicity of *Giardia duodenalis* in gnotoxenic mice. J Med Microbiol 2000; 49(3): 209-15. [http://dx.doi.org/10.1099/0022-1317-49-3-209] [PMID: 10707940]

[333] Gerbaba TK, Gupta P, Rioux K, Hansen D, Buret AG. *Giardia duodenalis*-induced alterations of commensal bacteria kill *Caenorhabditis elegans*: a new model to study microbial-microbial interactions in the gut. Am J Physiol Gastrointest Liver Physiol 2015; 308(6): G550-61. [http://dx.doi.org/10.1152/ajpgi.00335.2014] [PMID: 25573177]

[334] Koehler AV, Jex AR, Haydon SR, Stevens MA, Gasser RB. *Giardia*/giardiasis - a perspective on diagnostic and analytical tools. Biotechnol Adv 2014; 32(2): 280-9. [http://dx.doi.org/10.1016/j.biotechadv.2013.10.009] [PMID: 24189092]

[335] Jokipii AM, Jokipii L. Prepatency of giardiasis. Lancet 1977; 1(8021): 1095-7. [http://dx.doi.org/10.1016/S0140-6736(77)92345-5] [PMID: 68190]

[336] Verweij JJ, Schinkel J, Laeijendecker D, van Rooyen MA, van Lieshout L, Polderman AM. Real-time PCR for the detection of *Giardia lamblia.* Mol Cell Probes 2003; 17(5): 223-5. [http://dx.doi.org/10.1016/S0890-8508(03)00057-4] [PMID: 14580396]

[337] Verweij JJ, Blangé RA, Templeton K, *et al.* Simultaneous detection of *Entamoeba histolytica, Giardia lamblia*, and *Cryptosporidium parvum* in fecal samples by using multiplex real-time PCR. J Clin Microbiol 2004; 42(3): 1220-3. [http://dx.doi.org/10.1128/JCM.42.3.1220-1223.2004] [PMID: 15004079]

[338] ten Hove R, Schuurman T, Kooistra M, Möller L, van Lieshout L, Verweij JJ. Detection of diarrhoea-causing protozoa in general practice patients in The Netherlands by multiplex real-time PCR. Clin Microbiol Infect 2007; 13(10): 1001-7. [http://dx.doi.org/10.1111/j.1469-0691.2007.01788.x] [PMID: 17714523]

[339] Watkins RR, Eckmann L. Treatment of giardiasis: current status and future directions. Curr Infect Dis Rep 2014; 16(2): 396. [http://dx.doi.org/10.1007/s11908-014-0396-y] [PMID: 24493628]

[340] Ansell BR, McConville MJ, Maayeh SY, *et al.* Drug resistance in *Giardia duodenalis.* Biotechnol Adv 2015; 33(6 Pt 1): 888-901. [http://dx.doi.org/10.1016/j.biotechadv.2015.04.009] [PMID: 25922317]

[341] Gardner TB, Hill DR. Treatment of giardiasis. Clin Microbiol Rev 2001; 14(1): 114-28. [http://dx.doi.org/10.1128/CMR.14.1.114-128.2001] [PMID: 11148005]

[342] Rossignol JF, Maisonneuve H, Cho YW. Nitroimidazoles in the treatment of trichomoniasis, giardiasis, and amebiasis. Int J Clin Pharmacol Ther Toxicol 1984; 22(2): 63-72. [PMID: 6698665]

[343] Requena-Méndez A, Goñi P, Lóbez S, *et al.* A family cluster of giardiasis with variable treatment responses: refractory giardiasis in a family after a trip to India. Clin Microbiol Infect 2014; 20(2): O135-8. [http://dx.doi.org/10.1111/1469-0691.12327] [PMID: 23926944]

[344] Escobedo AA, Almirall P, Robertson LJ, *et al.* Giardiasis: the ever-present threat of a neglected disease. Infect Disord Drug Targets 2010; 10(5): 329-48. [http://dx.doi.org/10.2174/187152610793180821] [PMID: 20701575]

[345] United States Environment Protection Agency (USEPA). Final Long Term 2 Enhanced Surface Water Treatment Rule, EPA815-R-06-002. Ohio: Office of Water (4606-M) 2005.

[346] Medema G, Teunis O, Blokker M, Deere D, Charles P, Loret J-F. Risk assessment of *Cryptosporidium* in drinking water. Geneva: World Health Organization 2009.

[347] Drinking Water Inspectorate for England and Wales. Water Supply (Water Quality) (Amendment) Regulations (S.I. No. 1524 of 1999), 1999.

[348] Health Canada. Guidelines for Canadian Drinking Water Quality: Guideline Technical Document - Enteric Protozoa: *Giardia* and *Cryptosporidium*. Water, Air and Climate Change Bureau, Healthy Environments and Consumer Safety Branch, Health Canada, Ottawa, Ontario 2012. Available from http://healthycanadians.gc.ca/publications/healthy-living-vie-saine/water-protozoa-eau/alt/water-protozoa-eau-eng.pdf.

[349] Ministry of health. Guidelines for Drinking-water Quality Management for New Zealand (2nd edn), Chapter 8: Protozoa Compliance. Wellington: Ministry of health 2016. Available from http://www.health.govt.nz/publication/guidelines-drinking-water-quality-management-new-zealand.

[350] National Health and Medical Research Council (NHMRC). Australian Drinking Water Guidelines Paper 6 National Water Quality Management Strategy. National Health and Medical Research Council, National Resource Management Ministerial Council, Commonwealth of Australia, Canberra 2011. Available from https://www.clearwater.asn.au/user-data/resource-files/Aust_drinking_water_guidelines.pdf.

[351] US Department of the Interior Bureau of Reclamation. Managing Water in the West, Desalination Series Report 68, Water Treatment primer for communities in Need. Denver: Bureau of Reclamation 2010. Available from https://www.usbr.gov/research/AWT/reportpdfs/report068.pdf.

[352] Centers for Disease Control and Prevention (CDC) [homepage on Internet]. Atlanta: *Giardia* control and Prevention; [updated 18th July 2012]. Available from http://www.cdc.gov/parasites/giardia/prevention-control-general-public.html.

[353] Betancourt WQ, Rose JB. Drinking water treatment processes for removal of *Cryptosporidium* and *Giardia.* Vet Parasitol 2004; 126(1-2): 219-34.
[http://dx.doi.org/10.1016/j.vetpar.2004.09.002] [PMID: 15567586]

[354] Mujeriego R, Asano T. The role of advanced treatment in wastewater reclamation and reuse. Water Sci Technol 1999; 40: 1-9.
[http://dx.doi.org/10.1016/S0273-1223(99)00479-5]

[355] Hoff JC. Inactivation of microbial agents by chemical disinfectants, Report No. EPA-600/2-86-067. Cincinnati: US Environmental protection Agency 1986.

[356] Wickramanayake GB, Rubin AJ, Sproul OJ. Inactivation of *Giardia lamblia* cysts with ozone. Appl Environ Microbiol 1984; 48(3): 671-2.
[PMID: 6497374]

[357] Linden KG, Shin GA, Faubert G, Cairns W, Sobsey MD. UV disinfection of *Giardia lamblia* cysts in water. Environ Sci Technol 2002; 36(11): 2519-22.
[http://dx.doi.org/10.1021/es0113403] [PMID: 12075814]

[358] Mofidi AA, Meyer EA, Wallis PM, *et al.* The effect of UV light on the inactivation of *Giardia lamblia* and *Giardia muris* cysts as determined by animal infectivity assay (P-295101). Water Res 2002; 36(8): 2098-108.
[http://dx.doi.org/10.1016/S0043-1354(01)00412-2] [PMID: 12092585]

[359] Levy DA, Bens MS, Craun GF, Calderon RL, Herwaldt BL. Surveillance for waterborne-disease outbreaksUnited States, 19951996. MMWR CDC Surveill Summ 1998; 47(5): 1-34.
[PMID: 9859954]

[360] Joce RE, Bruce J, Kiely D, *et al.* An outbreak of cryptosporidiosis associated with a swimming pool. Epidemiol Infect 1991; 107(3): 497-508.
[http://dx.doi.org/10.1017/S0950268800049190] [PMID: 1752299]

[361] Centers for Disease Control and Prevention (CDC). Violations identified from routine swimming pool inspections selected states and counties, United States, 2008. MMWR Morb Mortal Wkly Rep 2010; 59(19): 582-7.
[PMID: 20489681]

[362] Centers for Disease Control and Prevention (CDC). Model Aquatic Health Code. Atlanta: US Department of Health and Human Services 2014. Available from http://www.cdc.gov/mahc/index.html.

CHAPTER 8

Microsporidia

Nadia A. El Dib[1,*] and **Thomas Weitzel**[2]

[1] *Faculty of Medicine, Cairo University, Cairo, Egypt*

[2] *Clínica Alemana School of Medicine, Universidad del Desarrollo, Santiago, Chile*

Abstract: Microsporidia the tiny unicellular eukaryotes are intracellular parasites of almost all animals. The diverged and specialized nature of these organisms, show some similarity to fungi. They cause opportunistic infections in animals and humans ranging between asymptomatic and severe life-threatening infections in immunocompromised individuals. Transmission occurs mainly by oral route, but other methods of transmission include inhalation, sexual contact, ocular mucosa, wounds, and insect bites. Food and water are relevant vehicles of infection. Animals act as reservoirs as they harbor most of the species that can also infect man and might contaminate water and environment with spores expelled in feces and/or urine. Clinical presentation is mainly intestinal with chronic diarrhea, mal-absorption, and loss of weight in immunocompromised persons, and self-limiting diarrhea in the immunocompetent individuals. Dissemination to other organs, may threaten the life of patients. Clinical picture of disseminated infection includes fever, cerebral manifestations or some other unexplained symptoms. Diagnosis of spores in feces, urine, CSF, sputum and in tissue is difficult and necessitates the use of special stains. Other methods of laboratory diagnosis include immunofluorescence, Electron Microscopy, and DNA detection. Treatment with Albendazole is effective for intestinal and other deep infections of various species of microsporidia except *E. bieneusi*, where fumagillin, can be considered. This drug is also used as topical treatment for eye infections by *E. hellem* and other species. Trials to produce vaccine against microsporidia are still under study. The increasing awareness will lead to a better understanding of the epidemiology, clinical relevance and control of microsporidiosis in humans and animals.

Keywords: AIDS, Dissemination, *Encephalitozoon*, *Enterocytozoon*, HIV, Microsporidia, Microsporidiosis, Parasitophorous vacuole, Protista, Polar tubule, *Septata*, Spores.

* **Corresponding author Nadia A. El Dib:** Faculty of Medicine, Cairo University, Cairo, Egypt; Tel; 002 01001171606: E-mail: naeldib@hotmail.com

Fabrizio Bruschi (Ed.)

INTRODUCTION

Microsporidia are unicellular, obligate intracellular spore formers of eukaryotic origin. They parasitize almost all animals. Understanding the basic biology of Microsporidia, have taken almost about 150 years of scientific research. The identification of DNA of organisms, created a new era of molecular phylogeny. Microsporidia, which were considered as protozoa (Kingdom Protista), are now considered as highly specialized fungi [1, 2]. There are more than 170 genera and approximately 1300 species of microsporidian organisms that parasitize a wide variety of vertebrates and invertebrates with at least 14 species and 8 genera known to infect humans [3 - 5]. The genera of microsporidia that cause human diseases are: *Nosema* [6, 7], *Brachiola* [8, 9], *Vittaforma* [10], *Pleistophora* [11, 12], *Trachipleistophora* [13], *Enterocytozoon* [14, 15], *Encephalitozoon* [16 - 18], *Septata* [19], and *Anncaliia* [20]. Microsporidia have been known to cause seriously damaging diseases in honeybees and silk worms, that consequently led to a serious economic loss [21]. Infection also was detected in different animals as rabbits, laboratory rodents and furred animals [22]. However microsporidia were considered as opportunistic pathogens in humans after the emergence of AIDS pandemic [23], and also have been detected in immunocompetent persons [24]. Infected cases may be asymptomatic or they may suffer severe life threatening disease according to the tissue or organs affected, as microsporidia may infect almost any part of the body. The most common site of infection is the intestine, which may account for up to 50% of all infections, with chronic diarrhea and wasting as the predominant manifestations [23 - 26].

HISTORY

There was an important economic problem in the year 1850, due a decline in the European silk industry as a result of a disease that affected the silk worms. This disease was called the pepper-disease (pébrine). Investigations were carried out in scientific centers in order to identify the microbial causative agents of the disease. There was some association between the disease and characteristic globular organisms, which were described, later by the Swiss microbiologist Karl Wilhelm von Nägeli in 1857 as the first microsporidium and he gave them the name *Nosema bombycis* [27]. Nägeli described *N. bombycis*, as a yeast-like fungus and included it in the Schizomycetes, which fits into the tree of eukaryotes according to the recent classification [27]. In 1870, Louis Pasteur incriminated microsporidia as a cause of infection of silkworm and cause of decimation of the silk industry. Aided by his colleagues, they could identify the nature of this parasite [28], with subsequent improvement of the European silk industry [29].

Further studies by Edouard-Gérard Balbiani by the year 1882, has created a new group for *Nosema* organisms, gave them the name 'microsporidies' and included them in the group of sporozoa within the Kingdom Protozoa [30]. Sporozoa is an old group of pathogens united together in an assemblage based on their similarity as spore formers. Recently, studies showed that they have distant relations and were subgrouped as members of Apicomplexes, haplosporidians and the Cnidosporidia. The last subgroup included Myxsosporidia (affecting animals), *Actinomyxidae* (of unknown origin), *Helicosporidia* (green algae) and the Microsporidia [31].

In the year 1976 Sprague created the Phylum *Microspora*, which was later included in the subkingdom Protozoa, a subdivision of the Kingdom Protista created in 1980 by Levine [30 - 33]. Shortly after, Sprague and Bencil changed the name of the phylum to Microsporidia, Balbiani 1882 [34]. This was in honor of Balbiani, who has created the order Microsporidia in 1882 [30].

Phylogeny and Taxonomy Considerations

Species of microsporidia have been classified according to studies based on their habitat, morphological and ultrastructural details. However the most important was the recent molecular phylogenetic classification [35, 36]. The most spectacular features of the spores of microsporidia have been explored by electron microscopy [37, 38],

Electron microscopy studies showed that the microsporidial spores lack some important structures of the Eukaryotic cells as mitochondria, peroxomes, Golgi apparatus, flagella and microtubules [39].

Intimate resemblance to fungi was proved by ultrastructural morphological studies based on the spore size, number of coils of the polar tube inside the spore as well as the life cycle and the host parasite relationship [40 - 43]. On the other hand, the resemblance to prokaryotes was revealed by biochemical analysis, after detecting that microsporidia include 70S ribosomes as in case of prokaryotes [44, 45]. Study of the rDNA sequences for the phylogenetic classification of microsporidia, suggested that they were among the deep-branching early eukaryotes [46]. This is based on lacking mitochondria, Golgi bodies and peroxisomes and having the small ribosomes of the prokaryotes as mentioned before [46].

Phylogenetic study of the sequence of the small subunit rRNA gene of *Variamphora necatrix*, one of the species of microsporidia, showed that there is closer resemblance to prokaryotes than to eukaryotes, suggesting that they have an ancient origin [46]. Furthermore, microsporidia showed that they possess fast evolving genes that make them closer to prokaryotes. At present, microsporidia are considered highly specialized, well-adapted and diverged organisms, that are either belonging to fungi or a near relative to them [47 - 50]. Studies on the genomic molecular sequence of the *Encephalitozoon cuniculi*, supported its relationship to fungi [51 - 54].

Nevertheless, in most medical textbooks microsporidias are still discussed within the parasite section and also the life cycle still uses the terminology of parasitic pathogens. The first human infection with microsporia species was detected in a 9 year-old Japanese boy that suffered from fever, vomiting and spastic convulsions due to dissemination of infection with *Encephalitozoon* [55].

Until 1985, there were only few detected cases with microsporidiosis, when a new species "*Enterocytozoon bieneusi*" was diagnosed in another case with AIDS from Haiti and subsequently other cases of intestinal microsporidiosis were detected in HIV positive patients in France [56, 14]. A wide range of studies have reported infection with microsporidiosis in non-HIV persons. However there is lacking of data concerned with parasitological detection of spores *versus* serology [57].

Over the last 25 years, there have been improvements of diagnostic methods and equipment, which eventually led to the identification of many other species of microsporidia, some of which can disseminate to different organs and show unexplained symptoms [58].

MORPHOLOGY AND LIFE CYCLE

Microsporidia are named for their small, resistant spore stage. The spores that infect humans are ovoid in shape, around 1.5 to 5 μm in length and ~1 μm in width. The spore coat is composed of an outer cover of a proteinacious electron dense material, a median endospore made of chitin and protein and a plasmalemma or an inner membrane [59]. The spore has a membrane-bound nucleus and an intra-cytoplasmic membrane system [60]. Nuclear configuration differs among genera of microsporidia. In the spores of some genera, two nuclei are arranged as a tightly joined pair (called a diplokaryon), whereas in others the nucleus is single [22]. In *Enterocytozoon*, diplokarya may occur early in the life cycle [56], while single nuclei occur at later stages. In other genera, the nuclear configuration is constant (either single, *e.g. Encephalitozoon*, or double, *e.g. Nosema*) throughout the life cycle [57]. When organisms with diplokarya divide, each diplokaryon divides producing "double diplokarya" [9]. The potential link to fungi has been proposed, based on the presence of some important features, mainly the presence of chitin in the wall of the microsporidial spores, identifiable Golgi organelles [58 - 60], the microtubule gene data [61, 62] as well as several enzyme processes [5]. Studies showed that many characteristics of microsporidia are similar to prokaryotes, however their 70S ribosomes make them different from prokaryotes. They also contain 16S and 23S ribosomal ribonucleic acids (RNAs) similar to prokaryotes with the smallest genome of any eukaryote thus far reported [51].

All microsporidian spores contain a single long coiled structure called the polar filament, a unique structure attached at the anterior end by a large, mushroom-shaped anchoring disk [63]. Electron microscopy reveals that this structure coils around the single or double-nucleated sporoplasm inside the thick, resistant and refractile spore coat. The host-parasite interface may involve: 1) Direct interaction with the cytoplasm of the host cell, 2) Indirect contact in a parasite-secreted envelope (sporophorous vesicle, SPOV), 3) Indirect contact by production of a parasite- induced, host-produced envelope "parasitophorous vacuole" [64], or 4) Indirect contact by producing a host- produced "parasitophorous vacuole" and parasite-induced secretions [3, 15, 65].

The life cycle of microsporidia has three phases: infective, proliferative and sporogenic. The majority of microsporidian infections are initiated in the susceptible host *via* oral ingestion, with the spores gaining access to the digestive tract. This has led to the discovery that spores germinate in response to stimuli such as: pH, ion concentration, osmolarity, digestive enzymes, redox potential, and/or digestive products [66]. The stimulus changes the spore's permeability, triggering the eversion of the polar filament resulting in the projection of a long hollow polar tubule that jumps out from the anchoring disk coiling several times inside the posterior part of the spore (Fig. **1**).

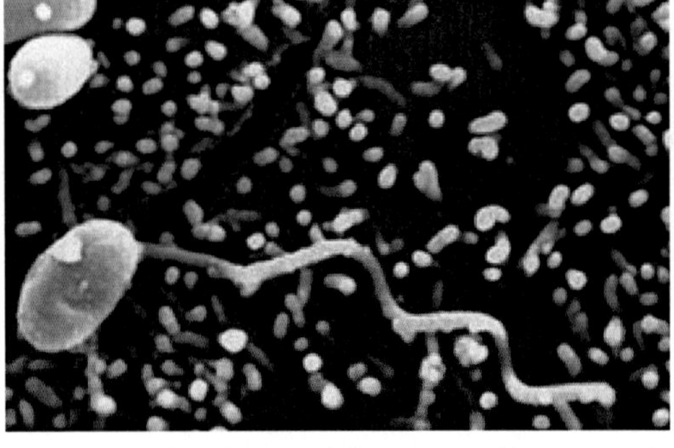

Fig. (1). Scanning electron micrograph of a microsporidian spore with an extruded polar tubule.

It emerges with sufficient speed and force to penetrate the host cell and transfer the sporoplasmic material directly into the host cell through its 50-500 μM long polar tube [67, 68] initiating a new infection in less than a second [69]. Some spores evert their polar tubules, releasing sporoplasms within the same host, thus establishing a cycle of autoinfection, which leads to chronicity and/or additional sites of infection [4]. The whole process from the beginning of germination, protrusion of the polar tube and the inoculation of the sporoplamic material into the host cell was described to have a resemblance to a hypodermic needle [70, 71]. In the proliferative phase, extensive multiplication begins when the injected sporoplasm proliferates to meronts. The injected sporoplasm grows and divides either by merogony, by just by simple binary fission or by schizogony, which occurs by multiple fission producing multinucleate plasmodial forms inside the cytoplasm of the infected host cell [22].

The sporogenic cycle is signaled by one or more changes: secretions deposited on the surface membrane of the meront and/or formation of an isolating envelope called sporophorous vesicle (SPOV). Sporonts divide one or more times then become sporoblasts, which mature into spores [4]. Once the cell becomes full with spores, it bursts into the surroundings releasing the new-formed spores that continue their cycle into new cells (Fig. **2**).

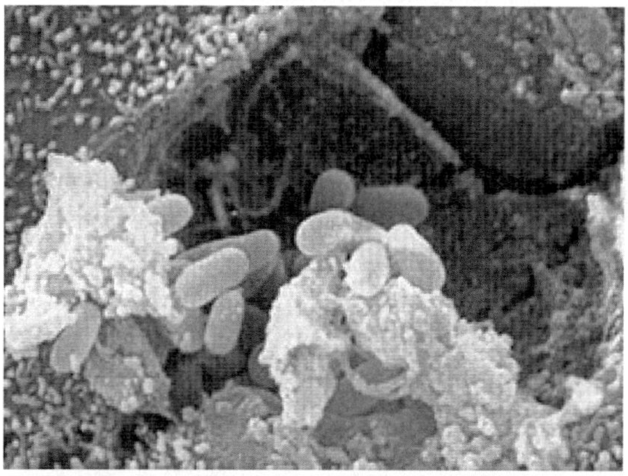

Fig. (2). Scanning electron micrograph showing an infected cell bursting and releasing spores of *Encephalitozoon hellem* into the surroundings.

Multiplication of spores inside the infected cells by both merogony and sporogony produce an enormous number of organisms [72]. While the spore's structures are characteristic of microsporidia, the number of spores produced in sporogony, the manner in which they are produced, and the host-parasite interface vary among different genera [4].

Nosema and *Anncaliia* (*Brachiola*) spores, members of the families *Nosematidae* and *Tublinomatidae*, are approximately 4 µm with paired abutted nuclei (diplokarya). There are over 100 species of *Nosema*, most of which are parasites of insects; however, few *Nosema* species have been described from human ocular infections [73]. After several nuclear and cellular divisions, large clusters develop. Each sporont produces two sporoplast cells that develop into two spores [3]. *Nosema* form thickened plasmalemma in the sporogenic phase, and *Ancaliia* (*Brachiola*) spp. in all developmental stages [74]. The family *Pleistophoridae* was named in 1893 by Gurly in relation to the fish parasite *Pleistophora typicalis* Gurley, 1893 [75]. Species of the parasite have been identified in human muscle infection in immunodeficient persons [11]. Infection has also been diagnosed from an HIV negative case in the USA [12]. Spores of *Trachipleistophora hominis* are approximately 4 x 2 µm, and develop by multiple fragmentations inside sporophorous vesicles (SPOV). *T. anthropophthera* is dimorphic. There are 2 types of SPOVs, one contains approximately 8 thick-walled spores, measuring 3.7 x 2.0 µm and the other type contains 2 thin- walled spores with 3-5 polar filament coils and measuring 2.2-5 x 1.8-2.0 µm [76]. *Enterocytozoon*, family *Enterocytozoonidae*, was the first genus of microsporidia identified from human infections [14, 56]. Spores are 1.3 µm x 0.8 µm, and contain a single nucleated sporoplasm surrounded by approximately 6 polar tubule coils arranged in a double row [15, 77]. Development of the parasite occurs inside the host cell cytoplasm. Finding several polar filaments within a multinucleate plasmodium is diagnostic for *Enterocytozoon*. The plasmodium divides by multiple fission producing a dozen or more of sporoblasts, which mature into spores [15] (Fig. **3**).

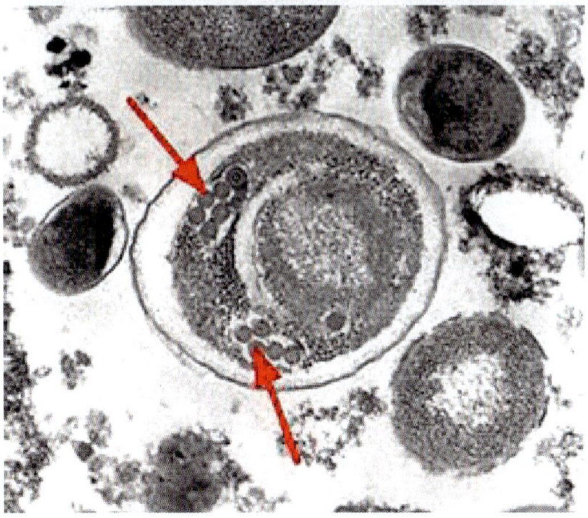

Fig. (3). Electron micrograph of *Enterocytozoon bieneusi*. The arrows are pointing to double rows of polar coils cut in cross section.

E. cuniculi, Family *Encephalitozoonidae*, was first discovered in rabbits in 1924 [78] and was considered as a synonym of the genus *Nosema* [79]. In 1987, the parasite was reported from over 30 different mammalian hosts including man [16]. Later it was classified as a genus Microsporidia and the family was established in 1989 [80]. This genus is characterized by a phagosome-like parasitophorous vacuole surrounding the 6 developing parasites, isolating them from the host-cell cytoplasm. During development, the parasite may contain one or more separate nuclei and the proliferative cells usually depend on the vacuole membrane, which ruptures to free the spores. Within the parasitophorous vacuole, each sporont elongates, divides and produces spores.

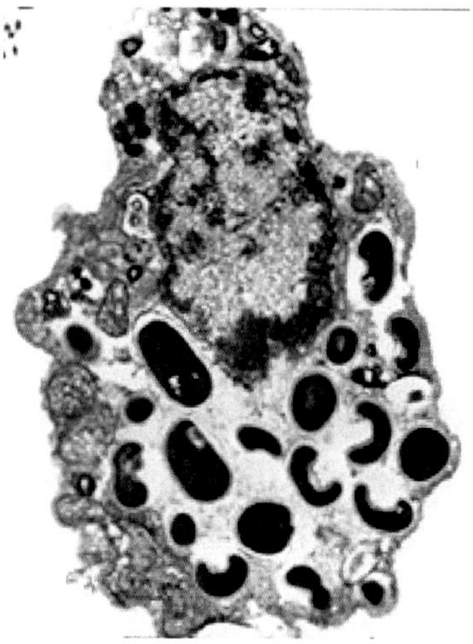

Fig. (4). Electron micrograph of a cell infected with *Encephalitozoon intestinalis* spores developing inside a septated parasitophorous vacuole.

A spore is about 1-1.5 x 0.5 µm, and contains a sporoplasm with a single nucleus, besides approximately 6 polar tubular coils arranged in a single row [81, 57]. *Septata* was the second microsporidial genus created for human infection in 1991, and was considered as a new genus within its family, based on the similarity of some morphological features [19, 65]. *S. intestinalis* species is characterized by parasite secreted material surrounding the developing stages and spores inside the parasitophorous vacuole. The proliferative and sporogenic stages have 1 to 4 nuclei. Cells are rounded at first, but they elongate when containing 2 or 4 nuclei. In sporogony, there is thickening of the plasmalemma and elongation of sporonts. Each sporont divides into 4 single-nucleate sporoblasts. Each develops the polar filament complex and matures into a spore. *S. intestinalis* cells are tightly packed in clusters (Fig. **4**) while some cells condense leaving a space between individual developing forms.

Early and late forms develop asynchronously, with the parasite secretions surrounding individual cells within the parasitophorous vacuole. Spores are 2 x 1.2 μm with a single nucleus and 4-7 (approximately 5) polar tubular coils, in a single row [18, 19]. Based on molecular data, *S. intestinalis* is moved into the genus *Encephalitozoon* and is given the name *E. intestinalis* [4].

GLOBAL EPIDEMIOLOGY AND RISK FACTORS

Although microsporidia prevail among almost all members of the animal kingdom from honeybees, silk worms, and mosquitoes to mammals and birds, the prevalence is thought to be underestimated due to difficulty and unreliability in detection [57]. The small size of these organisms makes their identification in specimens very difficult. The epidemiological evaluations mainly depend on the geographical area and the method of diagnosis used. Serological prevalence in different localities reported rates ranging from 0 to 42%. There was a high rate of infection in homosexual males in Sweden as well as in cases infected with other parasitic infections [22, 82, 83]. The prevalence of intestinal microsporidiosis showed no significant seasonal variation among HIV seropositive patients [84].

Until the mid 1980s, microsporidia were not recognized as human pathogens [22, 77]. The identification of the new microsporidian *E. bieneusi,* was reported in 1985 from an AIDS Haitian patient that complained from diarrhea and wasting. *E. bieneusi* is still considered the commonest species associated with HIV positive individuals [14].

Before the introduction of HAART (Highly Active Anti Retoviral Therapy), the identification of intestinal microsporidiosis in HIV-positive cases showed wide variability ranging from 2-50% depending on geographical location, laboratory methods and personal experience in diagnosis [57].

The prevalence of systemic microsporidiosis is difficult to estimate, due to absence of clinical signs or nonspecific symptoms that is why microsporidiosis is usually missed in clinical practice. Chronic diarrhea and wasting were primarily associated with microsporidiosis in HIV-positive cases in the early reports. *E. bieneusi* and *E. intestinalis* were the predominant species infecting the small intestine.

In Sub-Saharan Africa, infection with microsporidia in HIV positive cases is a cause of high morbidity and mortality [85, 87]. In other countries, especially in Asia (India, Thailand), the Middle East (Turkey), Europe, Africa (Tunisia, Mali, Uganda, Senegal, Zimbabwe), and Latin America (Brazil, Peru) HIV positive cases account for high number of cases of microsporidiosis [20, 85 - 102]. Although infection prevails among HIV-infected patients, they have also been reported in HIV-negative individuals [72], such as travellers [103, 104], malnourished children [105 - 107], recipients of organ transplantation [108, 109], wearers of contact lens [110] and old individuals [90] among immunocomptent individuals [73].

People having organ transplantation are recently considered a risk group for microsporidiosis [111], since cases recognized in transplantation of solid organs and bone marrow, were all negative for HIV [96, 108, 111 - 124]. In this group of patients, diarrhea was the most common complaint and *E. bieneusi* was the predominating species of microsporidia [125 - 127]. Transmission of microsporidiosis transplacentally from a mother to the offspring was reported in animals as non-human primates, carnivores, rabbits and rodents [128 - 130]. This finding suggests that the same mode of infection may occur in humans; however, it is not proved till now.

Spores can be excreted in sputum of cases with respiratory infection and also can be expelled in stools or urine indicating a horizontal transmission and cause contamination of the environment. Oral-fecal or oral-oral transmission, inhalation of aerosols or ingestion of food and water contaminated with spores are methods of transmission [57, 131 - 133]. It was reported that mice experimentally infected with *E. cunniculi* as neonates didn't show clinical symptoms or developed mortality to experimental infection than adults as they developed cell-mediated and humoral immunity [134]. Experimental infection of animals orally, intra-rectally or by ocular inoculation with the *E. bieneusi, Encephalitozoon* species or *B. algeriae* supported the idea that horizontal transmission might occur between humans [135 - 137].

Intrarectal transmission hints to the possibility of sexual transmission among humans [135]. Since many human species of microsporidia also infect a wide range of animals, zoonotic infection may play a role. Moreover, *E. cuniculi* spores have been recovered from carnivores, rabbits and rodents [22, 129, 133]. The same spores have also been reported from foxes, goats, horses, and from non-human primates [138 - 143]. *Encephalitozoon hellem* spores were identified from birds [133] and *Encephalitozoon* (syn. *Septata*) *intestinalis* were also identified from donkeys, dogs, pigs, cows, goats, and gorillas [144, 145].

Spores of *E. bieneusi* were detected in domestic animals as dogs and cats, farm animals as rabbits, goats, pigs and cattle and in wild animals as llama, raccoons, muskrats, beavers, foxes and otters [90, 146 - 152].

Nosema species have been identified from insects [21], and *Pleistophora* species were reported from fish [153]. Microsporidiosis has been reported in a child that showed a sero-conversion after exposure to an animal that was infected with *E. cuniculi* [154]. In 2001, Weitzel and others reported a dual microsporidial infection with *E. cuniculi* and *E. bieneusi* in an HIV-positive patient with a high occupational zoonotic risk (dog hairdresser), giving further evidence for zoonotic transmission [155]. Depression of cell-mediated immunity, as occurring in HIV infection, is considered the main risk factor of human microsporidiosis [156, 157]. In HIV patients infection occurs in high degree of immunedepression with CD4+ counts <100 cells/ μL [92, 95, 96, 101, 156 - 158], although it is of importance to consider that seroprevalence is unreliable in HIV cases, due to their lowered immunity [132].

Since the identification of *E. bieneusi* as a human pathogen, it was incriminated as a cause of chronic diarrhea in HIV positive cases and studies were done to recognize the risk factors in humans. HIV/AIDS immunodeficiency, low CD4+ counts (<50 cells/μL), and young age, were identified as risk factors in cases infected with *E. bieneusi* in their intestine [157 - 162].

Internal transcribed spacer (ITS) could help to identify more than 90 genotypes of *E. bieneusi* in humans and animals. *E. bieneusi* is known to be the only and most common species of the genus *Enterocytozoon* that was reported to infect man [163, 172].

A study carried out on *E. intestinalis* isolates from animals and human showed no molecular differences, which may explain that there is no transmission barrier between different host species [164]. In constrast, antigenic diversity has been demonstrated among isolates of *E. cuniculi* and *E. hellem* from human infections [165] and *E.cunniculi* is considered the species that has the widest distribution within the genus *Encephalitozoon*, among mammals including humans [172].

Water is an important source of transmission of microsporidiosis. Spores of *E. intestinalis* were detected in almost all types of water including surface and ground water as well as sewage treated water [166].

Researchers identified spores of *E. bieneusi* in surface water and were able to identify microsporidial spores in water of swimming pools [167, 168]. *Vittaforma corneae* (syn. *Nosema corneum*) spores were isolated from river Seine [167, 169] and spores of *Nosema* species from ditch water [170] and tertiary effluent [166]. A survey comparing human microsporidiosis in two Mexican villages with different water sources found that people in a village receiving piped, untreated water from a spring had a significantly higher incidence of spores of *Encepalitozoon* in their stool samples than people in a village with a well water supply (40% *vs* 15%), highlighting the potential of transmission by contaminated water [171]. Water supplies contaminated by feces and urine of animals infected with microsporidia may infect humans, as most of microsporidia lack host specificity [172].

Microsporidian spores are environmentally resistant and live for considerably prolonged durations. Under experimental conditions, some of spores of *E.cuniculi* remained viable and infectious in the medium 199 (M99) of tissue culture system for a duration of 16 days, when incubated at 22°C and 98 days when incubated at 4°C [284]. Their small size allows spores to escape filtration. Besides, the availability in various water sources, favor the role of water as a vehicle for transmission [20, 172]. Microsporidia could live in fresh or salt water, in tissue culture, or after dehydration for prolonged periods at suitable temperatures [174]. *E. cuniculi* were able to survive when incubated in distilled water or freezed and thawed for 24 hours after incubation at pH4 and pH9 [173].

Spores of *E. intestinalis* and *E. hellem*, still had the ability to infect cells of tissue culture for weeks to months after incubation in water at 10-30°C [174]. Some epidemiological studies reported that the recreational water as well as hot tubs, occupational water and drinking water constitute a risk factor for human infection [175, 161] and increased rates of infection were in the vicinity of distribution subsystems [176].

A study carried out in Peru on HIV/AIDS patients showed that there were risk factors for *E. bieneusi* when getting in contact with excreta of ducks and chicken with lack of fresh and clean water, in addition to the flush toilets and collection of garbage [96].

Studies depending on molecular epidemiology have created a better understanding of the geographic, zoonotic, demographic and the environmental outlines of microsporidia that infect man. The role of water in the transmission of infection has led to the consideration of microsporidia into the NIH category B list of Biodefense pathogens, and the important contaminant organisms transmitted by water by EPA "Environmental Protection Agency" [88].

The globalization of food, the increasing travel of consumers and the change in food consumption patterns have created a concern about the role of food in the transmission of microsporidial infections [177, 178]. Investigations showed that eating undercooked beef was associated with microsporidiosis in HIV positive cases, although adequate cooking of infected meat can avoid infection [175]. *Trachiopleistophora hominis*, which is similar to that of fish, was recognized to cause myositis in patients with AIDS [179]. These organisms could grow actively in culture temperature of 32-34°C, which may give in idea that this species is not completely adapted to human infection [180]. Further phylogenetic studies raised the concern that human infection may develop after eating improperly cooked fish or by transmission by mosquito bite [180].

Vector-borne transmission has been studied by trying to inoculate *Brachiola* (syn. *Nosema*) *algerae*, a natural pathogen of mosquitoes, into rats or athymic rats. There was failure in the inoculation of the organisms by oral or intravenous routes, however local infection occurred after subcutaneous inoculation [181, 182]. Inoculation of the spores of *B. algerae* into the eyes of SCID mice (severely deficient in functional T or B lymphocytes), failed to develop ocular signs, but microorganisms appeared in the liver after 60 days [137].

Additional evidence for vector-borne transmission wasexplained when infection with *T. hominis* was transmitted to *Anopheles quadrimaculatus* and *Culex quadrimaculatus*. The infection developed in the muscle bundles of the insect's abdominal segments. Spores of *T. hominis* were later identified in the sugar solution used for feeding mosquitoes, and also from the proventricului and the rest of the gut of mosquito, indicating that mammalian transmission may occur during blood meals of infected mosquitoes [183].

Brachiola (syn. *Nosema*) *algerae* spores were also detected in mosquitoes, and it was suggested that there might be a risk of transmission to HIV infected patients, if they were stung by bees, wasps or hornets [184, 162]. Infection has been traced among asymptomatic apparently healthy individuals

The indirect immunofluorescence assay (IFA), was used in the Czech Republic for the detection of specific antibodies against micrsporidia. The result showed unexpected higher incidence of infection than reported before, raising a doubt of the possibility of reactivation of previous latent infection in immunocompromised cases leading to serious results [185].

PATHOPHYSIOLOGY

The pathophysiology of microsporidiosis is not adequately known. The pathogenesis and clinical picture depend mainly on the species of microsporidia causing infection, the site of infection, and most of all on the immune status of the infected host [20, 99, 186]. Although infection is mostly diagnosed in patients with impaired immunity, microsporidia have also been detected from immunocompetent individuals [187, 188]. Infection occurs mainly by ingestion or inhalation of the environmentally resistant spores of microsporidia, in addition to other routes of infection including their passage through injured skin or ocular surface, trauma, sexual route [88], as well as possibility of transmission by insect bites [88, 162]. *E. bieneusi* and *Encephalitozoon spp.* are considered the most common microsporidia recovered from human infections [88, 132, 189]. They tend to infect the gastrointestinal tract. *E. bieneusi* locates in the apical part of the intestinal villi, while *Encephalitozoon intestinalis* infects not only the intestinal villi, but also the cryptic cells, thus reaching and invading macrophages, fibroblasts, and endothelial cells [20, 23, 190, 191]. The infection of intestinal villi results in flattening of the epithelium and leads to subsequent atrophy of the villi of the brush border with compensatory elongation and hyperplasia of the crypts reducing the absorption surface area up to 40%. There will be lymphoid exocystosis with edema as well as vesiculation and necrosis of the enterocytes [20, 23, 88, 89]. *E. intestinalis* can induce extensive ulcerations with subsequent mucosal atrophy, acute and chronic inflammation, and zones of submucosal macrophage infiltration [24, 192], resulting in malabsorption of lipids, vitamin B12, and D-xylose as well as electrolyte imbalance (especially potassium and magnesium) and decreased level of serum bicarbonate [89, 193, 194]. Infection of the intestinal epithelium with *E. bieneusi* is restricted to the enterocytes at the tip of the villi leading to villous atrophy, cellular degeneration, necrosis and sloughing. The preferred site of infection is the jejunum and the duodenum is less commonly affected while the large intestine is not included in infection [195, 56]; *E. intestinalis* causes granulomatous interstitial enteritis accompanied by severe diarrhea and may disseminate to lungs and sinuses [19].

Theoretically all organs could be infected during dissemination of *Encepalitozoon* spesies [196]. Clinically, disseminated microsporidiosis might present as encephalitis, keratoconjunctivitis, sinusitis, pneumonia, myositis, peritonitis, nephritis, and hepatitis [77]. Dissemination of *E. intestinalis* induces inflammatory reactions in infected organs such as the liver, the pancreas, the lungs, and the kidneys [89]. The functional impairment was attributed to decline in the enzymatic activities at the basal portion of the intestinal villi [20, 193].

Hepatic infections with *Encephalitozoon spp.* can cause a granulomatous necrosis with the presence of microsporidia disseminated within the hepatic parenchyma or a non-granulomatous inflammatory reaction [23, 20].

Bile duct infection may be complicated by papillary stenosis, alithiasic cholecystitis, bile duct dilatation, and sclerosing cholangitis [20, 23]. Muscle infections with *Pleistophora spp.* lead to muscle atrophy and diffuse degenerative lesions with numerous microsporidia spores infiltrating the muscle fibers [20, 23]. *E. cuniculi* may infect the kidneys and CNS in a big variety of mammals including man. The infected tissue generally exhibits minimal inflammatory reactions, but changes might range from normal tissue architecture to severe degenerative lesions of the epithelium [20, 23].

Kidney lesions are seen as a tubulointerstitial granulomatous nephritis with an inflammatory infiltration composed of macrophages, lymphocytes, plasma cells, and Langhans type multinucleated giant cells [23]. Infection of the ureters can lead to a granulomatous inflammatory reaction. In the bladder, the lesions produce an ulcerated cystitis with lympho-histiocyte infiltration [23]. *Encephalitozoon spp.* infects the genitourinary system in most mammals, including humans [186, 197 - 199].

Granulomatous interstitial nephritis composed of infiltration by plasma cells and lymphocytes are the main pathological finding. Usually it is associated with necrosis of the kidney tubules, with their lumens full of amorphous granular material. Spores of microsporidia are detected in the necrotic tubules and the sloughing tubular cells [105, 109, 116, 119]. Upon their shedding into the urinary bladder, they are able to infect other epithelial cells in their way causing ureteritis, prostatitis, and cystitis [197]. Infection also affects the muscle cells, fibroblasts and macrophages of the affected mucosa and is often associated with the shedding of spores in urine. However sometimes spores of microsporidia may not be detected in urine in cases with renal failure and intestinal microsporidiosis, indicating that dissemination does not always occur [200].

Erosive tracheitis, bronchitis, and bronchiolitis have been reported in microsporidial infection. Typically, organisms are found in intact or sloughed epithelial cells [201, 202].

Biopsies taken from AIDS cases complaining of chronic sinusitis and microsporidiosis showed the presence of spores in the epithelial cells of sinuses as well as in supporting structures [203 - 205].

Deep inflammatory keratitis may occur in microsporidial ocular infection with inflammatory cellular infiltration and zones of necrosis associated with thickening of the cornea [123, 206]. This reaction is however generally moderate or even absent in the event of superficial keratoconjunctivitis where the inflammatory infiltration is made up of neutrophil polymorphonuclears and mononuclears (punctate keratopathy) [23].

The relation between microsporidiosis and diarrhea in HIV-positive cases is complicated, as there is always the possibility of the effect of other intestinal pathogens in cases with declining immune status plus the direct effect of HIV on the gut. In case of mono-infection with microsporidia, intestinal biopsies showed villus atrophy and crypt hyperplasia where as cases without other intestinal pathogens didn't show the same changes [77].

IMMUNOLOGY AND IMMUNOPATHOLOGY

The immune response against microsporidia and the receptors involved in recognition by the infected host are not well known. Toll-like receptors (TLR) are receptors that can recognize and bind to certain specific molecules on the surface of pathogens and stimulate a variety of inflammatory reactions [207].

TLR2 can recognize *E. cuniculi* and *E. intestinalis* on primary human infected macrophages which activate nuclear factor, kappa- light-chain-enhancer of activated B cells (NF- κB), releasing some inflammatory cytokines, mainly interlukin-8 (IL-8) and tumor necrosis factor alpha (TNF- α) [207]. About an hour after infection of macrophages, they develop nuclear translocation of NF-kB with production of TNF-α and IL-8. To test the role of TLR2, small interfering RNA was used to knock down the receptors of the primary human macrophages. After challenge with spores there will be an increased nuclear translocation of NF-14kB and the levels of TNF- α and IL-8 [207]. Infection leads to activation of antibody production by the host. Persistence of these specific antibodies is associated with latent infection and they play an important in resistant to infections. However specific antibodies fail to act as a barrier against infection when acting alone [208]. When hyperimmune serum was injected into athymic mice previously infected with *E. cuniculi*, it failed to improve their survival rate, which was explained by the possibility of having another ancillary defense mechanism [208]. Macrophage-mediated phagocytosis could be facilitated by an opsonic function of specific antibodies against microsporidia. Sometimes uncontrolled antibody response was associated with disease, in such cases there is hypersensitivity reactions and hypergammaglobulinaemia resulting in the formation of immune-complexes and renal failure. This type of reaction has been reported in some arctic foxes and dogs infected with microsporidiosis [209 - 211].

Immunologically competent animals develop IgG antibody response; two weeks after infection, which peaks at week 5-6 and in most of the cases, it persists lifelong. The long-term presence of high levels of specific antibodies is used in diagnostics to isolate seropositive animals from others [69, 212].

Antibodies against *E. cuniculi* were also detected in immunologically competent people, but the authors did not directly observe microsporidia [83, 213].

Specific antibodies against microsporidia with variable levels; have been detected in HIV positive cases with confirmed microsporidiosis and in HIV negative cases with previous history of microsporidiosis. It is thought that the variability depends mainly on the immune condition of the person at time of microsporidial infection [17, 214].

Macrophages are part of the primary response against pathogens as they reside at the site of their entry, so they are considered a link between innate and adaptive immunity [215]. Some pattern recognition receptors (PRR) on the surface of local macrophages can recognize foreign pathogens resulting in a circle of host defense mediators as cytokines, chemokines, nitric oxide (NO), nitric oxide synthase (iNOS) and radical oxygen species.

Activated T cells secrete IFN that initiate respiratory burst to kill the phagocytized intracellular pathogens [216]. Microsporidia can evade these protective immune responses by using macrophages as their "Trojan horses" that carry them to different organs of the body initiating a disseminated infection [217].

Dissemination is thought to occur in two steps. The first is the initial infection in the intestine by species like *E. cuniculi* or *E. intestinalis* invading the resident macrophages [218]. These macrophages secrete chemo-attractants in response to infection, to recruit new cells as monocytes to help resolving the infection. In patients with multifocal involvement of the organs with microsporidia, the lesions appear as micro-abscesses and granulomas.

During the second phase, macrophages, which failed to kill the cellular intruder, migrate into the lymphatic system, blood and tissues. Microsporidia gain access to the host cells through eversion of their filaments and penetration of the cell membrane or by phagocytosis of the released spores [218].

Studies showed that microsporidia could inhibit the process of fusion of phagosome and lysosome, thus affecting the ability of the parasite to survive inside the macrophages [219]. This finding can explain the theory that the microsporidia remaining inside the primary phagolysosomes can evade the immune responses of macrophages and continue their life inside the cells. Cell-mediated immunity has its role in the prevention of severe infection with

Encephalitozoon. The Th-lymphocytes with CD4+ receptors, Tc-lymphocytes with CD8+ receptors and a few populations of the TCR-associated with CD3+ trigger the reaction. The adoptive transfer of sensitized T-splenocytes could protect athymic BALB/c and SCID mice infected with microsporidia *E.cuniculi* from death [208, 217, 220].

CD8 T-cells participate in the pro-inflammatory response by the production of cytokines such as interferon gamma (INF-α) and their direct cytotoxic effect [190]. They also contribute to the regulation of the immune response by secreting interleukin-10 (IL10) [23, 85]. Recent studies have identified the significance of pro-inflammatory cytokines as INF-α, tumor necrosis factor and interleukin-12 (IL12) in resistance against *Encephalitozoon* infections [190]. Studies showed that IL-10 blocked the effect of INF-α in controlling the cellular immunity [223]. This implicates IL-10 in preventing early dissemination of microsporidia as observed in SCID mice, which do not produce IL-10 [222]. In immunodeficient mice without T or B-lymphocytes [223], even fully functional macrophages could not produce IL-10 [221]. The higher levels of IL-12 found in this situation stimulates the production of INF-α, which is the main cytokine of macrophages that helps in the process of phagocytosis and elimination of spores.

In vitro, injection with antibodies against INF-α or IL-12 was able to neutralize the resistance to the parasite [216]. Aged mice infected with *E. cuniculi* developed unusual priming of the T-cells by dendritic cells and could restore their adaptive immunity when injected with dendritic cells (DCs) extracted from younger mice [224]. This observation is in accordance with clinical data from humans, since the elderly are more susceptible to microsporidiosis.

Encephalitozoon, Trachipleistophora and *Pleistophora* species could disseminate causing systemic disease with the affection of the sinuses, eyes, liver, muscles, kidneys, peritoneum, CNS and respiratory tract in immunodifficient individuals. *E. cuniculi* can cause serious disease due to development of immune complexes and renal disease in carnivores such as domestic dogs, blue foxes and mink [225, 129, 130].

As a final result, both hyper and hypo-immune responses to microsporidia can be a cause of disease and only the well-regulated immune response in a host can control these pathogens resulting in suclinical infection [225].

SYMPTOMS

Clinical manifestations of microsporidiosis mainly depend on both the immunological response of the host and the site of infection, which shows great variability as the organisms can almost infect every tissue and organ of the affected hosts [172]. In immunocompromised patients, microsporidia most likely develop disease, manifesting as severe opportunistic infections, often with fatal outcome [128 - 130]. In AIDS patients with CD4+ counts <100 cells/μL microsporian species *E. bieneusi* and *E. intestinalis* were the most common causes of diarrhea with fever, lack of appetite and loss of weight [77, 226, 227, 132].

Most of the reported cases are confined to the intestine, however there are wide spectrum of diseases ranging from asymptomatic carrier to dissemination to different organs leading to keratoconjunctivitis, hepatitis, cholangitis, ascites, myositis, sinusitis, kidneys and urogenital infections [3, 23, 193]. Infection with *E. intestinalis* and *E. bieneusi* usually manifests as chronic diarrhea of 3-10 motions/day, bloating, lack of appetite and loss of weight not associated with fever [25, 132, 186, 228, 229]. Diarrhea is often associated with mal-absorption, weight loss and wasting [228]. The mortality rate was reported to be more than 50% among cases with wasting and advanced HIV infection [25]. Organ transplant recipients under immunosuppressive therapy were reported to have fever, fatigue, nausea and diarrhea when they develop infection with *E. bieneusi* or *Encephalitozoon spp*. [77, 108, 226]. Children in tropical countries infected by microsporidia, primarily *E. bieneusi*, might suffer from persistent diarrhea, malnutrition and lowered immunity [106, 107, 161]. Recently, it was suggested that the elderly might have decreased immune competency, therefore become more susceptible to infection with microsporidia [90].

The association of microsporidiosis with human disease was discovered in the mid 1980s, when the organism was detected in stool samples of cases with HIV/AIDS and chronic diarrhea [195]. There was controversy on the pathogenicity of microsporidia, as they have been reported in persons without diarrhea [216, 230], until the opportunistic nature of these microorganisms became more clear [88, 216]. In experimental infection of immunocompetent laboratory animals, early acute stage of infection showed clinical signs, followed by asymptomatic shedding, whereas the same infection caused death of immunodeficient athymic and SCID mice [85]. Asymptomatic chronic infection was observed in immunocompetent hosts, which were infected with *E. cuniculi*, naturally or experimentally [128, 129]. In some cases, mild clinical signs are developed early after infections. An example is the formation of ascites in some mice experimentally infected with *E. cuniculi,* which resolved 2 weeks after inoculation, with development of motor paralysis, convulsions and torticollis in infected rabbits [128, 129]. Self-limiting traveller's diarrhea, with a duration of about 2-3 weeks may develop in healthy individuals after getting infection with microsporidia [104, 231, 232].

Intestinal and biliary infections: The detected pathogens in these sites in immunocompromised cases were mainly *E. bieneusi* and less frequently *E. intestinalis.*

Infection usually causes severe non-bloody, non-mucoid intermittent diarrhea with gradual onset and months duration. Cases develop malabsorption of nutrients with progressive weight loss. There is an association between intestinal infection and lactase deficiency, decreased activity of alkaline phosphatase and α-glucosidase at the base of villi with atrophy of villi and reduction in their height. The patients may have nausea and loss of appetite. Reports show that microsporidial spores may be excreted in the diarrheic or normal stool. In patients with chronic diarrhea, without any other known intestinal pathogens, *E. bieneusi* have been detected in 7- 50% of the study cases depending upon the methods of diagnosis and the group of study [77, 193]

Hepatitis and Peritonitis

Encephalitozoon spp. is able to cause hepatitis and/or peritonitis. *E. cuniculi* was identified on the basis of ultrastructure in two HIV-infected patients at autopsy [16, 233]. In these cases, infection was diagnosed on ultrastructure basis without exact species identification. Another case of infection with *Encephalitozoon spp.* was in a patient with AIDS that suffered from diarrhea for 2 months before he died from fulminant hepatitis. The autopsy specimen examinations showed disseminated microsporidiosis in the liver, gall bladder and mediastinal lymph nodes [234]. *T. anthropophthera* have been reported in a 8-year-old HIV-infected girl with disseminated infection in the liver and pancreas [235]. *E. bieneusi* and *E. intestinalis* were detected from the non-parenchymal cells of the liver in some cases with HIV infections without signs of hepatitis.

Ocular Infections

Ocular infection is considered the second most common manifestation of microsporidiosis after gastrointestinal infection [73]. Keratoconjunctivitis may be caused by all *Encephalitozoon spp.* (*E. intestinalis, E. cuniculi* and *E. hellem*) in HIV-infected cases. Most cases complained of bilateral conjunctival inflammation and bilateral punctuate keratopathy with subsequent decreased visual acuity. Keratoconjunctivitis is often asymptomatic or moderate, but also could be severe ending in corneal ulcers. Other species of microsporidia (*V. cornea, N. ocularum,*

T. hominis, M. ceylonensis and *M. africanum*) have also been reported as single case reports [73].

From 1989 to 1991, six cases of microsporidian keratoconjunctivitis were reported in patients with AIDS, four from New York, one from Texas, and one from Ohio [6, 17, 18, 236 - 238]. All had conjunctivitis, blurred vision, and photophobia. By 1999 over 20 cases were characterized, reported and reviewed [73]. Organisms were observed in corneal epithelial cell scrapings examined by light and electron microscopy [18, 239]. The organisms were morphologically similar to *E. cuniculi*, but a 19 clearly defined parasitophorous vacuole surrounding the organisms, was not always visible [81]. *Encephalitozoon hellem* was identified as morphologically identical to *E. cuniculi*, but was serologically different [17]. Topical steroid treatment was thought to promote a localized immunosuppression of the eye with exacerbation of ocular microsporidial infection in some cases [240, 241].

Sinusitis

It is one of the common manifestations of microsporidiosis in humans [203]. *Encephalitozoon* species (*E. hellem*, *E. cuniculi* and *E. intestinalis*) have the a ability to cause rhinosinusitis in many HIV-infected patients, while other species *E. bieneusi* and *T. hominis* caused less frequent infections in patients with nasal polyps and severe rhinitis [203].

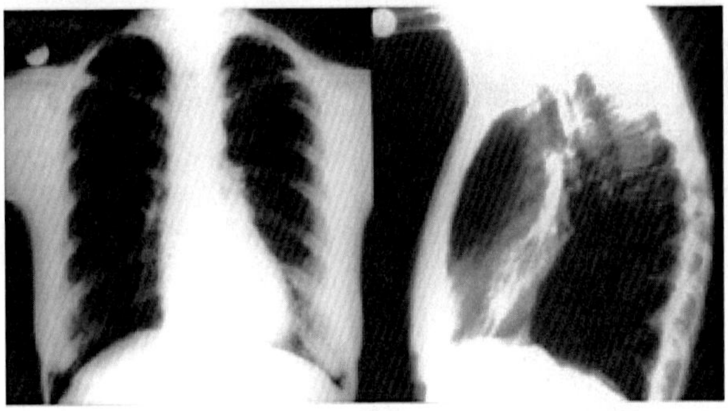

Fig. (5). Chest X-rays of a female HIV patient with left-sided pneumonia caused by *Enterocytozoon cuniculi*.

Lower Repiratory Tract Infections

It is less frequent than other microsporidial infections. It may show asymptomatic infection or could be associated with bronchiolitis. Pneumonia and respiratory failure might be the main manifestation of systemic infection in HIV positive patients [155] (Fig. **5**).

All species of *Encephalitozoon* have been reported to infect the bronchial epithelial cells in cases with disseminated microsporidiosis in HIV-infected patients. Pulmonary *E. bieneusi* was only detected sporadically [77].

Urinary Tract Infections

Urinary tract infection usually occurs with disseminated *Encephalitozoon* infections in HIV positive cases. Infection may be asymptomatic or may present with cystitis or nephritis with dysuria and haematuria, or may be the cause of progressive renal failure [20].

Myositis

This type of infection has been described in few immunocompromised cases and infection was diagnosed to be due to *Pleistophora*-like microsporidia and *Trachipleistophora spp*. [8, 11, 179, 242]. Patients presented with fever with generalized muscle weakness. Spores of microsporidia were detected in muscle biopsies.

Cerebral Infections

Involvement of the CNS was reported in two HIV-positive children with disseminated *Encephalitozoon* infection [55, 213]. Both patients suffered from signs of intracranial affection *e.g.* headache, vomiting, seizures and spastic convulsions. Diagnosis by immunohistochemistry and molecular analysis, was described in the mentioned cases [243]. Affection of the CNS with *T. anthropophthera* was also diagnosed in cases presented with seizures and cerebral manifestations. Autopsy specimen examination showed disseminated infection including the brain [253].

Rare Manifestations

Urethritis

Microsporidiosis was reported in two AIDS patients suffering from urethritis, sinusitis and diarrhea. One of the patients had *Encephalitozoon*-like spores in his nasal discharge, stool samples, urine, and urethral pus and the stool samples of the other case [236, 237].

Cutaneous Microsporidiosis

A nodular cutaneous lesion has been reported as due to infection with *Encephalitozoon intestinalis* in the leg of an HIV positive patient [4].

Vocal Cord Infection

Vocal cord infection with microsporidia has been reported in a patient with lymphocytic leukemia, who had received chemotherapy. The patient complained of hoarseness and shortness of breath. A biopsy of the area of the false vocal cord nodules was examined by Electron Microscopy and confirmed by molecular technique, showed infection with *Anncalila algerae*, that is reported as an insect pathogen [4].

Systemic Infections

The first reported human case with microsporidial infection in 1959, was a case of disseminated infection with *Encephalitozoon* in a 9-year-old Japanese child. The patient presented with intermittent fever and signs of CNS affection in the form of: headache, vomiting and spastic convulsions [55]. CSF and urine samples showed the presence of *Encephalitozoon*-like organisms. In the year 1984, a new similar case has been reported in a 2-year-old Colombian boy living in Sweden. The child complained of convulsive seizures and *Encephalitozoon*-like organisms had been recovered from his urine. The patient's serum samples had IgG and IgM against *E. cuniculi* [213]. Disseminated microsporidiosis with all *Encephalitozoon* species have been reported in immunosuppressed HIV-positive cases [23]. The possible manifestations in such cases include: ocular lesions in the form of keratoconjunctivitis, respiratory tract lesions in the form of bronchiolitis, and pneumonia, urogenital and gastrointestinal lesions. However there were significant distribution pattern for each species of microsporidia [77, 186]. *E. hellem* was identified as a cause of keratoconjunctivitis, sinusitis, bronchial disease and urinary tract infection. *E. intestinalis* was mainly affecting the gastrointestinal and biliary system. It disseminates to the eyes, nasal sinuses, respiratory tract and kidneys. *E. cuniculi* was identified as a cause of wide dissemination in all organs, with clinical symptoms varying from no symptom to severe disease [20, 77, 186]. There was also a single case report for disseminated microsporidiosis due to infection with other species (*N. connori, V. corneae, T. hominis* and *T. anthropophthera).*

DIAGNOSIS AND DETECTION METHODS

Microsporidiosis is probably overlooked as a disease because the detectable elements, "the microsporidian spores", are very small and microscopical diagnosis therefore requires expertise [88]. The index of suspicion for microsporidiosis should be highest in patients with cellular immunosuppression such as HIV or transplant recipients. Intestinal infection with microsporidia should be included in differential diagnosis in any patient with unexplained chronic diarrhea or hepatobiliary disease [72]. Some authors suggest including microsporiadia in the differential diagnosis of travel-associated diarrhea; although in our experience such cases are extremely rare [unpublished data]. Infection with microsporidia should probably be considered in cases of unexplained keratoconjunctivitis or corneal ulcers, in unexplained renal insufficiency or in cases of myositis. Since dissemination can occur, microsporidiosis may affect virtually any organ system, including bone and central nervous system. Therefore the identification of microsporidia in any specimen should prompt a thorough search in all other readily available sources, including stool, urine, sputum, nasal and conjunctival swabs, and possibly cerebrospinal fluid, with consideration of more invasive approaches for other sites of infections *e.g.* myositis [72]. Since microsporidial spores can occur in virtually any clinical sample, microbiologists and pathologist should be familiar with their appearance. In most stains such as routine Gram or Giemsa stain, they are visible as oval structures resembling yeast cells (Fig. **6** and **7**) and microsporidial infection has to be considered in samples with such "yeast-like" cells, which are negative in fungal cultures.

Most detection methods of microsporidia were developed to diagnose infections in immunocompromised patients with a higher load of microorganisms. As awareness of microsporidiosis increased and more sensitive (molecular) techniques became available, more infections in immunocompetent individuals were reported. They might be more frequent than previously expected [185], but if those positive cases represent true infections or only temporary shedding, requires further studies.

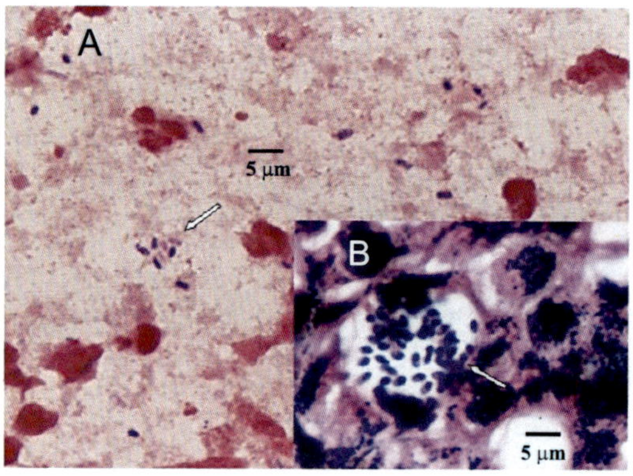

Fig. (6). Sample of HIV patient with pneumonia caused by *Encephalitozoon cuniculi*. Routine Gram stain of broncho-alveolar lavage sample showed Gram-positive oval structure (arrow), initially misidentified as yeast cells (A). Tissue gram stain of transbronchial biospies revealed typical intracellular spores, which were identified with monoclonal antibodies and electrone microscopy as *E. cuniculi*.

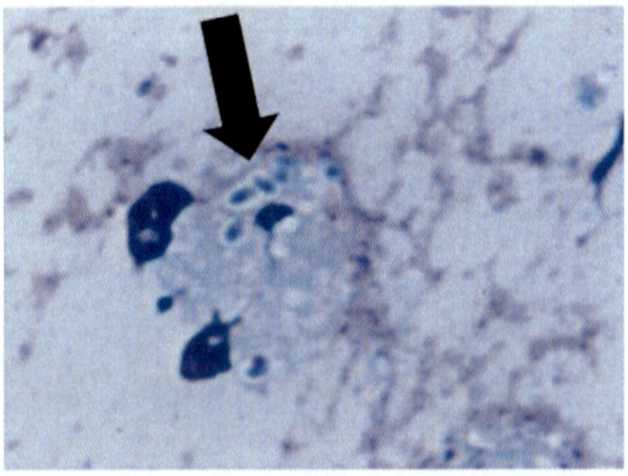

Fig. (7). Giemsa stain of a broncho-alveolar lavage of an HIV patient with pneumonia caused by *Encephalitozoon cuniculi.*

As soon as microsporidia are detected in a clinical specimen, examination of other body tissues and fluids should be considered. Urine must be examined as a routine in all cases suspected to have microsporidial dissemination. This regimen is thought to have a therapeutic implication because the most common microsporidia causing dissemination, *Encephalitozoon spp.,* is sensitive to albendazole, whereas *E. bieneusi,* which usually does not disseminate, is resistant to this drug [72].

Specimen Collection

Spores of microsporidia can resist environmental conditions for years keeping their infectivity, if they are protected from from excessive desiccation. Stool samples sent to the laboratory should be preserved in 5% or 10% formalin or in sodium acetic acid-formalin. In case of suspected dissemination, urine, sputum, bronchoalveolar lavage, nasal secretion, cerebrospinal fluid (CSF), conjunctival smears and corneal scrapings are submitted to the laboratory. They are submitted in formalin for ordinary microscopic examination, fixed in glutaraldehyde for electron microscopy and fresh for cell culture or molecular studies [60].

Stool Examination

Detection of intestinal microsporidiosis by light microscopy in cases with chronic diarrhea requires sufficient experience, as spores have a size similar to bacterial and yeast cells and can easily be missed within the sample's microflora and debris. In stool specimens (and other samples contaminated by other microorganisms) it is therefore necessary to use special stains to identify microsporidia. The most common stain for stool is Weber's chromotrope stain (Fig. **8**), which is considered practical [228].

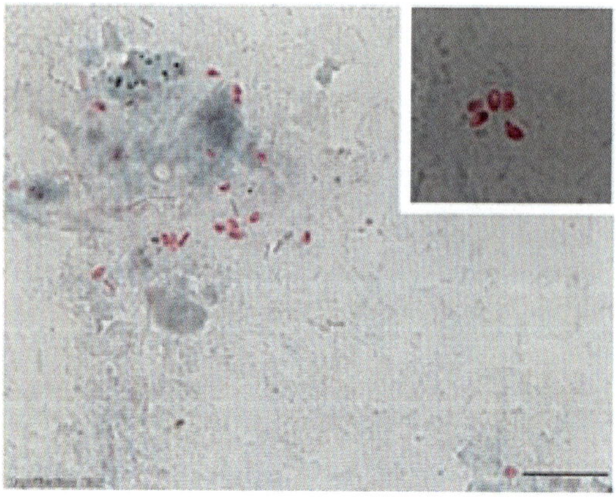

Fig. (8). Stool sample of a male HIV patient with chronic diarrhea and wasting stained with Weber's stain. Microsporidial spores (in this case *Enterocytozoon bieneusi*) can be identified as multiple red structures with oval shape of 2-3 x 0.8 μm (arrow). The main characteristic is that spores are not homogenously stained but show typical vacuoles (insert).

The use of positive control material is highly recommendable, especially if positive samples are rare. Spore detection requires a total magnification of ×1000 with oil immersion. The spore wall stains with variable degrees of red, while its interior shows a characteristic inhomogeneous pattern "vacuoles" (see Fig. **8**).

The counterstain gives the background a blue or green color, according to the type of stain used. At least 100 fields should be examined under oil immersion (1000 x) and the size of spores should be measured. Bacterial spores, as well as other findings of yeast cells and debris could be stained in red color.

Therefore morphological identification has to be performed by an experienced microscopist considering the size and staining pattern. For quality control reasons, positive control smears have to be included [60]. Modifications of Weber's method have been described and include modified trichrome stain (Ryan's stain) and the Gram chromotrope method [172, 244]. Optical brightener binds to the chitineous wall of microsporidian spores and can be used to visualize them under UV light. Commonly used agents are Calcofluor white M2R (Fig. **9**) and Uvitex 2B (Fungiqual A), others such as fluorescent brightner 28 and Fungi-Flour [245] can also be used in the same manner. Since these reagents bind to chitin, they also stain fungal structures and many fibers.

Therefore these stains can only serve as screening tool and do not allow the differentiation of microsporidia from other structures such as yeast cells. They might increase sensitivity since they allow screening a higher number of microscopical fields. If elements compatible with microsporidial spores are visualized with these stains, confirmation by specific stains (*e.g.* Weber's stain) or other methods has to be performed.

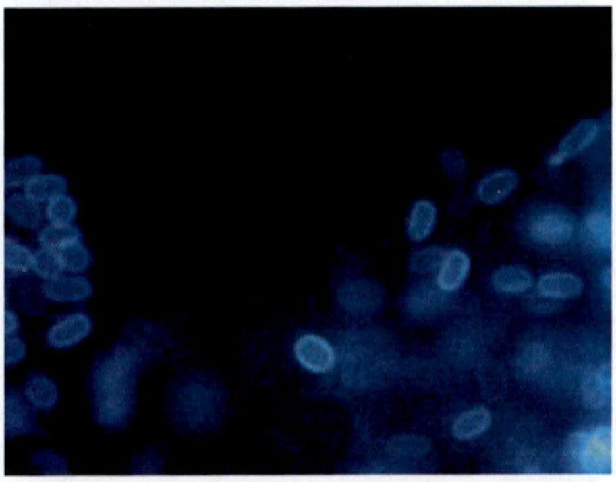

Fig. (9). Encephalitozoon intestinalis stained with Calcoflour white.

Direct Immunofluorescence

Direct immunofluorescence using monoclonal antibodies for the diagnosis of infection and species identification is effective [246, 247]. It is technically simple and rapid, and does not require costly reagents or equipment (Fig. **10**).

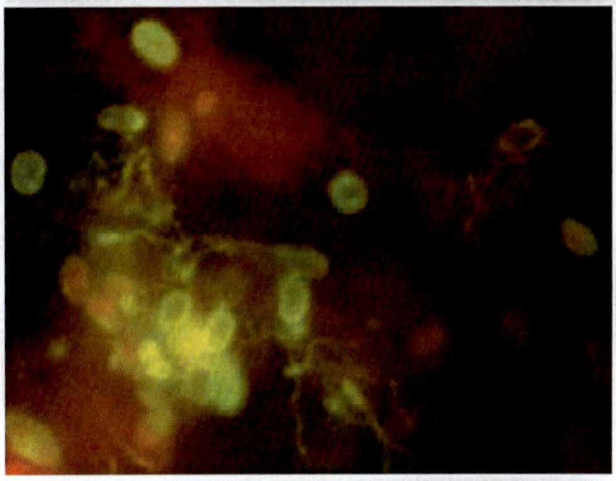

Fig. (10). Monoclonal antibody-based immunofluorescence identification of *Encephalitozoon hellem.*

High sensitivity and specificity values have been reported [246]. Some authors find the technique comparable to PCR, which is more complex and costly and is therefore mainly used in research laboratories [247]. Thus, the use of monoclonal antibodies might become the routine method for the specific diagnosis of microsporidiosis [247]. Monoclonal antibodies kits are currently available (*e.g.* Bordier Laboratories). Still, these kits are not widely available and do not always include positive control slides [1, 246, 247].

Electron Microscopy

Before the widespread use of molecular techniques in the diagnosis of microsporidosis, EM was used as the gold standard for confirmation of diagnosis and for species identification of microsporidia.

Electron microscopy is a very reliable approach in the diagnosis of microsporidia. One of its limitations is the difficulty to have an electron microscope in most laboratories. Another limitation depends on its restricted ability of morphological identification to the species level, which will need antigenic characterization or molecular identification [60].

Although electron microscopy was used in the diagnosis of microsporidiosis in body fluid specimens with success, it showed difficulty to differentiate species in the tissue biopsy specimens due to absence of proliferative stages and the small amount of sample to be examined renders this technique of low sensitivity [193]. It is advised to always place the body fluid sediments or the tissue biopsies in case of suspected infection in the fixatives used for Electron Microscopy for further examinations [60]. EM is important for the diagnosis of the details of the spores of microsporia. Slides should be ultrathin (1 micron) to identify the internal structures. Preparation and examination of the specimens are also time consuming and need trained personnel.

Molecular Methods

The molecular analysis of microsporidia can provide a highly sensitive and specific tool for detecting and differentiating species in biological samples. It could also explain a wide range of geographic distribution of microsporidia that may infect man, as well as their demographic data, zoonotic relationship and their survival in the environment [146, 193, 248].

Conventional PCR is a sensitive, specific and reproducible method that is considered an alternative to electron microscopy. The detection threshold for microsporidia is 102spores/g fecal samples, much lower than detected by light microscopy, where the cutoff is around 104-106spores/g [20, 156, 172, 193, 249]. Conventional PCR has however several drawbacks. First, it is a long, expensive technique performed by specialized laboratories. There is also risk of contamination and the parasite load cannot be quantified [249]. Since small number of spores might be ingested with food or water, false positive results are possible.

Quantitative PCR: Over the last few years, quantitative PCR, particularly with the advent of real- time procedures, has revolutionized the diagnosis of certain infectious diseases, including microsporidiosis.

The Quantitative PCR is considered one of the most reliable methods of detection and identification of microsporidia in stools up to <40 spores/ml suspensions [248]. Real-time PCR offers the advantage of eliminating all post-PCR manipulation, reducing the reaction time and limiting the risk of contamination, and consequently of false positives [249, 250].

Multiplex PCR: This technique is designed to use two couples of probes specific for two species or more simultaneously. Multiplex PCR is a sensitive specific technique for the diagnosis of *E. beineusi* and *Encephalitozoon spp.* [251, 252]. Also multiplex PCR could be used, if the specimen is thought to be containing more than one pathogen [253]. Cell culture: The first attempt to culture microsporidia was made in 1937 by Trager, who was partially successful in establishing *Nosema bombycis* infection of a cell culture developed from the ovarian tube lining cells of the silkworm (*Bombyx mori*) [254].

Culture

In vitro culture has also been implicated to determine the species as well as the antimicrosporidial effect of several drugs on *E. cuniculi, E. hellem* and *E. intestinalis* [255, 256]. Microsporidia couldn't be grown well in axenic cultures. However cell culture were used to *e.g.* monkey and rabbit kidney cell lines (Vero and RK-13), human fetal lung fibrobalsts cell line (MRC-5) and Madin-Darby canine kidney cell line (MDCK) were used to grow *Encephalitozoon spp.Trachipleistophora hominis,V. corneae* and *B. algerae. E. bieneusi* which is an important human pathogen, grew only in short term cultures [255, 256].

Only a few attempts have been made to isolate microsporidia into culture from feces because enteric bacteria and yeasts usually overgrow the rich culture medium that is used, which impedes isolation of the fastidious microsporidia, especially *E. bieneusi* [257].

Nucleospora salmonis, a fish microsporidia was successfully cultured on salmonid leukocytes for a relatively long duration [258]. This cell line is promising for growing *E. beineusi* in culture. Little success was achieved in growing *Pseudoloma neutriphilia* on different cell lines of fish [258]. Although *in vitro* culture of microsporidia has a limited diagnostic value, however it may provide important diagnostic information [255].

Serological Methods

A variety of serological methods have been used to detect the antibodies IgG and IgM against microsporidial antigens, particularly *E. cuniculi* in experimental animals. Antibodies against *Encephalitozoon* species (*E. cuniculi* and *E. hellem*) have been identified in both HIV and non-HIV positive cases. The presence of these antibodies does not explain if there is a true infection or not, due to possibility of cross reaction with other species or non-specific reactions.The lack of long-term culture made it difficult to prepare suitable antigens for proper serological studies [60].

TREATMENT

Many drugs have been tried for the management of intestinal microsporidiosis. Efficacy has been variable depending on the causal species. The criteria of therapeutic success are the resolution of the clinical manifestations and the absence of spores from samples [191, 259]. At the present time, albendazole and fumagillin are considered the most effective compounds against *Encephalitozoon intestinalis* and *E. bieneusi*, respectively. Other therapeutic alternatives are either less effective or under trial [172, 260 - 263]. Albendazole, is a benzimidazole derivate used for the treatment of a variety of helminthic infestations and giardiasis, that has been tested against microsporidia *in vitro* and *in vivo* with a considerably a good therapeutic effect against *Encephalitozoon spp.* especially *E. hellem* [88, 172, 191, 259, 264 - 267].

The mechanism of action of albendazole consists in the inhibition of microsporidial division by blocking the synthesis of tubulin, a major constituent of the mitosis spindle [89, 191, 260]. Thus, it causes inhibition of the microtubule assembly of microsporidia, including the *Encephalitozoon spp* [268, 269]. Electron microscopical study showed that albendazole affects mainly the developmental stages leading to partial inhibition of the reproduction of the microsporidia affecting the small intestine [270] and was suggested as a suitable treatment for systemic parasitic disease [271]. Clinical studies have demonstrated the efficacy of albendazole against species of the *Encephalitozoon* genus in HIV-infected patients for whom it is the treatment of choice for intestinal, ocular and disseminated microsporidiosis [172, 259]. On the other hand, it exhibits a less effective action against *E. bieneusi* since it yields only a decline in the parasite load and degenerative alterations of the spores. Clinically, diarrhea might become less severe and the body weight might stabilize. Relapse is however common after treatment withdrawal [260, 279]. Consequently, albendazole has a parasitostatic effect on *E. bieneusi* by incomplete inhibition of replication; stool and duodenal biopsy samples remain positive [172, 189, 191, 194, 259, 260, 269, 278]. Albendazole is absorbed well after oral intake when associated with a fat-rich meal. It is metabolized in the liver where a sulfoxide metabolite is formed and is more active and less toxic than albendazole itself [191, 194, 261, 269]. The oral dose of albendazole is 400 mg b.i.d. for adults and 7.5 mg/kg b.i.d for children with total daily dose of 15 mg/kg for 2-4 weeks [89, 194, 260]. Severely immunosuppressed individuals might require longer periods of treatment or maintenance therapy. Albendazole is well tolerated and does not require special surveillance. Rare adverse effects have been described in the form of abdominal pain and diarrhea. There also may be minor elevation of the serum transaminases, which is reversible at withdrawal as well as proteinuria and/or neurological manifestations. Exceptional cases of reversible alopecia, malaise with vertigo, skin rash, fever, pruritis, and less commonly, hematological disorders (neutropenia, pancytopenia) have been reported [194, 260, 262]. It is contraindicated for patients with known hypersensitivity to albendazole as well as pregnant or lactating women [194, 262]. Drug interactions exist with cimetidine, dexamethasone and praziquantel, leading to increased serum levels of albendazole [194, 262].

Other Benzimidazole Derivates

Certain benzimidazole derivatives have been studied in terms of efficacy for the treatment of microsporidiosis. Mebendazole has been found to be active against *E. intestinalis in vitro* but is poorly absorbed after oral administration. Nocodazole and parbendazole also showed anti microsporidial effect, but presented toxic effects limiting their use. Thiabendazole is well absorbed but poorly active [191, 262]. Fenbendazole may be of potential interest because of its rapid absorption after oral intake and its metabolism into oxfendazole. Fenbendazole and oxfendazole are very active against *E. intestinalis* and are non-toxic *in vitro*. These compounds appear to be promising for the treatment of microsporidiosis [191, 260].

Fumagillin: (Fumidil B, Fumadil, Fugillin, Fumagillin, Flesint), a known antibiotic, anti-angiogenic substance and a product of *Aspirigillus fumigatus* has demonstrated a good anti-*E. bieneusi* activity [191, 272], although various adverse side effects have been reported [272, 273]. Fumagillin was identified in 1949 and was used in 1953 by beekeepers against encephalitozoonosis in bees caused by *Nosema apis* and as a human drug for the treatment of amoebiasis prior to the development of more effective amoebicidal agents [88, 89, 191, 260, 272]. The target of fumagillin is a cellular metalloprotease, methionine aminopeptidase-2

(MetAP2). This enzyme is indispensable for microsporidia metabolism and survival. It is essential for eliminating methionine on the terminal end of proteins, necessary for post-translational and functional modifications [88, 273]. The fumagillin acts by inhibition of the replication of microsporidia by blockage of the site of action of MetAP2 and by the inhibition of RNA synthesis and consequent death of the organism [88, 191, 171, 260, 273]. Fumagillin has been used successfully against species 28 of the *Encephalitozoon* genus and against *Vittaforma cornea in vitro* and in humans for the treatment of ocular infections caused by *E. hellem* and intestinal infection by *E. bieneusi* [88, 273]. Topical fumagillin is effective as eye drops in a concentration of 70 µg/mL in saline. Microsporidial eye infections might require long-term and may be lifelong therapy [263, 274]. Systemic side effects are negligible. Since in such cases systemic spread is possible, a combination with albendazole should be considered. The drug is prescribed for oral intake 20 mg t.i.d for a total dose of 60 mg/ d for 14 days. The efficacy of systemic fumagillin is counter balanced by its adverse effects. When administered orally, the drug exhibits bone marrow toxicity by its direct effect on the megakaryocytic line and myeloid progenitors [272]. Thrombocytopenia and neutropenia are the most common adverse effects requiring regular medical follow up for the entire duration of treatment [88, 89, 189, 191, 194, 271, 273]. In addition, abdominal pain, diarrhea, vomiting, and hyperlipasemia have been noted during the use of fumagillin. This drug is contraindicated in the event of hypersensitivity [194, 261]. A fumagilin analog, TNP-470, with fewer side effects could replace fumagillin for the systemic treatment of *E. bieneusi* infection in the future [274].

Other drugs used: **Nitazoxanide** (Cryptaz): Is a broad-spectrum oral anti-parasite agent against protozoa such as amoeba, nematodes, cestodes, and trematodes [260, 275]. It is also used for the treatment of cryptosporidiosis [260, 262, 275]. The drug inhibits the action of pyruvate ferrodoxine oxydoreductase of the electron transport system [262]. It is prescribed at the dose of 1g b.i.d for 60 days [260, 262]. Nitazoxanide has proven efficacy *in vivo* on cell cultures of *E. intestinalis* and *Vittaforma cornea.* A clinical effect has been reported in a single case report of an AIDS patient with *E. bieneusi* infection [275].

***Antiretroviral Therapy (HAART)*:** These combination therapies aim to suppress viral replication and to restore cell-mediated immunity. As immunodeficiency is the main factor promoting micrsporidiosis, HAART is the most effective treatment for microsporidisis in HIV patients [265]. Countries where HAART has been used showed a dramatic decline in the microsporidial incidence among HIV cases.

Apart from the restoration of cell-mediated immunity, some antiretroviral drugs might also exhibit direct antiparasitic activity (*i.e.*, protease inhibitors) [276]. In situations where there is difficulty to get HAART, there will be no improvement in the incidence of microsporidiosis [57]

With antiretroviral therapy, HIV-infected patients have a lower viral load and improved CD4 counts with reconstitution of their immune defense. Consequently, anti-retroviral therapy reduces the prevalence of opportunistic infections, including microsporidiosis, and reduces the morbidity and mortality related to HIV infection [84, 95, 172, 189, 259, 277]. It also enables the eradication of microsporidia infection without use of a specific treatment. It considerably reduces the risk of recurrent microsporidiosis observed after treatment withdrawal in subjects with severe immune deficiency [277].

PERSPECTIVES OF CONTROL

In the management system for the control of microsporidiosis, it is of great importance to view the epidemiological extent of the infection. The potential role of water and food in the transmission of microsporidia is of importance and the usual measures and precautions should be taken to prevent their contaminations with the urine or feces of infected animals or humans.

Unfortunately, in many countries there is insufficient epidemiological data on the clinical and environmental magnitude of infection and contamination with these organisms.

Microsporidial spores are able to survive for a considerably long durations and remain infective in the environment [278]. Spores of *E.cuniculi* could be affected by exposure to bleach, ethanol, and other disinfectants leading to decrease in the infectivity of the organisms in tissue culture models [21, 279 - 281]. Physicians should contribute in diagnosis and treatment of infection by including microsporidiosis in the differential diagnosis of chronic diarrhea, fever of unknown origin, myosistis, renal disease, biliary-hepatic disease, and other unusual symptoms in HIV patients, organ transplant recipients, and other patients at risk. Laboratories should also include the methods for the detection of microsporidia *e.g.* staining and DNA identification of these parasites. Besides, the laboratory staff should be trained for different techniques and the morphological features of microsporidia in specific and unspecific stains.

The Center of Disease Control and the HIV Medicine Association of the Infectious Diseases, provide a comprehensive guideline for the treatment and control of the opportunistic infections detected in HIV cases including microsporidiosis [282]. There are new guidelines are available in: http://aidsinfo.nih.gov/contentfiles/lvguidelines/adult_oi.pdf [282]. For immuno-compromised patients, as the organ transplant recipients or AIDS, microsporidiosis could be life threatening. The recommendations are similar to other opportunistic food- or water-borne and zoonotic infections. They include proper hand hygiene and drinking bottled or at least boiled water [57]. Meat, fish, and seafood should be properly cooked and fruits and vegetables washed before consuming. Avoiding exposure to animals suspected to be carriers of microsporidia or handing them with great care. Boiling of water for not less than 5 minutes can kill spores of *E. cuniculi.* The application of some disinfectants as Quaternary Ammonium, ethyl alcohol (70%), Formaldehyde (0.3-1%), phenol, Hydrogen peroxide (1%), chloramine or Sodium hydroxide for 30 minutes completely destroyed *E. cuniculi* spores [173, 276]. Treatment with Ozone, ultraviolet, gamma rays and chlorination at pH 7 reduced the infectivity of *Encephalitozoon spp.* [277, 283 - 286]. Antiparasitic prophylactic agent has not be mentioned, however the improvement of immunity in the immunocompromised patients, as in case of using HAART is effective in remission of microsporidiosis. [287 - 289].

Ocular infection, which may occur as a result of conjunctival inoculation with fingers contaminated with organisms from body secretions, could be prevented by hand washing. Some instructions should be given to infected people to avoid dispersal of spores in their sputum or respiratory secretions.

In hospitals, spores of *E. cuniculi* have the ability to survive and keep their infectivity to a period of about one month, however they can be rendered noninfectious by a 30-minute exposure to the most common disinfectants and by the methods employed for sterilization. Therefore, it is of utmost importance to limit infection in hospital rooms by sufficient cleaning with proper disinfectants.

In experiments with the *E. intestinalis*, a 16-minute exposure to a 2.0 mg/l chlorine treatment was needed to achieve a three-log reduction (99.9%) of viable spores as determined by infection of cell cultures [290]. Treatment with chlorine at concentrations of approximately 2.5 mg/l for a minimum of 4 minutes produced a 3.3 log inactivation of *E. cuniculi,* but only a 0.70 log reduction of *E. hellem* [285]. Scientists developed a very efficient and rapid extraction free, filter-based method to prepare DNA template for the use in PCR, to identify microorganisms including microsporidia spores. The method could be adapted for detection of parasites from clinical and environmental samples using multiplex PCR with great sensitivity and minimal preparation [291].

The United States Environmental Protection Agency prepared some instructions modified in 2005, These instructions were given in response to the Safe water Amendments of 1999 to identify and determine the water-borne parasites: *Cryptosporidium parvum* and *Giardia intestinalis*. The recommended methods are also applied to detect and identify microsporidia (http://www.epa.gov/-nerlcwww/). The main steps of these approaches depend on filtration followed by separation of spores by an immune-magnetic bead separation (IMS) assay. Immunofluorescence antibody (FA) staining is used to identify the separated microsporidia. Another variation in the methods of detection uses the separation of spores on an immune-magnetic separation assay (IMS) followed by PCR [167] or water filtration followed by PCR [169].

VACCINES

Little has been done and published in the field of preparation and use of vaccines against microsporidia. Some of the main difficulties are establishing and maintaining microsporidial culture, development of resistance associated with cellular immune response and the choice of the biological product that can serve as a vaccine. However the main few successful trials carried out were directed towards the protection of farm fish from serious infections with microsporidia. Spores of a low virulence strain of the microsporidian *Loma salmonae* were used to vaccinate trout against the gill disease by intraperitoneal route and showed success in protecting fish against infection with the virulent strain [292]. In an experiment concerned with *Loma salmonae*, Rodriguez-Tavor *et al.*, reported that the main argument that favors the production of a vaccine was the development of resistance associated to cellular immune response [293]. In another study, the juvenile rainbow trout were vaccinated against microsporidiosis using whole viable spores of the microsporidian *Glugea anomala* and *Glugea hertwigi* by intraperitoneal route. The results showed reduction in the numbers of branchial xenomas by 80% and 91%, respectively, after a standard experimental infection with the microsporidian *Loma salmonae*. Significant protection was obtained when killed-spore preparations were used [294]. Although these studies are limited, yet they may direct the attention towards the success that may be obtained with vaccination in animals or humans.

CONCLUSIONS

Microsporidia are tiny unicellular organisms that parasitize cells of almost all living creatures particularly the immunocompromised subjects. Infection occurs mainly by oral route, however inhalation, sexual relations, direct inoculation into ocular mucosa and wounds and insect bites are also described. Animals are important reservoirs of infection to humans. Food and water are relevant vehicles of infection and could be contaminated by animal and human excreta. Most human infections affect the intestine, however dissemination to other organs may occur with serious clinical manifestations. Proper diagnosis is needed for proper management of cases. The increased awareness of the modes of infection, pathogenesis and diagnosis can improve the understanding of the epidemiology and management of microsporidiosis in humans and animals.

CONFLICT OF INTEREST

The authors confirm that they have no conflict of interest to declare for this publication.

ACKNOWLEDGEMENTS

The authors are grateful to the team of CDC for allowing the use of figure no. 1,2,3,4,9,10.

These photos are used after permission from DPDX team: CDC http://www.dpd.cdc.gov/dpdx/HTLM/microsporidiosis.html.

REFERENCES

[1] Keeling PJ, Fast NM. Microsporidia: biology and evolution of highly reduced intracellular parasites. Annu Rev Microbiol 2002; 56: 93-116. [http://dx.doi.org/10.1146/annurev.micro.56.012302.160854]

[2] Weiss LM. The First United Workshop on Microsporidia from Invertebrate and Vertebrate Hosts. Folia Parasitol (Praha) 2005; 52: 1-7. [http://dx.doi.org/10.14411/fp.2005.001]

[3] Wittner M, Weiss LM. The microsporidia and microsporidiosis. Washington, DC: American Society, for Microbiology Press 1999.

[4] Cali A, Neafie RC, Takvorian PM. Microsporidiosis Topics on the Pathology of Protozoan Arthropod Disreases. Washington, DC: Armed Forces Institute of Pathology Press 2011.

[5] Stark D, Barratt JL, van Hal S, *et al.* Clinical significance of enteric protozoa in immunosuppressed human population. 2009. Clin Microbiol Rev 2009; 634-50. [http://dx.doi.org/10.1128/CMR.00017-09]

[6] Bryan R, Cali A, Owen R, *et al.* Microsporidia: newly recognized opportunistic pathogens in patients with AIDS. In: Sun T, Ed. Progress in Clinical Parasitology. Blue Bell, PA: Field and Wood, Inc Pub 1990; Vol. 2: pp. 1-26.

[7] Pinnolis M, Egbert PR, Font RL, *et al.* Nosematosis of the cornea: case report including electron microscopic studies. Arch Ophthalmol 1981; 99: 1044-14. [http://dx.doi.org/10.1001/archopht.1981.03930011044012]

[8] Cali A, Takvorian PM, Lewin S, *et al. Brachiola vesicularum*, n. g., n. sp., a new microsporidium associated with AIDS and myositis. J Eukaryot Microbiol 1998; 45: 240-51. [http://dx.doi.org/10.1111/j.1550-7408.1998.tb04532.x]

[9] Margileth AM, Strano AJ, Chandra R, *et al.* Disseminated nosematosis in an immunologically compromised infant. Arch Pathol 1973; 95: 145-50.

[10] Silveira H, Canning EU. *Vittaforma corneae* n. comb. for the human microsporidium *Nosema corneum* Shadduck, Meccoli, Davis & Font, 1990, based on its ultrastructure in the liver of experimentally infected athymic mice. J Eukaryot Microbiol 1995; 42: 158-65. [http://dx.doi.org/10.1111/j.1550-7408.1995.tb01557.x]

[11] Ledford DK, Overman MD, Gonzalvo A, *et al.* Microsporidiosis myositis in a patient with the acquired immunodeficiency syndrome. Ann Intern Med 1985; 102: 628-30. [http://dx.doi.org/10.7326/0003-4819-102-5-628]

[12] Cali A, Takvorian PM. Ultrastructure and development of *Pleistophora ronneafiei* n. sp., a Microsporidium (Protista) in the skeletal muscle of an immunecompromised individual. J Eukaryot Microbiol 2003; 50: 77-85. [http://dx.doi.org/10.1111/j.1550-7408.2003.tb00237.x]

[13] Hollister WA, Canning EU, Anderson CL. Identification of microsporidia causing human disease. J Eukaryot Microbiol 1996; 43: 104S-5S. [http://dx.doi.org/10.1111/j.1550-7408.1996.tb05026.x]

[14] Desportes I, Le Charpentier Y, Galian A, *et al.* Occurrence of a new microsporidan: *Enterocytozoon bieneusi* n.g., n. sp., in the enterocytes of a human patient with AIDS. J Protozool 1985; 32: 250-4. [http://dx.doi.org/10.1111/j.1550-7408.1985.tb03046.x]

[15] Cali A, Owen RL. Intracellular development of *Enterocytozoon*, a unique microsporidian found in the intestine of AIDS patients. J Protozool 1990; 37(2): 145-55. [http://dx.doi.org/10.1111/j.1550-7408.1990.tb05885.x]

[16] Terada S, Reddy KR, Jeffers LJ, *et al.* Microsporidan hepatitis in the acquired immunodeficiency syndrome. Ann Intern Med 1987; 107: 61-2. [http://dx.doi.org/10.7326/0003-4819-107-1-61]

[17] Didier ES, Didier PJ, Friedberg DN, *et al.* Isolation and characterization of a new human microsporidian, *Encephalitozoon hellem* (n. sp.), from three AIDS patients with keratoconjunctivitis. J Infect Dis 1991; 163: 617-21. [http://dx.doi.org/10.1093/infdis/163.3.617]

[18] Cali A, Meisler DM, Rutherford I, *et al.* Corneal microsporidiosis in a patient with AIDS. Am J Trop Med Hyg 1991; 44: 463-8.

[19] Cali A, Kotler DP, Orenstein JM. *Septata intestinalis* N. G., N. Sp., an intestinal microsporidian associated with chronic diarrhea and dissemination in AIDS patients. J Eukaryot Microbiol 1993; 40: 101-12. [http://dx.doi.org/10.1111/j.1550-7408.1993.tb04889.x]

[20] Franzen C, Muller A. Molecular techniques for detection, species differentiation and phylogenetic analysis of microsporidia. Clin Microbiol Rev 1999; 12: 243-85.

[21] Becnel JJ, Andreadis TG. Microsporidia in insects. In: Wittner M, Ed. The Microsporidia and Microsporidiosis. Washington, D.C: ASM Press 1999; pp. 447-501. [http://dx.doi.org/10.1128/9781555818227.ch14]

[22] Canning EU, Lom J. The microsporidia of vertebrates. London, United Kingdom, Orlando, Fla.: Academic Press 1986.

[23] Weber R, Bryan RT, Schwartz DA, *et al.* Human microsporidial infections. Clin Microbiol Rev 1994; 7: 426-61.
[http://dx.doi.org/10.1128/CMR.7.4.426]

[24] Bakowski MA, Priest M, Young S. Genome sequence of the microsporidian species Nematocida sp1 strain ERTm6 (ATCC PRA-372). Genome announcements 2014; 2(5) e00905-14.

[25] Molina JM, Sarfati C, Beauvais B, *et al.* Intestinal microsporidiosis in human immunodeficiency virus-infected patients with chronic unexplained diarrhea—Prevalence and clinical and biologic features. J Infect Dis 1993; 167: 217-21.
[http://dx.doi.org/10.1093/infdis/167.1.217]

[26] Verre J, Harkness J, Marriot D. Prevalence of *Enterocytozoon bieneusi* in faecal specimens from non-HIV-infected patients with diarrhoea [abstract]. In: Abstract book of the Workshop on Microsporidiosis and Cryptosporidiosis in Immunodeficient Patients. Ceske Budejovice, 1993.

[27] Nageli KW. 1857. Uber die neue Krankheit der Seidenraupe und verwandte Organismen. Bot. Z. 15: 760–761, Quoted from: Korrady, N. and Keeling, P J: Microsporidia, a journey through radical taxonomical revision. Fungal Biol Rev 2009; 23: 1-8.

[28] Pasteur L. 1870. Etudes sur la maladie des vers a Soie. Gauthier–Villars, Imprimeur–Libraire, Paris. p. 322. Quoted from: Korrady, N and Keeling, P J. Microsporidia, a journey through radical taxonomical revision. Fungal Biol Rev 2009; 23: 1-8.

[29] Parry W. Simply strange, The story of tiny parasite, Microsporidia. www.livescience.com/- 13237-strange-fungi-microsporidia-nosema-colon. 2011.

[30] Balbiani G. Sur les microsporidies ou psorospermies des articules. CR Acad. Sci. 95: 1168–1171. Quoted from: Korrady, N and Keeling, P J: Microsporidia, a journey through radical taxonomical revision. Fungal Biol Rev 1882; 2009(23): 1-8.

[31] Kudo RR. Protozoology, third ed., Springfield, Illinois. . Quoted from: Korrady, N. and Keeling, P J: Microsporidia, a journey through radical taxonomical revision. Fungal biology reviews. 2009; 23:1-8

[32] Sprague V. Classsification and phylogeny of the microsporidia. In: Bulla LA, Cheng TC, Eds. Comparative pathology. New York, NY: Plenum Press 1976; Vol. 2: pp. 1-30.

[33] Levine ND, Corliss JO, Cox FE, *et al.* A newly revised classification of the Protozoa. J Protozool 1980; 27(1): 37-58.
[http://dx.doi.org/10.1111/j.1550-7408.1980.tb04228.x]

[34] Sprague VV, Becnel JJ. Note on the name–author–date combination for the taxon Microsporidies Balbiani, 1882, when ranked as a phylum. J Invertebr Pathol 1998; 71: 91-4.
[http://dx.doi.org/10.1006/jipa.1997.4702]

[35] Larsson JI. Molecular *versus* morphological approach to microsporidian classification. Folia Parasitol (Praha) 2005; 52: 143-4.
[http://dx.doi.org/10.14411/fp.2005.018]

[36] Vossbrinck CR, Debrunner-Vossbrinck BA. Molecular phylogeny of the Microsporidia: ecological, ultrastructural and taxonomic considerations. Folia Parasitol (Praha) 2005; 52: 130-42.
[http://dx.doi.org/10.14411/fp.2005.017]

[37] Kudo RR, Daniels EW. An electron microscope study of the spore of a microsporidian, *Thelohania californica*. J Protozool 1963; 10: 112-20.
[http://dx.doi.org/10.1111/j.1550-7408.1963.tb01645.x]

[38] Weiser J. *Nosema laphygmae n. sp.* and the internal structure of the microsporidian spore. J Insect Pathol 1959; 1: 52-9.

[39] Vavra J, Larsson JT. Structure of the microsporidia. In: Wittner M, Weiss LM, Eds. The Microsporidia and Microsporidiosis. Washington, DC: American Society for Microbiology 1999; pp. 7-84. [http://dx.doi.org/10.1128/9781555818227.ch2]

[40] Tuzet O, Maurand J, Fize JA, *et al.* Proposition d'un nouveau cadre systimatique pour les genres de Microsporidies. CR Acad Sci (Paris) 1971; 272: 1268-71.

[41] Sprague V, Vavra J. Systematics of the Microsporidia. In: Bulla LA, Cheng TC, Eds. Comparative Pathobiology. New York: Plenum Press 1977; Vol. 2.

[42] Issi IV. Microsporidia as a phylum of parasitic protozoa. Protozoology (Leningrad) 1986; 10: 1-136.

[43] Larsson JI. Identification of microsporidian genera- A guide with comments on the taxonomy. Arch Protistenkd 1988; 136: 1-37. [http://dx.doi.org/10.1016/S0003-9365(88)80032-0]

[44] Curgy JJ, Vavra J, Vivares CP. Presence of ribosomal RNAs with prokaryotic properties in Microsporidia, eukaryotic organisms. Biol Cell 1980; 38: 49-51.

[45] Ishihara R, Hayashi Y. Some properties of ribosomes from the sporoplasm of *Nosema bombycis.* J Invertebr Pathol 1968; 11: 377-85. [http://dx.doi.org/10.1016/0022-2011(68)90186-9]

[46] Vossbrinck CR, Maddox JV, Friedman S, *et al.* Ribosomal RNA sequence suggests microsporidia are extremely ancient eukaryotes. Nature 1987; 326: 411-4. [http://dx.doi.org/10.1038/326411a0]

[47] Fedorov A, Hartman H. What does the microsporidian *E-cuniculi* tell us about the origin of the eukaryotic cell? J Mol Evol 2004; 59: 695-702. [http://dx.doi.org/10.1007/s00239-003-0085-1]

[48] Keeling PJ, Slamovits CH. Simplicity and complexity of microsporidian genomes. Eukaryot Cell 2004; 3: 1363-9. [http://dx.doi.org/10.1128/EC.3.6.1363-1369.2004]

[49] Thomarat F, Vivare`s CP, Gouy M. Phylogenetic analysis of the complete genome sequence of *Encephalitozoon cuniculi* supports the fungal origin of microsporidia and reveals a high frequency of fast-evolving genes. J Mol Evol 2004; 59: 780-91. [http://dx.doi.org/10.1007/s00239-004-2673-0]

[50] Gill EE, Fast NM. Assessing the microsporidia-fungi relationship: combined phylogenetic analysis of eight genes. Gene 2006; 375: 103-9. [http://dx.doi.org/10.1016/j.gene.2006.02.023]

[51] Katinka MD, Duprat S, Cornillot E, *et al.* Genome sequence and gene compaction of the eukaryote parasite *Encephalitozoon cuniculi.* Nature 2001; 414: 450-3. [http://dx.doi.org/10.1038/35106579]

[52] Vivarès D, Belloni L, Tardieu A, *et al.* Catching the PEG-induced attractive interaction between proteins. Eur Phys J E 2002; 9: 15-25.

[53] TA revised six-kingdom system of life. Biol Rev Camb Philos Soc 1998; 73: 203-66. [http://dx.doi.org/10.1017/S0006323198005167]

[54] Keeling PJ, Luker MA, Palmer JD. Evidence from beta-tubulin phylogeny that microsporidia evolved from within the fungi. Mol Biol Evol 2000; 17: 23-31. [http://dx.doi.org/10.1093/oxfordjournals.molbev.a026235]

[55] Matsubayashi H, Koike T, Mikata T, *et al.* A case of *Encephalitozoon*-like-body infection in man. Arch Pathol 1959; 67: 181-7.

[56] Modigliani R, Bories C. Diarrhea and malabsorption in acquired immune deficiency syndrome: a study of four cases with special emphasis on opportunisyic protozoan infestation. Gut 1985; 26: 179-81.
[http://dx.doi.org/10.1136/gut.26.2.179]

[57] Bryan RT, Schwartz DA. Epidemiology of microspridiosis. In: Wittner M, Weiss L, Eds. The Microsporidia and Microsporidiosis. Washington, DC: American Society of Microbiology 1999; pp. 502-16.
[http://dx.doi.org/10.1128/9781555818227.ch15]

[58] Didier ES, Weiss LM. Microsporidiosis: Not just in AIDS patients. Curr Opin Infect Dis 2011; 24(5): 490-5.
[http://dx.doi.org/10.1097/QCO.0b013e32834aa152]

[59] Vavra J. Structure of the microsporidia. Comparative Pathology. New York, NY: Plenum Press 1976; pp. 1-85.
[http://dx.doi.org/10.1007/978-1-4684-3114-8_1]

[60] Garcia LS. Laboratory identification of the microsporidia. J Clin Microbiol 2002; 40(6): 1892-901.
[http://dx.doi.org/10.1128/JCM.40.6.1892-1901.2002]

[61] Edlind T, Katiyar S, Visvesvara G, Li J. Evolutionary origins of Microsporidia and basis for benzimidazole sensitivity: an update. J Eukaryot Microbiol 1996; 43: 109S-S.
[http://dx.doi.org/10.1111/j.1550-7408.1996.tb05029.x]

[62] Edlind TD, Li J, Visvesvara GS, *et al.* Phylogenetic analysis of beta-tubulin sequences from amitochondrial protozoa. Mol Phylogenet Evol 1996; 5: 359-67. [b].
[http://dx.doi.org/10.1006/mpev.1996.0031]

[63] Cali A. General microsporidian features and recent findings on AIDS isolates. J Protozool 1991; 38: 625-30.

[64] Cali A. Comparison of the biology and pathology of microsporidia from different host groups. Fundamental and Applied Aspects of Invertebrate Pathology. In: Fourth International Colloquium of Invertebrate Pathology; Netherlands. 1986; pp. 356-9.

[65] Cali A, Takvorian PM. Developmental morphology and life cycles of the Microsporidia. In: Wittner M, Weiss LM, Eds. The Microsporidia and microsporidiosis. Washington, DC: ASM Press 1999; pp. 85-128.
[http://dx.doi.org/10.1128/9781555818227.ch3]

[66] Jaronski ST. Role of the larval mosquito midgut in determining host susceptibility to *Nosema algerae* (Microsporida). Ithaca: Cornell University 1979; pp. 1-141.

[67] Weidner E, Manale SB, Halonen SK, *et al.* Microsporidian spore invasion tubes as revealed by fluorescentprobes. Biol Bull 1994; 187: 255-6.
[http://dx.doi.org/10.1086/BBLv187n2p255]

[68] Frixione E, Ruiz L, Cerbon J. Germination of *Nosema algerae* (Microspora) spores: conditional inhibition by D2O, ethanol and Hg2+ suggests dependence of water influxuponmembrane hydration and specific transmembrane path ways. J Eukaryot Microbiol 1997; 44: 109-16.
[http://dx.doi.org/10.1111/j.1550-7408.1997.tb05946.x]

[69] Cali A, Takvorian PM. *Brachiola algerae* sporoplasms. J Eukaryot Microbiol 2001; 81S-2S.

[70] Lom J. On the structure of the extruded microsporidian polar filament. Z Parasitenkd 1972; 38: 200-13.
[http://dx.doi.org/10.1007/BF00329598]

[71] Weinder E. Ultrastructural study of microsporidian invasion into cells. Zeitschrift fur Parasitenkitnde 1972; 40: 227-42.

[72] Weiss LM, Schwartz DA. Microsporidiosis. Tropical Infectious Diseases, Principles, Pathogenesis and Practice. 3rd ed. Elsevier 2011; pp. 1126-40.

[73] Friedberg DN, Ritterband DC. Ocular microsporidiosis. In: Wittner M, Weiss LM, Eds. The Microsporidia and Microsporidiosis. Washington, DC: ASM Press 1999; pp. 293-314. [http://dx.doi.org/10.1128/9781555818227.ch9]

[74] Cali A, Takvorian PM. The Microsporidia; Pathology in Man and occurance in nature. Southeast Asian J Trop Med Public Health 2004; 35 (Suppl. 1): 58-64.

[75] Canning EU, Nicholas JP. Genus Pleistophora (Phylum Microspora): Rediscription of the type species *Pleistophora typicalis* Gurley, 1893 and ultrastructural characterization of the genus. J Fish Dis 1980; 3(4): 317-83. [http://dx.doi.org/10.1111/j.1365-2761.1980.tb00402.x]

[76] Vavra J, Yachnis AT, Shadduck JA, *et al.* Microsporidia of the genus *Trachipleistophora* – causative agents of human microsporidiosis: description of *Trachipleistophora anthropopthera n.* sp. (Protozoa: Microsporidia). J Eukaryot Microbiol 1998; 45: 273-83. [http://dx.doi.org/10.1111/j.1550-7408.1998.tb04536.x]

[77] Kotler DP, Orenstein JM. Clinical syndromes associated with microsporidiosis. In: Wittner M, Weiss LM, Eds. The Microsporidia and Microsporidiosis. Washington, D.C.: American Society for Microbiology 1999; pp. 258-92. [http://dx.doi.org/10.1128/9781555818227.ch8]

[78] Levaditi C, Nicolau S, Schoen R. Etiology of epizootic rabbit encephalitis in reports with experimental studies of encephalitis lethargica, *Encephalitozoon cuniculi* (nov spec). Ann Inst Pasteur (Paris) 1923; 38: 675-711.

[79] Weiser J. Protozoan diseases of the gipsy moth. Proc Int Congr Protozool. 497-9.

[80] Voronin VN. Macrotaxonomy of the phylum Microsporidia. Parazitologiia 2001; 35(1): 35-44.

[81] Sprague V, Vernick SH. The ultrastructure of *Encephalitozoon cuniculi* (Microsporidia, Nosematidae) and its taxonomic significance. J Parasitol 1971; 18: 560-9.

[82] Hollister WA, Canning EU. An enzyme-Linked immunosorbent assay (ELISA) for detection of antibodies to *Encephalitozoon cuniculi* and its use in determination of infections in man. Parasitology 1987; 94: 209-19. [http://dx.doi.org/10.1017/S0031182000053890]

[83] Hollister WS, Canning EU, Willcox A. Evidence for widespread occurrence of antibodies to *Encephalitozoon cuniculi* (Microspora) in man provided by ELISA and other serological tests. Parasitology 1991; 102: 33-43. [http://dx.doi.org/10.1017/S0031182000060315]

[84] Conteas CN, Berlin OG, Speck CE, *et al.* Modification of the clinical course of intestinal microsporidiosis in acquired immunodeficiency virus therapy. Am J Trop Med Hyg 1998; 58: 555-8.

[85] Mathis A, Weber R, Deplazes P. Zoonotic potential of the microsporidia. Clin Microbiol Rev 2005; 18: 423-45. [http://dx.doi.org/10.1128/CMR.18.3.423-445.2005]

[86] Sarfati C, Bourgeois A, Menotti J, *et al.* Prevalence of intestinal parasites including microsporidia in human immunodeficiency virus-infected adults in Cameroon: a cross-sectional study. Am J Trop Med Hyg 2006; 74: 162-4.

[87] Raccurt CP, Agnamey P, Sarfati C, *et al.* Premier cas tchadien de microsporidiose intestinale due à *Enterocytozoon bieneusi* chez une patiente infectée par le virus de l'immunodéficience humaine VIH-1. Med Trop 2008; 68: 307-8.

[88] Didier ES, Weiss LM. Microsporidiosis: current status. Curr Opin Infect Dis 2006; 19: 485-92. [http://dx.doi.org/10.1097/01.qco.0000244055.46382.23]

[89] Desportes-Livage I, Darty A. Infections a microsporidies, Isospora et Sarcocystis. EMC (Elsevier SAS, Paris), Maladies infectieuses 2005. 8-503-A-10.

[90] Lores B, Lopez-Miragaya I, Arias C, *et al.* Intestinal microsporidiosis due to *Enterocytozoon bieneusi* in elderly human immunodeficiency virus-negative patients from Vigo, Spain. Clin Infect Dis 2002; 34: 918-21.
[http://dx.doi.org/10.1086/339205]

[91] Michiels JF, Saint Paul MC, Hofman P, *et al.* Caractéristiques histopathologiques des infections opportunistes intestinales au cours du SIDA. Ann Pathol 1992; 12: 165-73.

[92] Espern A, Morio F, Miegeville M, *et al.* Molecular study of microsporidiosis due to *Enterocytozoon bieneusi* and *Encephalitozoon intestinalis* among human immunodeficiency virus-infected patients from two geographical areas: Niamey, Niger and Hanoi, Vietnam. J Clin Microbiol 2007; 45: 2999-3002.
[http://dx.doi.org/10.1128/JCM.00684-07]

[93] Liguory O, David F, Sarfati C, *et al.* PCR (Polymerase Chain Reaction): intérêt pour l'identification d'*Enterocytozoon bieneusi* et *Encephalitozoon intestinalis* chez les patients infectés par le VIH atteints de microsporidiose intestinale. Eurobiologiste 1996; 30: 469-75.

[94] Lebbad M, Norrgren H, Nauclér A, *et al.* Intestinal parasites in HIV-2 associated AIDS cases with chronic diarrhea in Guinea –Bissau. Acta Trop 2001; 80: 45-9.
[http://dx.doi.org/10.1016/S0001-706X(01)00142-5]

[95] Ferreira FM, Bezerra L, Santos MB, *et al.* Intestinal microsporidiosis: a current infection in HIV-seropositive patients in Portugal. Microbes Infect 2001; 3: 1015-9.
[http://dx.doi.org/10.1016/S1286-4579(01)01465-4]

[96] Bern C, Kawai V, Vargas D, *et al.* The epidemiology of intestinal microsporidosis in patients with HIV/AIDS in Lima, Peru. J Infect Dis 2005; 191: 1658-64.
[http://dx.doi.org/10.1086/429674]

[97] Konate A, Minta D, Diarra M, *et al.* Parasitoses digestives au cours de la diarrhée du SIDA. Bull Soc Pathol Exot 2005; 98: 33-5.

[98] Dworkin MS, Buskin SE, Davidson AJ, *et al.* Prevalence of intestinal microsporidiosis in human immunodeficiency virus-infected patients with diarrhea in major United States cities. Rev Inst Med Trop Sao Paulo 2007; 49: 339-42.
[http://dx.doi.org/10.1590/S0036-46652007000600001]

[99] Billaud E, Miegeville M, Piolet D, *et al.* Microsporidiose au cours du SIDA, une infection opportuniste d'actualité. Med Mal Infect 1997; 27: 719-22.
[http://dx.doi.org/10.1016/S0399-077X(97)80182-5]

[100] Nkinin SW, Asonganyi T, Didier ES, *et al.* Microsporidian infection is prevalent in healthy people in Cameroon. J Clin Microbiol 2007; 45: 2841-6.
[http://dx.doi.org/10.1128/JCM.00328-07]

[101] Wumba R, Enache-Angoulvant A, Develoux M, *et al.* Prévalence des opportunistes digestives parasitaires à Kinshasa (république démocratique du Congo), résultats d'une enquête préliminaire chez 50 patients au stade SIDA. Med Trop 2007; 67: 145-8.

[102] Samie A, Obi CL, Tzipori S, *et al.* Microsporidia in South Africa: PCR detection in stool samples of HIV-positive and HIV-negative individuals and school children in Vhembe district. Limpopo province. Trans R Soc Trop Med Hyg 2007; 101: 547-54. [http://dx.doi.org/10.1016/j.trstmh.2007.02.005]

[103] Raynaud L, Delbac F, Brousolle V, *et al.* G., Vivares, CP., and Peyron, F. Identification of *Encephalitozoon intestinalis* in travelers with chronic diarrhea by specific PCR amplification. J Clin Microbiol 1998; 36: 37-40.

[104] Okhuysen PC. Traveler's diarrhea due to intestinal protozoa. Clin Infect Dis 2001; 1: 110-4. [http://dx.doi.org/10.1086/320894]

[105] Hautvast JL, Tolboom JJ, Derks TJ, *et al.* Asymptomatic intestinal microsporidiosis in a human immunodeficiency virus-seronegative immunocompetent Zambian child. Pediatr Infect Dis J 1997; 16: 415-6. [http://dx.doi.org/10.1097/00006454-199704000-00017]

[106] Desportes-Livage I, Doumbo O, Pichard E, *et al.* Microsporidiosis in HIV-seronegative patients in Mali. Trans R Soc Trop Med Hyg 1998; 92: 423-4. [http://dx.doi.org/10.1016/S0035-9203(98)91076-8]

[107] Valperga SM, de Jogna Prat SA, de Valperga GJ, *et al.* Microsporidian spores in the stool specimens of toddlers, with or without diarrhea, from Tucuman. Argentina. Rev Argent Microbiol 1999; 31: 157-64.

[108] Gumbo T, Hobbs RE, Carlyn C, *et al.* Microsporidia infection in transplant patients. Transplantation 1999; 67: 482-4. [http://dx.doi.org/10.1097/00007890-199902150-00024]

[109] Mohindra AR, Lee MW, Visvisvara G, *et al.* Disseminated microsporidiosis in a renal transplant recipient. Transpl Infect Dis 2002; 4: 102-7. [http://dx.doi.org/10.1034/j.1399-3062.2002.01011.x]

[110] Lewis NL, Francis IC, Hawkins GS, *et al.* Bilateral microsporidial keratoconjunctivitis in an immunocompetent non-contact lens wearer. Cornea 2003; 22: 374-6. [http://dx.doi.org/10.1097/00003226-200305000-00018]

[111] Lanternier F, Boutboul D, Menotti J, *et al.* Microsporidiosis in solid organ transplant recipients: two *Enterocytozoon bieneusi* cases and review. Transpl Infect Dis 2009; 11: 83-8. [http://dx.doi.org/10.1111/j.1399-3062.2008.00347.x]

[112] Carlson JR, Li L, Hilton CL, *et al.* Disseminated microsporidiosis in a pancreas/ kidney transplant recipient. Arch Pathol Lab Med 2004; 128: e41-3.

[113] Fogla R, Padmanabhan P, Therese KL, *et al.* Chronic microsporidial stromal keratitis in an immunocompetent, noncontact lens wearer. Indian J Ophthalmol 2005; 53: 123-5. [http://dx.doi.org/10.4103/0301-4738.16177]

[114] Gamboa-Dominguez A, De Anda J, Donis F, *et al.* Disseminated *Encephalitozoon cuniculi* infection in a Mexican kidney transplant recipient. Transplantation 2003; 75: 1898-900. [http://dx.doi.org/10.1097/01.TP.0000064623.57821.22]

[115] Goetz MS, Eichenlaub S, Pape GR, *et al.* Chronic diarrhea as a result of intestinal microsposidiosis in a liver transplant recipient. Transplantation 2001; 71: 334-7. [http://dx.doi.org/10.1097/00007890-200101270-00029]

[116] Guerard A, Rabodonirina M, Cotte L, *et al.* Intestinal microsporidiosis occurring in two renal transplant recipients treated with mycophenolate mofetil. Transplantation 1999; 68: 699-707. [http://dx.doi.org/10.1097/00007890-199909150-00017]

[117] Kakrania R, Joseph J, Vaddavalli PK, *et al.* Microsporidia keratoconjunctivitis in a corneal graft. Eye (Lond) 2006; 20: 1314-5.
[http://dx.doi.org/10.1038/sj.eye.6702178]

[118] Kelkar R, Sastry PS, Kulkarni SS, *et al.* Pulmonary microsporidial infection in a patient with CML undergoing allogeneic marrow transplant. Bone Marrow Transplant 1997; 19: 179-82.
[http://dx.doi.org/10.1038/sj.bmt.1700536]

[119] Latib MA, Pascoe MD, Duffield MS, *et al.* Microsporidiosis in the graft of a renal transplant recipient. Transpl Int 2001; 14: 274-7.
[http://dx.doi.org/10.1111/j.1432-2277.2001.tb00058.x]

[120] Mohammed H, Endeshaw T, Kebede A, *et al.* Comparison of Chromotrope 2R and Uvitex 2B for the detection of intestinal microsporidial spores in stool specimens of HIV patients attending Nekempte Hospital, West Ethiopia. Ethiop Med J 2009; 47: 233-7.

[121] Sax PE, Rich JD, Pieciak WS, *et al.* Intestinal microsporidiosis occurring in a liver transplant recipient. Transplantation 1995; 60: 617-8.
[http://dx.doi.org/10.1097/00007890-199509270-00018]

[122] Teachey DT, Russo P, Orenstein JM, *et al.* Pulmonary infection with microsporidia after allogeneic bone marrow transplantation. Bone Marrow Transplant 2004; 33: 299-302.
[http://dx.doi.org/10.1038/sj.bmt.1704327]

[123] Vemuganti GK, Garg P, Sharma S, *et al.* Is microsporidial keratitis an emerging cause of stromal keratitis? a case series study. BMC Ophthalmol 2005; 5: 19.
[http://dx.doi.org/10.1186/1471-2415-5-19]

[124] Rabodonirina M, Bertocchi M, Desportes-Livage I, *et al.* *Enterocytozoon bieneusi* as a cause of chronic diarrhea in a heart-lung transplant recipient who was seronegative for human immunodeficiency virus. Clin Infect Dis 1996; 23: 114-7.
[http://dx.doi.org/10.1093/clinids/23.1.114]

[125] Liguory O, Sarfati C, Derouin F, *et al.* Evidence of different *Enterocytozoon bieneusi* genotypes in patients with and without human immunodeficiency virus infection. J Clin Microbiol 2001; 39: 2672-4.
[http://dx.doi.org/10.1128/JCM.39.7.2672-2674.2001]

[126] Rabodonirina M, Cotte L, Radenne S, *et al.* Microsporidiosis and transplantation: a retrospective study of 23 cases. J Eukaryot Microbiol 2003; 50 (Suppl.): 583.
[http://dx.doi.org/10.1111/j.1550-7408.2003.tb00639.x]

[127] ten Hove RJ, Van Leishout L, Beadsworth MB, *et al.* Characterization of genotypes of *Enterocytozoon bieneusi* in immunosuppressed and immunocompetent patient groups. J Eukaryot Microbiol 2009; 56: 388-93.
[http://dx.doi.org/10.1111/j.1550-7408.2009.00393.x]

[128] Snowden KF, Didier ES, Orenstein JM, *et al.* Animal models of human microsporidial infections. Lab Anim Sci 1998; 48: 589-92.

[129] Snowden KF, Shadduck JA. Microsporidia of higher vertebrates. In: Wittner M, Weiss L, Eds. The Microsporidia and Microsporidiosis. Washington, DC: American Society of Microbiology 1999; pp. 393-419.
[http://dx.doi.org/10.1128/9781555818227.ch12]

[130] Didier ES, Maddry JA, Kwong CD, *et al.* Screening of compounds for antimicrosporidial activity *in vitro.* Folia Parasitol (Praha) 1998; 45: 129-39.

[131] Mota P, Rauch CA, Edberg SC. *Microsporidia* and *Cyclospora*: epidemiology and assessment of risk from the environment. Crit Rev Microbiol 2000; 26: 69-90. [http://dx.doi.org/10.1080/10408410091154192]

[132] Weber R, Deplazes P, Schwartz D. Diagnosis and clinical aspects of human microsporidiosis. Contrib Microbiol 2000; 6: 166-92. [http://dx.doi.org/10.1159/000060360]

[133] Deplazes P, Mathis A, Weber R. Epidemiology and zoonotic aspects of microsporidia of mammals and birds. Contrib Microbiol 2000; 6: 236-60. [http://dx.doi.org/10.1159/000060363]

[134] Liu JJ, Greeley EH, Shadduck JA. Murine encephalitozoonosis: the effect of age and mode of transmission on occurrence of infection. Lab Anim Sci 1988; 38: 675-9.

[135] Fuentealba IC, Mahoney NT, Shadduck JA, *et al.* Hepatic lesions in rabbits infected with *Encephalitozoon cuniculi* administered per rectum. Vet Pathol 1992; 29: 536-40. [http://dx.doi.org/10.1177/030098589202900608]

[136] Tzipori S, Carville A, Widmer G, *et al.* Transmission and establishment of a persistent infection of *Enterocytozoon bieneusi*, derived from a human with AIDS, in simian immunodeficiency virus-infected rhesus monkeys. J Infect Dis 1997; 175: 1016-20. [http://dx.doi.org/10.1086/513962]

[137] Koudela B, Visvesvara GS, Moura H, *et al.* The human isolate of *Brachiola algerae* (Phylum Microsporidia): development in SCID mice and description of its fine structure features. Parasitol 2001; 123: 153-62. [http://dx.doi.org/10.1017/S0031182001008228]

[138] Zeman DH, Baskin GB. Encephalitozoonosis in squirrel monkeys (*Saimiri sciureus*). Vet Pathol 1985; 22: 24-31.

[139] Van Rensburg IB, Volkmann DH, Soley JT, Stewart CG. *Encephalitozoon* infection in a still-born foal. J S Afr Vet Assoc 1991; 62: 130-2.

[140] Cislakova L, Literak I, Balent P, *et al.* Prevalence of antibodies to *Encephalitozoon cuniculi* (microsporidia) in angora goats—a potential risk of infection for breeders. Ann Agric Environ Med 2001; 8: 289-91.

[141] Akerstedt J. An indirect ELISA for detection of *Encephalitozoon cuniculi* infection in farmed blue foxes (Alopex lagopus). Acta Vet Scand 2002; 43: 211-20. [http://dx.doi.org/10.1186/1751-0147-43-211]

[142] Muller-Doblies UU, Herzog K, Tanner I, *et al.* First isolation and characterisation of *Encephalitozoon cuniculi* from a free-ranging rat (Rattus norvegicus). Vet Parasitol 2002; 107: 279-85. [http://dx.doi.org/10.1016/S0304-4017(02)00163-2]

[143] Guscetti F, Mathis A, Hatt JM, *et al.* Overt fatal and chronic subclinical *Encephalitozoon cuniculi* microsporidiosis in a colony of captive emperor tamarins (Saguinus imperator). J Med Primatol 2003; 32: 111-9. [http://dx.doi.org/10.1034/j.1600-0684.2003.00016.x]

[144] Bornay-Llinares FJ, Da Silva AJ, Moura H, *et al.* Immunologic, microscopic, and molecular evidence of *Encephalitozoon intestinalis* (*Septata intestinalis*) infection in mammals other than humans. J Infect Dis 1998; 178: 820-6. [http://dx.doi.org/10.1086/515356]

[145] Graczyk TK, Bosco-Nizeyi J, da Silva AJ, *et al.* A single genotype of *Encephalitozoon intestinalis* infects free-ranging gorillas and people sharing their habitats in Uganda. Parasitol Res 2002; 88: 926-31.
[http://dx.doi.org/10.1007/s00436-002-0693-5]

[146] Rinder H, Thomschke A, Dengjel B, *et al.* Close genotypic relationship between *Enterocytozoon bieneusi* from humans and pigs and first detection in cattle. J Parasitol 2000; 86: 185-8.
[http://dx.doi.org/10.1645/0022-3395(2000)086[0185:CGRBEB]2.0.CO;2]

[147] Reetz J, Rinder H, Thomschke A, *et al.* First detection of the microsporidium *Enterocytozoon bieneusi* in non-mammalian hosts (chickens). Int J Parasitol 2002; 32: 785-7.
[http://dx.doi.org/10.1016/S0020-7519(02)00045-0]

[148] Mansfield KG, Carville A, Shvetz D, *et al.* Identification of an *Enterocytozoon bieneusi*-like microsporidian parasite in simian-immunodeficiency-virus-inoculated macaques with hepatobiliary disease. Am J Pathol 1997; 150: 1395-405.

[149] Sestak K, Merritt CK, Borda J, Saylor E, *et al.* Infectious agent and immune response characteristics of chronic enterocolitis in captive rhesus macaques. Infect Immun 2003; 71: 4079-86.
[http://dx.doi.org/10.1128/IAI.71.7.4079-4086.2003]

[150] Dengjel B, Zahler M, Hermanns W, *et al.* Zoonotic potential of *Enterocytozoon bieneusi.* J Clin Microbiol 2001; 39: 4495-9.
[http://dx.doi.org/10.1128/JCM.39.12.4495-4499.2001]

[151] del Aguila C, Izquierdo F. *Enterocytozoon bieneusi* in animals: rabbits and dogs as new hosts. J Eukaryot Microbiol 1991; 46: 8S-9S.

[152] Sulaiman IM, Fayer R, Lal AA, *et al.* Molecular characterization of microsporidia indicates that wild mammals harbor host-adapted *Enterocytozoon* spp. as well as humanpathogenic *Enterocytozoon bieneusi.* Appl Environ Microbiol 2003; 69: 4495-501.
[http://dx.doi.org/10.1128/AEM.69.8.4495-4501.2003]

[153] Shaw RW, Kent ML. Fish microsporidia. In: Wittner M, Weiss LM, Eds. The Microsporidia and Microsporidiosis. Washington, DC: American Society of Microbiology 1999; pp. 418-46.
[http://dx.doi.org/10.1128/9781555818227.ch13]

[154] McInnes EF, Stewart CG. The pathology of subclinical infection of *Encephalitozoon cuniculi* in canine dams producing pups with overt encephalitozoonosis. J S Afr Vet Assoc 1991; 62: 51-4.

[155] Weitzel T, Wolff M, Dabanch J. Dual microsporidial infection with *Encephalitozoon cuniculi and Enterocytozoon bieneusi* in an HIV positive patient. Infection 2001; 29: 237-9.
[http://dx.doi.org/10.1007/s15010-001-1164-0]

[156] Sarfati C, Liguory O, Derouin F. Les microsporidies Press Med. 2001; 30 : 143-147.

[157] Hennequin C, Pialoux G, Bouree P, Dupont B. Microsporidiose et syndrome d'immunodéficience acquise (SIDA). Med Mal Infect 1992; 22: 487-92.
[http://dx.doi.org/10.1016/S0399-077X(05)80284-7]

[158] Endeshaw T, Kebede A, Verweij JJ, *et al.* Intestinal microsporidiosis in diarrheal patients infected with immunodeficiency virus in Addis Ababa, Ethiopia. Jpn J Infect Dis 2006; 59(5): 306-10.

[159] Pagornrat W, Leelayoova A, Rangsin R, *et al.* Carriage rate of *Enterocytozoon bieneusi* in an orphanage in Bangkok, Thailand. J Clin Microbiol 2009; 47(11): 3739-41.
[http://dx.doi.org/10.1128/JCM.01606-09]

[160] Lobo ML, Xiao L, Antunes F, *et al.* Microsporidia as emerging pathogens and the implication for public health: a 10-year study on HIV-positive and negative patients. Int J Parasitol 2012; 42: 197-205.
[http://dx.doi.org/10.1016/j.ijpara.2011.12.002]

[161] Tumwine JK, Kekitiinwa A, Bakeera-Kitaka S, *et al.* Cryptosporidiosis and microsporidiosis in Ugandan children with persistent diarrhea with and without concurrent infection with the human immunodeficiency virus. Am J Trop Med Hyg 2005; 73: 921-5.

[162] Dascomb K, Frazer T, Clark RA, *et al.* Microsporidiosis and HIV. J Acquir Immune Defic Syndr 2000; 24(3): 290-2.
[http://dx.doi.org/10.1097/00126334-200007010-00018]

[163] Sokolva OI, Demyanov AV, Bowers LC, *et al.* Emerging microsporidian infctions in Russian HIV-infected patients. J Clin Microbiol 2011; 49(6): 2102-8.
[http://dx.doi.org/10.1128/JCM.02624-10]

[164] Matos O, Lobo ML, Xiao L. Epidemiology of *Enterocytozoon bieneusi* Infection in Humans. J Parasitol Res 2012; 981424.

[165] Santín M, Fayer R. *Enterocytozoon bieneusi* genotype nomenclature based on the internal transcribed spacer sequence: a consensus. J Eukaryot Microbiol 2009; 56(1): 34.
[http://dx.doi.org/10.1111/j.1550-7408.2008.00380.x]

[166] Xiao L, Li L, Visvesvara GS, Moura II, *et al.* Genotyping *Encephalitozoon cuniculi* by multilocus analyses of genes with repetitive sequences. J Clin Microbiol 2001; 39: 2248-53.
[http://dx.doi.org/10.1128/JCM.39.6.2248-2253.2001]

[167] Dowd SE, Gerba CP, Pepper IL. Confirmation of the human-pathogenic microsporidia *Enterocytozoon bieneusi, Encephalitozoon intestinalis*, and *Vittaforma corneae* in water. Appl Environ Microbiol 1998; 64: 3332-5.

[168] Fournier S, Dubrou S, Liguory O, *et al.* Detection of microsporidia, cryptosporidia and *Giardia* in swimming pools: a one year prospective study. FEMS Immunol Med Microbiol 2002; 33: 209-13.
[http://dx.doi.org/10.1111/j.1574-695X.2002.tb00592.x]

[169] Sparfel JM, Sarfati C, Liguory O, *et al.* Detection of microsporidia and identification of *Enterocytozoon bieneusi* in surface water by filtration followed by specific PCR. J Eukaryot Microbiol 1997; 44: 78S.
[http://dx.doi.org/10.1111/j.1550-7408.1997.tb05792.x]

[170] Avery SW, Undeen AH. The isolation of microsporidia and other pathogens from concentrated ditch water. J Am Mosq Control Assoc 1987; 3: 54-8.

[171] Enriquez FJ, Taren D, Cruz-Lopez A, *et al.* Prevalence of intestinal encephalitozoon in Mexico. ClinIn. Infect Dis 1998; 26: 1227-9.
[http://dx.doi.org/10.1086/520278]

[172] Didier ES, Stovall ME, Green LC, *et al.* Epidemiology of microsporidiosis: sources and modes of transmission. Vet Parasitol 2004; 126: 145-66.
[http://dx.doi.org/10.1016/j.vetpar.2004.09.006]

[173] Shadduck JA, Polley MB. Some factors influencing the *in vitro* infectivity and replication of *Encephalitozoon cuniculi.* J Protozool 1978; 25: 491-6.
[http://dx.doi.org/10.1111/j.1550-7408.1978.tb04174.x]

[174] Li X, Palmer R, Trout JM, *et al.* Infectivity of microsporidia spores stored in water at environmental temperatures. J Parasitol 2003; 89: 185-8.
[http://dx.doi.org/10.1645/0022-3395(2003)089[0185:IOMSSI]2.0.CO;2]

[175] Hutin YJ, Sombardier MN, Ligouory O, *et al.* Risk factors for intestinal microsporidiosis in patients with human immunodeficiency virus infection: a case study. J Infect Dis 1998; 178: 904-7.
[http://dx.doi.org/10.1086/515353]

[176] Cotte L, Rabodonirina M, Chupis F, *et al.* Waterborne outbreak of intestinal microsporidiosis in persons with and without human immunodeficiency virus infection. J Infect Dis 1999; 180: 2003-8. [http://dx.doi.org/10.1086/315112]

[177] Slifko TR, Smith HV, Rose JB. Emerging parasite zoonoses associated with water and food. Int J Parasitol 2000; 30: 1379-93. [http://dx.doi.org/10.1016/S0020-7519(00)00128-4]

[178] Orlandi PA, Chu DT, Bier JW, *et al.* Parasites and the food supply. Food Technol 2002; 56: 72-81.

[179] Field AS, Marriott DJ, Milliken ST, *et al.* Myositis associated with a newly described microsporidian, *Trachipleistophora hominis*, in a patient with AIDS. J Clin Microbiol 1996; 34: 2803-11.

[180] Cheney SA, Lafranchi-Tristem NJ, Canning EU. Phylogenetic relationships of Pleistophora-like microsporidia based on small subunit ribosomal DNA sequences and implications for the source of *Trachipleistophora hominis* infections. J Eukaryot Microbiol 2000; 47: 280-7. [http://dx.doi.org/10.1111/j.1550-7408.2000.tb00048.x]

[181] Undeen AH, Alger N. *Nosema algerae*: infection of the white mouse by a mosquito parasite. Exp Parasitol 1976; 40: 86-8. [http://dx.doi.org/10.1016/0014-4894(76)90068-0]

[182] Trammer T, Dombrowski F, Doehring M, *et al.* Opportunistic properties of *Nosema algerae* (Microspora), a mosquito parasite, in immunocompromised mice. J Eukaryot Microbiol 1997; 44: 258-62. [http://dx.doi.org/10.1111/j.1550-7408.1997.tb05709.x]

[183] Weidner E, Findley AM, Dolgikh V, *et al.* Microsporidian biochemistry and physiology. In: Wittner M, Weiss LM, Eds. The Microsporidia and Microsporidiosis. Washington, DC: ASM Press 1999; pp. 172-95. [http://dx.doi.org/10.1128/9781555818227.ch5]

[184] Vavra J, Undeen AH. *Nosema algerae spp.* (Cnidospora, Microsporida) a pathogen in a laboratory colony of *Anopheles stephensi* Liston (Diptera, Culicidae). J Protozool 1970; 17: 240-9. [http://dx.doi.org/10.1111/j.1550-7408.1970.tb02365.x]

[185] Sak B, Brady D, Pelikanova M, *et al.* Unapparent microsporidial infection among immunocompetent humans in the Czech Republic. J Clin Microbiol 2011; 49: 1064-70. [http://dx.doi.org/10.1128/JCM.01147-10]

[186] Weber R, Bryan RT. Microsporidial infections in immunodeficient and immunocompetent patients. Clin Infect Dis 1994; 19: 517-21. [http://dx.doi.org/10.1093/clinids/19.3.517]

[187] Sandfort J, Hannemann A, Gelderblom H, *et al.* *Enterocytozoon bieneusi* infection in an immunocompetent patient who had acute diarrhea and who was not infected with the human immunodeficiency virus. Clin Infect Dis 1994; 19: 514-6. [http://dx.doi.org/10.1093/clinids/19.3.514]

[188] Sobottka I, Albrecht H, Schottelius J, *et al.* Self-limited traveller's diarrhea due to a dual infection with *Enterocytozoon bieneusi* and *Cryptosporidium parvum* in an immunocompetent HIV-negative Child. Eur J Clin Microbiol Infect Dis 1995; 14: 919-20. [http://dx.doi.org/10.1007/BF01691502]

[189] Didier ES. Microsporidiosis: an emerging and opportunistic infection in human and animals. Acta Trop 2005; 94: 61-76. [http://dx.doi.org/10.1016/j.actatropica.2005.01.010]

[190] Anane S, Attochi H. Microsporidiosis: Epidemiology, Clinical data and therapy. Gastroenterol Clin Biol 2010; 34: 450-64.
[http://dx.doi.org/10.1016/j.gcb.2010.07.003]

[191] Conteas CN, Berlin OG, Ash LR. Therapy of human gastrointestinal microsporidiosis. Am J Trop Hyg 2000; 63: 121-7.

[192] Dei-Cas E. Infections a microsporidies, Isospora et Sarcocystis. EMC (Elsevier SAS, Paris), Maladies infectieuses 1994. 8-503-A-10.

[193] Franzen C, Muller A. Microsporidiosis: human diseases and diagnosis. Microbes Infect 2001; 3: 389-400.
[http://dx.doi.org/10.1016/S1286-4579(01)01395-8]

[194] Chijide VM. Microsporidiosis in http://emedicine.medscape.com. 2014.

[195] Orenstein JM, Chiang J, Steinberg W, *et al.* Intestinal microsporidiosis as a cause of diarrhea in human immunodeficiency virus-infected patients: a report of 20 cases. Hum Pathol 1990; 21: 475-81.
[http://dx.doi.org/10.1016/0046-8177(90)90003-N]

[196] Barratt JL, Harkness J, Marriott D, *et al.* Importance of nonenteric protozoan infections in immunocompromised people. Clin Microbiol Rev 2010; 23: 795-836.
[http://dx.doi.org/10.1128/CMR.00001-10]

[197] Schwartz DA, Sobottka I, Leitch GJ, *et al.* Pathology of microsporidiosis: Emerging parasitic infections in patients with acquired immunodeficiency syndrome. Arch Pathol Lab Med 1996; 120: 173-88.

[198] Molina JM, Oksenhendler E, Beauvais B, *et al.* Disseminated microsporidiosis due to *Septata intestinalis* in patients with AIDS: Clinical features and response to albendazole therapy. J Infect Dis 1995; 171: 245-9.
[http://dx.doi.org/10.1093/infdis/171.1.245]

[199] Gunnarsson G, Hurlbut D, De-Girolami PC, *et al.* Multiorgan microsporidiosis: Report of five cases and review. Clin Infect Dis 1995; 21: 37-44.
[http://dx.doi.org/10.1093/clinids/21.1.37]

[200] Fadl HO, Khairy H, Hanafy NA, El-Dib NA. Detection of microsporidiosis in patients with chronic renal failiure. In: Moderated e-poster presentation in the 34 Congress of the SIU (Socite Internationale D'Urologie); October 12-15, 2014; Glasgow, Scotland. 2007.

[201] Schwartz DA, Bryan RT, Hewanlowe KO, *et al.* Disseminated microsporidiosis (*Encephalitozoon hellem*) and acquired immunodeficiency syndrome—Autopsy evidence for respiratory acquisition. Arch Pathol Lab Med 1992; 116: 660-8.

[202] Schwartz DA, Visvesvara GS, Diesenhouse MC, *et al.* Pathologic features and immunofluorescent antibody demonstration of ocular microsporidiosis (Encephalitozoon hellem) in seven patients with acquired immunodeficiency syndrome. Am J Ophthalmol 1993; 115: 285-92.
[http://dx.doi.org/10.1016/S0002-9394(14)73577-9]

[203] Franzen C, Muller A, Salzberger B, *et al.* Chronic rhinosinusitis in patients with AIDS: potential role of microsporidia. AIDS 1996; 10: 687-8.
[http://dx.doi.org/10.1097/00002030-199606000-00022]

[204] Didier ES, Rogers LB, Orenstein JM, *et al.* Characterization of *Encephalitozoon* (*Septata*) *intestinalis* isolates cultured from nasal mucosa and bronchoalveolar lavage fluids of two AIDS patients. J Eukaryot Microbiol 1996; 43: 34-43.
[http://dx.doi.org/10.1111/j.1550-7408.1996.tb02470.x]

[205] Pedro-de-Lelis FJ, Sabater-Marco V, Herrera-Ballester A. Necrotizing maxillary sinus mucormycosis related to candidiasis and microsporidiosis in an AIDS patient. AIDS 1995; 9: 1386-8. [http://dx.doi.org/10.1097/00002030-199512000-00019]

[206] Joseph J, Vemuganti GK, Garg P. Histopathological evaluation of ocular microsporidiosis by different stains. BMC Clin Pathol 2006; 6: 6. [http://dx.doi.org/10.1186/1472-6890-6-6]

[207] Fischer J, Suire C, Donze HH. Toll-Like Receptor and recognition of the microsporidiadia *Encephalitozoon spp.* induces nuclear translocation of NF-κB and Subsequent Inflammatory Responses. Infect Immun 2008; 76(10): 4737-44.

[208] Schmidt EC, Shadduck JA. Mechanisms of resistance to the intracellular protozoan *Encephalitozoon cuniculi* in mice. J Immunol 1984; 133: 2712-9.

[209] Arnesen K, Nordstoga K. Ocular encephalitozoonosis (nosematosis) in blue foxes, polyarteritis nodosa and cataract. Acta Ophthalmol 1977; 55: 641-51. [http://dx.doi.org/10.1111/j.1755-3768.1977.tb05662.x]

[210] Mohn SF, Nordstoga K, Moller OM. Experimental encephalitozoonosis in the blue fox: transplacental transmission of the parasite. Acta Vet Scand 1982; 23: 211-20.

[211] Moretto M, Casciotti L, Durell B, *et al.* Lack of CD^4 T cells does not affect induction of CD8 T- cell immunity against *Encephalitozoon cuniculi* infection. Infect Immun 2000; 68: 6223-32. [http://dx.doi.org/10.1128/IAI.68.11.6223-6232.2000]

[212] Cox JC, Gallichio HA, Pye D, *et al.* Application of immunofluorescence to the establishment of an *Encephalitozoon cuniculi* free rabbit colony. Lab Anim Sci 1977; 27: 204-9.

[213] Bergquist NR, Stintzing G, Smedman L, *et al.* Diagnosis of encephalitozoonosis in man by serological tests. Br Med J (Clin Res Ed) 1984; 288: 902. [http://dx.doi.org/10.1136/bmj.288.6421.902]

[214] Didier ES, Kotler DP, Dieterich DT, *et al.* Serological studies in human microsporidia infections. AIDS 1993; 7: 8-11.

[215] Buc M. Imunológia. Bratislava: Veda 2001; p. 463.

[216] Didier ES, Didier PJ, Snowden KF, *et al.* Microsporidiosis in mammals. Microbes Infect 2000; 2: 709-20. [http://dx.doi.org/10.1016/S1286-4579(00)00354-3]

[217] Valencakova A, Halanova M. Immune response to Encephalitozoon infection review. Comp Immunol Microbiol Infect Dis 2012; 35: 1-7. [http://dx.doi.org/10.1016/j.cimid.2011.11.004]

[218] Franzen C. Microsporidia: how can they invade other cells. Trends Parasitol 2004; 20: 275-9. [http://dx.doi.org/10.1016/j.pt.2004.04.009]

[219] Couzinet S, Cejas E, Schittny J, *et al.* Phagocytic uptake of *Encephalitozoon cuniculi* by nonprofessional phagocytes. Infect Immun 2000; 68: 6939-45. [http://dx.doi.org/10.1128/IAI.68.12.6939-6945.2000]

[220] Hermánek J, Koudela B, Kuˇcerová A, *et al.* Prophylactic and therapeutic immune reconstitution of SCID mice infected with *Encephalitozoon cuniculi.* Folia Parasitol (Praha) 1993; 40: 287-91.

[221] Braunfuchsova P, Kopecky J, Ditrich O, *et al.* Cytokine response to infection with the microsporidian, *Encephalitozoon cuniculi.* Folia Parasitol (Praha) 1999; 46: 91-5.

[222] Abbas AK, Lichman AH, Pober JS. Cellular and Molecular Immunology. Philadelphia: W.B. Saunders Company 1994; pp. 240-60.

[223] Bosma GC, Custer RP, Bosma MJ. A severe combined immunodeficiency mutation in mouse. Nature 1983; 301: 527-30.
[http://dx.doi.org/10.1038/301527a0]

[224] Moretto MM, Lawlor EM, Khan IA. Aging mice exhibit a functional defect in mucosal dendritic cell response against an intracellular pathogen. J Immunol 2008; 181: 7977-84.
[http://dx.doi.org/10.4049/jimmunol.181.11.7977]

[225] Shadduck JA, Orenstein JM. Comparative pathology of microsporidiosis. Arch Pathol Lab Med 1993; 117: 1215-9.

[226] Kotler DP, Orenstein JM. Clinical syndromes associated with microsporidiosis. 1998; 40: 321-49.

[227] Dascomb K, Clark R, Pulvierenti J, *et al.* Natural history of intestinal microsporidiosis among patients infected with human immunodeficiency virus. J Clin Microbiol 1999; 37: 3421-2.

[228] Weber R, Bryan RT, Owen RL, *et al.* Improved light-microscopal detection of microsporidia spores in stool and duodenal aspirates. N Engl J Med 1992; 326: 161-6.
[http://dx.doi.org/10.1056/NEJM199201163260304]

[229] Rijpstra AC, Canning EU, Van Ketel RJ, *et al.* Use of light microscopy to diagnose small-intestinal microsporidiosis in patients with AIDS. J Infect Dis 1988; 157: 827-31.
[http://dx.doi.org/10.1093/infdis/157.4.827]

[230] Rabeneck L, Genta RM, Gyorkey F, *et al.* Observations on the pathological spectrum and clinical course of microsporidiosis in men infected with the human immunodeficiency virus: follow-up study. Clin Infect Dis 1995; 20: 1229-35.
[http://dx.doi.org/10.1093/clinids/20.5.1229]

[231] Muller A, Bialek R, Kamper A, *et al.* Detection of microsporidia in travelers with diarrhea. J Clin Microbiol 2001; 39: 1630-2.
[http://dx.doi.org/10.1128/JCM.39.4.1630-1632.2001]

[232] Bryan RT, Cali A, Owen RL, *et al.* Microsporidia: opportunistic pathogens in patients with AIDS. New York: Field & Wood Medical Publishers 1991.

[233] Zender HO, Arrigoni E, Eckert J, *et al.* A case of *Encephalitozoon cuniculi* peritonitis in a patient with AIDS. Am J Clin Pathol 1989; 92: 352-6.
[http://dx.doi.org/10.1093/ajcp/92.3.352]

[234] Sheth SG, Bates C, Federman M, *et al.* Fulminant hepatic failure caused by microsporidial infection in a patient with AIDS. AIDS 1997; 11: 553-4.

[235] Yachnis AT, Berg J, Martinez-Salazar A. Disseminated microsporidiosis especially infecting the brain, heart, and kidneys. Report of a newly recognized pansporoblastic species in two symptomatic AIDS patients. Am J Clin Pathol 1996; 106: 535-43.
[http://dx.doi.org/10.1093/ajcp/106.4.535]

[236] Hartskeerl RA, Van Gool T, Schuitema AR, *et al.* Genetic and immunological characterization of the microsporidian *Septata intestinalis*. Reclassification to Encephalitozoon intestinalis Parasitology 1995; 110(Pt 3): 277-85.

[237] Friedberg DN, Stenson SM, Orenstein JM, *et al.* Microsporidial keratoconjunctivitis in acquired immunodeficiency syndrome. Arch Ophthalmol 1990; 108: 504-8.
[http://dx.doi.org/10.1001/archopht.1990.01070060052047]

[238] Lowder CY, Meisler DM, McMahon JT, *et al.* Microsporidia infection of the cornea in a man seropositive for human immunodeficiency virus. Am J Ophthalmol 1990; 109: 242-4.
[http://dx.doi.org/10.1016/S0002-9394(14)76005-2]

[239] Desser SS, Hong H, Yang YJ. Ultrastructure of the development of a species of *Encephalitozoon* cultured from the eye of an AIDS patient. Parasitol Res 1992; 78: 677-83. [http://dx.doi.org/10.1007/BF00931520]

[240] Silverstein B, Cunningham EJ, Margolis T, *et al.* Microsporidial keratoconjunctivitis in a patient without human immunodeficiency virus infection. Am J Ophthalmol Mol 1997; 124: 395-6. [http://dx.doi.org/10.1016/S0002-9394(14)70833-5]

[241] Chan CM, Theng JT, Li L, *et al.* Microsporidial keratitis in healthy individuals. Ophthalmology 2003; 110: 1420-5. [http://dx.doi.org/10.1016/S0161-6420(03)00448-2]

[242] Chupp GL, Alroy J, Adelman LS, *et al.* Myositis due to *Pleistophora* (Microsporidia) in a patient with AIDS. Clin Infect Dis 1993; 16: 15-21. [http://dx.doi.org/10.1093/clinids/16.1.15]

[243] Weber R, Deplazes P, Flepp M, *et al.* Cerebral microsporidiosis due to *Encephalitozoon cuniculi* in a patient with human immunodeficiency virus infection. N Engl J Med 1997; 336: 474-8. [http://dx.doi.org/10.1056/NEJM199702133360704]

[244] Moura H, Schwartz DA, Bornay-Linnares F, *et al.* A new and improved "Quick-Hot Gram-Chromotrope" technique that differentially stains microsporidian spores in clinical samples, including paraffin-embedded tissue sections. Arch Pathol Lab Med 1997; 121: 888-93.

[245] Vavra J, Dahbiova R, Hollister WS, *et al.* Staining of microsporidian spores by optical brighteners with remarks on the use of brighteners for the diagnosis of AIDS associated human microsporidioses. Folia Parasitol (Praha) 1993; 40: 267-72.

[246] Accoceberry I, Thellier M, Desportes-Livage I, *et al.* Production of monoclonal antibodies directed against the microsporidium *Enterocytozoon bieneusi.* J Clin Microbiol 1999; 37: 4107-12.

[247] Sheoran AS, Feng X, Singh I, *et al.* monoclonal antibodies against *Enterocytozoon bieneusi* of human origin. Clin Diagn Lab Immunol 2005; 12: 1109-13.

[248] Thellier M, Breton J. *Enterocytozoon bieneusi* in human and animals, focus on laboratory identification and molecular epidemiology. Parasite 2008; 15: 349-58. [http://dx.doi.org/10.1051/parasite/2008153349]

[249] Menotti J, Cassinat B, Sarfati C, *et al.* Development of a real-time PCR assay for quantitative detection of *Encephalitozoon intestinalis* DNA. J Clin Microbiol 2003; 41: 1410-3. [http://dx.doi.org/10.1128/JCM.41.4.1410-1413.2003]

[250] Menotti J, Cassinat B, Porcher R, *et al.* Development of a Real-time polymerase chain reaction assay for quantitative detection of *Enterocytozoon bieneusi* DNA in stool specimens from immunocompromised patients with intestinal microsporidiosis. J Infect Dis 2003; 187: 1469-74. [http://dx.doi.org/10.1086/374620]

[251] Verweij JJ, Ten Hove RE, Brienen AT, *et al.* Multiplex detection of *Enterocytozoon bieneusi* and *Encephalitozoon spp.* in fecal samples using real-time PCR. Diagn Microbiol Infect Dis 2007; 57(2): 163-7. [http://dx.doi.org/10.1016/j.diagmicrobio.2006.08.009]

[252] Rayan HZ, Ismail OA, El-Hamshary EM, *et al.* Contribution of Multiplex real time PCR for detection and differentiation of *Enterocytozoon bieneusi* and *Encephalitozoon intestinalis* in fecal samples of immunocompromized patients. Parasitol United J 2013; 6(1): 51-8. [PUJ].

[253] Lee S-H, Joung M, Yoon S, *et al.* Multiplex PCR detection of waterborne intestinal protozoa; *Microsporidia, Cyclospora* and *Cryptosporidium.* Korean J Parasitol 2010; 48(4): 297-301. [http://dx.doi.org/10.3347/kjp.2010.48.4.297]

[254] Joseph J, Sharma S. *In vitro* culture of various species of microsporidia causing Keratitis: evaluation of three immortalized cell lines. Indian J Med Microbiol 2009; 27(1): 35-9.

[255] Garcia LS. Diagnostic medical parasitology. 4th ed. Washington, D.C: ASM Press 2001; pp. 87-97.

[256] Weber R, Swartz DA, Deplazes P. Laboratory diagnosis of microsporidiosis. The microsporidia and microsporidiosis. Washington, DC: ASM Press 1999; pp. 315-61.

[257] Visvesvara G, Moura H, Leitch G. Culture and Propagation of Microsporidia. The microsporidia and microsporidiosis. Washington, DC: ASM Press 1999; pp. 363-92.

[258] Monaghan SR. Use of fish cell cultures for the study and cultivation of microsporidia. A thesis presented to the University of Waterloo in fulfillment of the requirement for the degree of Doctor of Philosophy in Biology Waterloo, Ontario, Canada 2011; 1-167.

[259] Bettens S, Gigi J, Delmée M. Les infections à microsporidies. Louv Med 1999; 118: 446-56.

[260] Costa SF, Weiss LM. Drug treatment of microsporidiosis. Drug Resist Updat 2000; 3: 384-99. [http://dx.doi.org/10.1054/drup.2000.0174]

[261] Didier ES, Bowers L, Stovall ME, *et al.* Antimicrosporidial activity of (fluoro) quinolones *in vitro* and *in vivo*. Folia Parasitol (Praha) 2005; 52: 173-81. [http://dx.doi.org/10.14411/fp.2005.022]

[262] Grob U. Treatment of microsporidiosis including albendazole. Parasitol Res 2003; 90: s14-8.

[263] Sobottka I, Bartscht K, Schäfer P, *et al. In vitro* activity of polyoxin D and nikkomycin Z against *Encephalitozoon cuniculi.* Parasitol Res 2002; 88(5): 451-3. [http://dx.doi.org/10.1007/s00436-002-0592-9]

[264] Farthing MJ. Treatment options for the eradication of intestinal protozoa. Nat Clin Pract Gastroenterol Hepatol 2006; 3: 436-45. [http://dx.doi.org/10.1038/ncpgasthep0557]

[265] Gazzard B. AIDS and the gastrointestinal tract. Medicine 2005; 33: 24-6. [http://dx.doi.org/10.1383/medc.33.6.24.66000]

[266] Leder K, Hellard ME, Sinclair MI, *et al.* No correlation between clinical symptoms and *Blastocystis hominis* in immunocompetent individuals. J Gastroenterol Hepatol 2005; 20: 1390-4. [http://dx.doi.org/10.1111/j.1440-1746.2005.03868.x]

[267] Kelly P, Baboo KS, Wolf M, *et al.* The prevalence nada etiology of persistent diarrhea in adults in urban Zambia. Acta Trop 1996; 61: 183-90. [http://dx.doi.org/10.1016/0001-706X(95)00142-2]

[268] MacDonald LM, Armson A, Thompson AR, *et al.* Characterisation of benzimidazole binding with recombinant tubulin from *Giardia duodenalis, Encephalitozoon intestinalis*, and *Cryptosporidium parvum.* Mol Biochem Parasitol 2004; 138: 89-96. [http://dx.doi.org/10.1016/j.molbiopara.2004.08.001]

[269] Tremoulet AH, Avila-agüero ML, Paris MM, *et al.* Albendazole therapy for Microsporidium diarrhea in immunocompetent Costa Rican children. Pediatr Infect Dis J 2004; 23: 915-8. [http://dx.doi.org/10.1097/01.inf.0000141724.06556.f9]

[270] Blanshard C, Ellis DS, Dowell SP, *et al.* Electron microscopic changes in *Enterocytozoon bieneusi* following treatment with albendazole. J Clin Pathol 1993; 46: 898-902. [http://dx.doi.org/10.1136/jcp.46.10.898]

[271] Horton J. Albendazole: abroad spectrum anthelminthic for treatment of individuals and populations. Curr Opin Infect Dis 2002; 15: 599-608. [http://dx.doi.org/10.1097/00001432-200212000-00008]

[272] Molina JM, Tourneur M, Sarfati C, *et al.* Fumagillin treatment of intestinal microsporidiosis. N Engl J Med 2002; 346: 1963-9.
[http://dx.doi.org/10.1056/NEJMoa012924]

[273] Zhang H, Huang H, Cali A, *et al.* Investigations into microsporidian methionine aminopeptidase type 2: a therapeutic target for microsporidiosis. Folia Parasitol (Praha) 2005; 52: 182-92.
[http://dx.doi.org/10.14411/fp.2005.023]

[274] Gross U. Treatment of microsporidiosis including albendazole. Parasitol Res 2003; 90 (Suppl. 1): S14-8.

[275] Bicart-Sée A, Massip P, Linas MD, *et al.* Successful treatment with nitrazoxanide of *Enterocytozoon bieneusi* microsporidiosis in a patient with AIDS. Antimicrob Agents Chemother 2000; 44: 167-8.
[http://dx.doi.org/10.1128/AAC.44.1.167-168.2000]

[276] Pozio E. Highly active antiretroviral therapy and opportunistic protozoan infections. Parassitologia 2004; 46: 89-93.

[277] Martins SA, Muccioli C, Belfort R, *et al.* Resolution of microsporidial keratoconjunctivitis in an AIDS patient treated with highly active antiretroviral therapy. Am J Ophthalmol 2001; 131: 378-9.
[http://dx.doi.org/10.1016/S0002-9394(00)00810-2]

[278] Waller T. Sensitivity of *Encephalitozoon cuniculi* to various temperatures, disinfectants and drugs. Lab Anim 1979; 13: 227-30.
[http://dx.doi.org/10.1258/002367779780937753]

[279] John DE, Haas CN, Nwachuku N, *et al.* Chlorine and ozone disinfection of *Encephalitozoon intestinalis* spores. Water Res 2005; 39: 2369-75.
[http://dx.doi.org/10.1016/j.watres.2005.04.013]

[280] Jordan CN, Zajac AM, Holliman D, *et al.* Effects of high-pressure processing on *in vitro* infectivity of *Encephalitozoon cuniculi.* J Parasitol 2005; 91: 1487-8.
[http://dx.doi.org/10.1645/GE-568R.1]

[281] Jordan CN, Dicristina JA, Lindsay DS. Activity of bleach, ethanol and two commercial disinfectants against spores of *Encephalitozoon cuniculi.* Vet Parasitol 2006; 136: 343-6.
[http://dx.doi.org/10.1016/j.vetpar.2005.11.005]

[282] Panel on Opportunistic Infections in HIV-Infected Adults and Adolescents. Guidelines for the prevention and treatment of opportunistic infections in HIV-infected adults and adolescents: recommendations from the Centers for Disease Control and Prevention, the National Institutes of Health, and the HIV Medicine Association of the Infectious Diseases Society of America. Available at http://aidsinfo.nih.gov/contentfiles/lvguidelines/adult_oi.pdf. Accessed 27 Apr 2015.

[283] Khalifa AM, El Temsahy MM, Abou El Naga IF. Effect of ozone on the viability of some protozoa in drinking water. J Egypt Soc Parasitol 2001; 31: 603-16.

[284] Li X, Trout JM, Jenkins MC, *et al.* Effects of gamma radiation on viability of *Encephalitozoon* spores. J Parasitol 2002; 88: 812-3.
[http://dx.doi.org/10.1645/0022-3395(2002)088[0812:EOGROV]2.0.CO;2]

[285] Johnson CH, Marshall MM, DeMaria LA, *et al.* Chlorine inactivation of spores of *Encephalitozoon spp*. Appl Environ Microbiol 2003; 69: 1325-6.
[http://dx.doi.org/10.1128/AEM.69.2.1325-1326.2003]

[286] Marshall MM, Hayes S, Moffett J, *et al.* Comparison of UV inactivation of spores of three *Encephalitozoon species* with that of spores of two DNA repair-deficient *Bacillus subtilis* biodosimetry strains. Appl Environ Microbiol 2003; 69: 683-5.

[287] Foudraine NA, Weverling GJ, van Gool T, *et al.* Improvement of chronic diarrhoea in patients with advanced HIV-1 infection during potent antiretroviral therapy. AIDS 1998; 12: 35-41. [http://dx.doi.org/10.1097/00002030-199801000-00005]

[288] Goguel J, Katlama C, Sarfati C, *et al.* Remission of AIDS-associated intestinal microsporidiosis with highly active antiretroviral therapy. AIDS 1997; 11: 1658-9.

[289] Maggi P, Larocca AM, Quarto M, *et al.* Effect of antiretroviral therapy on cryptosporidiosis and microsporidiosis in patients infected with human immunodeficiency virus type 1. Eur J Clin Microbiol Infect Dis 2000; 19: 213-7. [http://dx.doi.org/10.1007/s100960050461]

[290] Wolk DM, Johnson CH, Rice EW, *et al.* A spore counting method and cell culture model for chlorine disinfection studies of *Encephalitozoon syn. Septata intestinalis.* Appl Environ Microbiol 2000; 66: 1266-73. [http://dx.doi.org/10.1128/AEM.66.4.1266-1273.2000]

[291] Orlandi PA, Lampel KA. Extraction-free, filter-based template preparation for rapid and sensitive PCR detection of pathogenic parasitic protozoa. J Clin Microbiol 2000; 38: 2271-7.

[292] Speare DJ, Markham JF, Guesella NJ. Development of effective whole-spore vaccine to protect against microsporidial gill disease in rainbow trout (*Onchorhynchus mykiss*) by using a low-virulence strain of *Loma salmonae.* Clin Vaccine Immunol 2007; 14(12): 1652-4.

[293] Rodriguez-Tovar LE, Speare DJ, Markham RJ. Fish microsporidia: immune response, immunomodulation and vaccination. Fish Shellfish Immunol 2011; 30(4-5): 999-1006. [http://dx.doi.org/10.1016/j.fsi.2011.02.011]

[294] Harkness JE, Guselle NJ. And Speare, DJ. Development of a pilot Heterologous whole-spore vaccine against microsporidial gill disease in rainbow Trout. Clin Vaccine Immunol 2013; 20(9): 1483-4. [http://dx.doi.org/10.1128/CVI.00340-13]

SUBJECT INDEX

A

B

C

D

E

H

I

N

O

P

Q

R

S

T

U

V

W

Made in United States
North Haven, CT
04 November 2026

21727475R00166